WESTMAR COLLEGE

W9-BMI-157

The Hollywood Musical

The Hollywood Musical

John Russell
Taylor /
Arthur Jackson

McGraw-Hill Book Company
New York   St. Louis   San Francisco

782.81
T243

Copyright © 1971 by McGraw-Hill Book
Company Limited. All Rights Reserved. No
part of this publication may be reproduced,
stored in a retrieval system, or transmitted, in
any form or by any means, electronic,
mechanical, photocopying, recording, or
otherwise, without the prior written permission
of the publisher.

Library of Congress Catalog Card Number:
70-139565

07-062953-6

Printed in Great Britain

ML
2075
.T4

Acknowledgements

Stills by courtesy of:
Metro-Goldwyn-Mayer: *An American in
Paris, Anchors Aweigh, Annie Get Your Gun,
Babes in Arms, The Band Wagon, Born To Dance,
The Broadway Melody of 1938, Gigi, High Society,
It's Always Fair Weather, Jupiter's Darling,
Kiss Me Kate, Lovely To Look At, Love Me or
Leave Me, Meet Me in St Louis, Naughty
Marietta, A Night at the Opera, On the Town,
Pagan Love Song, The Pirate, Silk Stockings,
Singin' in the Rain, Summer Holiday, Summer
Stock, Three Little Words, The Wizard of Oz,
The Ziegfeld Follies*, portraits of Johnny Green,
Jane Powell. **Twentieth Century-Fox**: *The
Best Things in Life are Free, Centennial Summer,
Daddy Long Legs, Gentlemen Prefer Blondes,
Little Miss Broadway, Say One for Me, Sing Baby
Sing, Star!, Thin Ice*, portraits of Vivian Blaine,
Celeste Holm, Cesar Romero. **Paramount**:
*The Belle of the Nineties, High, Wide and Hand-
some, Lady in the Dark, The Love Parade, One
Hour With You, Paramount on Parade, Red
Garters, White Christmas*. **Warner Brothers**:
*Dames, Finian's Rainbow, Footlight Parade, The
Golddiggers of Broadway, Gypsy, The Jazz Singer,
The Music Man, The Pajama Game, The Singing
Fool, A Star is Born, Young at Heart*. **United
Artists**: *How to Succeed in Business Without
Really Trying, Some Like It Hot, West Side Story*,
portraits of Ethel Merman, Phil Silvers.
Columbia: *Cover Girl, Jolson Sings Again,
Three for the Show*. **Universal**: *100 Men and a
Girl, The King of Jazz, Sweet Charity, Thoroughly
Modern Millie*. **RKO-Radio**: *Follow the Fleet,
The Gay Divorcee, Hans Christian Andersen,
Shall We Dance?, Top Hat*. **First National**:
Flirtation Walk, The Golddiggers of 1933. **Walt
Disney Productions**: *Babes in Toyland*.
RCA-Victor Records: *Xavier Cugat,
Henry Mancini, André Previn, Artie Shaw*.
EMI Records: Nelson Riddle, David Rose.
Columbia Records: Leonard Bernstein.
Pye Records: Duke Ellington. **Capitol
Records**: Carmen Dragon. **MGM Records**
Kurt Weill.
Color material by courtesy of
The John Kobal Collection, London

85590

Frontispiece:
Gene Kelly / Singin' in the Rain

Contents

The Hollywood Musical
John Russell Taylor

Reference Section
Arthur Jackson

Fred Astaire/*Follow the Fleet*

The Show and the Film

What is a musical? All talking, all singing, all dancing? Well, yes, preferably. But a lot more, and sometimes a lot less. A film musical is, essentially, just that: a film which, in whole or in part, has its shape, its movement, its whole feeling dictated by music. Not every film with a single song or dance qualifies: there are hundreds of films which import a guest star for a number in a night-club which is really no more than a slight extension of background music, making no real modification in the way the film works, even momentarily. There are even films with a lot of music in them which do not really, except formally, qualify as musicals, because they have no real feeling for their musical materials (like most of the Elvis Presley films), or because they are deliberately treated non-musically. To take two examples, the secret of the success of Gian-Carlo Menotti's *The Medium* or Jacques Demy's *Les Parapluies de Cherbourg*, both of which are completely sung dramas, lies in their complete refusal to let that fact make any difference to the way they are made: both are essentially realistic pieces of film-making which carefully avoid any momentary lift into another world of fantasy such as characterizes the true musical. The truth of this in the case of Jacques Demy's work was indeed uncomfortably demonstrated when, after *Les Parapluies*, he set out to make a proper, Hollywood-type musical (with Gene Kelly in attendance) in *Les Demoiselles de Rochefort*, and proved in the process that he had no feeling at all for the variegated pattern of the true musical, with its imaginative leaps, its manœuvrability, its lightning changes of pace and word.

Not every musical approximates throughout to this ideal. From a perfection-ist point of view, many of the films I shall discuss are radically defective. But if every film cannot be a city built to music, then at least it should have the odd square or side street, perhaps just a turret or two, which would not have been shaped quite as they are if it were not for the effect of music on their creation. The musical genre is unique in its combination of two distinct difficulties. If it is virtually impossible to make the perfect musical, it is also virtually impossible to make a musical which is totally bad. However awful a musical may be, there is nearly always something to catch the imagination, to make the film endurable and enjoyable in spite of its faults. Almost any vehicle of a major musical star will at least have that star, however inferior the material he or she has to work with. On the other hand, even if you hate the star you still have something to watch and listen to apart from him or her. It might be a marvellous score, like Harold Arlen's for *Casbah*, an otherwise dire semi-musical remake of *Pépé-le-Moko*, or, more improbable yet, Jerome Kern's for the Abbott and Costello vehicle *One Night in the Tropics*. It might be a solitary dance number, like Cyd Charisse's 'Desert Song' piece in Stanley Donen's otherwise fairly undistinguished bio-graphical tribute to Sigmund Romberg, *Deep in My Heart*, or her contribution to Nicholas Ray's mainly non-musical thriller *Party Girl*, or for that matter the magical dance interlude by Donald O'Connor in a Ma and Pa Kettle movie *Feudin', Fussin' and a-fightin'*. It could even be an art director's field-day like that of Walter Holscher (or was it that of set decorator William Kiernan?) for the fantastically transformed 'Swan Lake' sequence in *Three for the Show*, or some-thing so seemingly peripheral as the off-screen contribution of Lisa Kirk, who sang most of the songs for Rosalind Russell in the generally lack-lustre screen version of *Gypsy*.

Always there is something. And the reason for this is not, after all, so very far to seek. It lies in the extraordinary complexity of even the most apparently simple musical. All films, whoever may be ultimately regarded as the artist-creator, must necessarily be collaborative efforts. Musicals, obviously, are likely to involve more prime collaborators than any other sort of film, in the way of choreographers, musical arrangers and others. But the essential difference lies not so much in the quantity as in the quality of the collaboration. Everyone involved is likely to be stretched to the utmost by the demands of the medium. Stars, ideally, should be able to sing and dance as well as to act, or to act at least a little as well as sing or dance or both. Music in most films is merely background, even if as such it may play an important role. But in the musical music comes to the fore, as in a sense the whole film is built round it. Sets and costumes are generally more prominently displayed in a musical than elsewhere. And the director has to meet demands far more varied and complex than for any other type of film, and to possess certain specific types of sensitivity which straight dramatic films would never call on, as well as being able to fuse the contributions of the others involved into some sort of coherent whole.

Naturally, from the point of view of the ideal musical this organizational complexity hardly makes things easier. But then the ideal musical may not exist, and the one true formula for it, like the Philosopher's Stone or the secret of perpetual motion, is one of those things which will always be hopefully sought for and never found. Yet in the search all sorts of unexpected, unsought-for things may turn up: they usually do. And if we are content to enjoy whatever is there to be enjoyed, without constantly measuring it against some impossible ideal, we are likely to find the musical the most continually satisfying of all film genres. And I mean continually. I think no film can be seen and re-seen more frequently than a musical. With even the best films sooner or later one reaches saturation-point, sometimes temporarily, sometimes permanently. Possibly it is because you have managed to get out of the film all that is in it, or have discerned that there was never as much in it as you first suspected anyway. Maybe it is the unpredictable effect of time passing, which not only makes the fashions dowdy and the ideas old-fashioned, but makes you a different person and what you bring to the film different. Fashion, of course, never stands still, and one may recognize that what looks dowdy and unappealingly dated today may well tomorrow have taken on a charming period quality. But then again it may not: how many masterpieces of an hour do not sink without trace when their hour is over, beyond all hope of recovery and revaluation?

Possibly it is only personal prejudice which makes me believe that these processes have less effect on the musical than on any other type of film. But I do not think so. Some filmgoers, and some critics, are just allergic to the genre. Many more enjoy it, but slightly shamefacedly: it is not, after all, 'serious' in itself, though it may be redeemed for serious attention by the seriousness of its themes. But once one gets round to those silly parlour games which so often, unexpectedly, reveal the truth, it nearly always turns out that the one film one would take to the moon, or choose to see immediately before going in solitary confinement for life, is *Singin' in the Rain*, or *Meet Me in St Louis*, or *Funny Face*, or *Love Me Tonight*, rather than *Battleship Potemkin*, or *Intolerance* or *Kameradschaft*, or *Le Jour se Lève*. And that, after all, must mean something.

What it does mean, it seems to me, is that the musical, whether thematically serious or not, is aesthetically as serious as any other kind of film. More so, perhaps, in that it uses all sorts of techniques, anything really that comes to hand, in order to work on the spectator at every level of his being. Its effect on us is unpredictable, undefinable, often downright sneaky. If we are moved by the sufferings of the downtrodden poor or excited by the triumph of right over wrong, good over evil in a film, we know more or less what is getting at us and why. But why are we so moved by Judy Garland singing 'Have yourself a merry little Christmas' in *Meet Me in St Louis*? Why are we moved at all by Audrey Hepburn singing 'How long has this been going on' in *Funny Face*? The song? The singer? Yes, and all those other imponderables: the relation between the emotion of the song and its placing in the film; the rhythmic and melodic structure of the song in relation to the precise way it is shot, the speed and direction of the camera's movement, the placing of cuts in the montage, the warmth or coldness of the designer's colours, the hardness or softness of the cameraman's lighting. And, of course, the overall control exerted by the director in fusing all these elements together into a perfectly integrated whole. We need not be solemn about the musical, but we should be very careful before we refuse to accept that on its own terms it can be taken as seriously as any experience the cinema has to offer.

But then there is not just '*the* musical'. There are all kinds of musical, all, within the basic requirements of the medium, working on us in different ways. There is one great divide, though even there not unbridgeable, in that the two types are every now and then mixed and mingled: between the singer's musical and the dancer's musical. Most real musical *aficionados* tend to prefer the dancer's musical, perhaps because, at its best, it brings so many more talents obviously into play, perhaps because on the whole the dancing stars have been more compelling than the purely singing stars. Despite which, there is no reason why a film in which no one dances a step should not be just as much a musical, in our definition, as one in which no one walks if he can possibly dance. The most obvious line between the two sorts of musical, I suppose, would be the one we might draw in the 1930s between the Astaire-Rogers films and the Jeanette Mac-Donald-Nelson Eddy films. And in many ways it would be a fair one. In neither series did really first-rate directors play much part, except perhaps for George Stevens directing the Astaire-Rogers *Swing Time* – and he, whatever his talents, could hardly be counted as a great *musical* director. Dancers' musicals are more likely, like the majority of the Astaire-Rogers vehicles, to be light, fast and modern in their setting; singers' musicals are more likely, like the majority of the MacDonald-Eddy vehicles, to look back nostalgically to glamorous and romantic periods and to the more expansive world of the operetta as opposed to the slick and jazzy musical comedy.

Here, however, we have come on a question so fundamental that it must be gone into rather more deeply before we move on. That is, the relations of the film musical to its parent forms in the theatre. The film musical undoubtedly arose out of the two allied forms of theatrical entertainment which were most popular at the time of the sound film's advent: the romantic operetta and the musical comedy. These were precisely the areas in which, up to then, the stage had a clear advantage over the screen: whatever substitutes might be found for stage dialogue, none could really be found for stage music, for what good was it to see dancing without hearing the right music rightly synchronized; what was the use of a singer opening and closing his mouth if the voice did not come out of it? It was not for nothing, therefore, that when the screen first found its voice, it was not to speak, but to sing. *The Jazz Singer* was conceived in 1927 as a desperate plunge into novelty in the hope of·thus restoring the failing fortunes of Warner Brothers, and the idea was to insert synchronized songs into a silent drama. A few impromptu words crept in, and then a whole scene of dialogue, but that was not the point of the exercise. Primarily, what was intended was the birth of the musical film, and to begin with that is what was achieved. Virtually every film in production in 1928 had, if at all possible, at least one song worked in somehow, and at that time it seemed more important for silent stars bent on proving their staying power to be able to sing than to be able to speak.

Of course that phase was soon over. Audiences became glutted with music, music everywhere, and before long some bright showmen had the idea of billing talkies as decidedly *not* musicals; for a couple of years, even, musicals were being prematurely written off as box-office poison. Until the tremendous success of *Forty-Second Street* gave them a shot in the arm. So it might seem to be one of those meaningless accidents of history that the whole of the sound cinema sprang from a musical rather than a straight drama. But I doubt it. Though the musical form has been written off again and again as played out, outdated, finished, somehow it always manages to spring back as full of life as ever. If it is a complete accident that the first talkie was a musical, then it is also a complete accident that the largest money-making film to date in the history of the cinema is another musical, *The Sound of Music*. And two accidents of that size are rather much to accept as simple coincidence. (1976)

But central as the musical has certainly been to the whole development of the sound film, the connection between the film musical and the stage musical still persists. *The Jazz Singer* was based on a stage success, though made over with new songs to suit the talents of its new star, Al Jolson. *The Sound of Music* came to the screen straight from a vast success on the stage as the last of Rodgers and Hammerstein's record-breaking musicals: indeed, in London it was still running on the stage, after nearly four years, when the film version opened. And it has been a constant cause of complaint, not always without some colouring of justification, that the film musical, despite its several masterpieces of original creation, has always remained tied to the stage, a subordinate, bastard form with little independent life of its own.

Against such an attack there are two lines of argument. One is the classic argument for any adaptation from one medium to another: that it does not matter at all where the material comes from, all that matters is what the new medium makes of it. The second is a counter-attack: for does not Robert Bresson have some reason when he says that after all it is the live theatre, not the cinema, which is the bastard art? It might be argued that what is true of all stage productions – that they are the results of an uneasy and unreliable collaboration of equal talents, often in conflict, rather than the clear and decisive creation of one single, indisputable artist – is true even more acutely of the stage musical, while the film musical always has some chance of pulling all the disparate elements together under the overriding direction of one man who is ultimately in control of everything.

13

But arguments on principle seldom get us very far. Let us agree that all the best screen musicals are original screen creations, in fact or (like *On the Town*) in effect. Of adaptations from the stage the most successful are as a rule those which, however faithful they may be to the spirit of the original, treat its letter with considerable freedom. Sometimes film versions are signally better than their stage originals, for one reason or another – *The Sound of Music*, for instance, or *Half a Sixpence* – sometimes about as good, remaining at least, like *Kiss Me Kate*, a very fair if not specially cinematic record, and sometimes, like *West Side Story*, markedly inferior, usually because they throw away the specific coherence of the original without substituting any new overall pattern of their own. Between the best and the worst of film musicals there is an infinite gradation, and far be it from me to exclude any possible source of enjoyment on the grounds of mere principle, so I shall try to deal with each particular example that comes up strictly on its own merits, referring to stage originals only when such a reference seems to clarify understanding.

All that is somewhat by the way, however, and has anyhow carried us a long way beyond the film musical's dual origin in stage operetta and stage musical comedy. We have to look a little further into this if we are to understand the conflicting influences in the film musical's later development. Historians of the musical stage generally say blandly that really operetta, comic opera, *opéra comique* and musical comedy are all the same thing, or at least that operetta derived from opera via light or comic opera and that musical comedy evolved from operetta. But in most ways that is a drastic over-simplification. Operetta as it was understood by the Viennese from Strauss to Lehar had nothing

May McAvoy and Al Jolson | The Jazz Singer

historically or actually to do with the English light or comic opera of Gilbert and Sullivan, though it is true they shared approximately the same relationship with grand opera, both featuring a lot of spoken dialogue interrupted every now and then by songs and concerted numbers (though seldom importantly by dances). Operetta as, say, Victor Herbert understood it in New York in the 1900s, the time of *The Red Mill* and *Naughty Marietta*, developed from the romantic Viennese school, and this particular torch was passed on to Rudolf Friml and Sigmund Romberg in the 1920s. They wrote shows essentially for singers, musicians of at least semi-operatic standard, and comedy was kept in the background as humble relief.

But other forces were at work in the American musical theatre, and other kinds of performer were coming to the fore, song-and-dance men (and women) with a vaudeville background whose popularity could be turned to the legitimate theatre's advantage only in a new sort of show: musical comedy. In musical comedy the demands made on the performers purely as singers were considerably less and different: instead of singing a full-throated romantic melody they had to be able to put over a number, to point witty words and make sure they could be heard. (Here the connection is much closer with Gilbert and Sullivan than with Lehar, and in fact a surprisingly strong influence from Gilbert and Sullivan can be detected in a lot of early Gershwin, Rodgers and Hart, etc.) They also tended to shine in dance as much as, perhaps even more than, song, and vehicles for the Astaires and others had to be designed to make full use of this. It must be emphasized, though, that this sort of show did not evolve from the romantic operetta, and though eventually, for a while at least, musical comedy superseded operetta, at first they co-existed quite happily. 1924 might be the year of Friml's *Rose Marie* and Romberg's *The Student Prince*, but it also brought forth the Gershwins' *Lady Be Good*. 1925 gave audiences Friml's *The Vagabond King*, but also Youmans's *No, No, Nanette*, the Gershwins' *Tip Toes* and Kern's *Sunny*. In 1926 there was Romberg's *The Desert Song* and the Gershwins' *Oh, Kay!*, Rodgers and Hart's *The Girl Friend*. In 1927 musical comedy carried off most of the laurels with Youmans's *Hit the Deck* and the Gershwins' *Funny Face*, but in 1928 operetta hit back with the enormous success of Romberg's *The New Moon*.

And this was precisely the confused situation into which the sound film, and so the film musical, was born. Which way should it lean? Where should it try to recruit its new talent: from among the singing stars of operetta or from the song-and-dance men of musical comedy? Ideally, of course, the answer was both: absorb everyone and everything that could possibly be of use. But for the moment that was easier said than done. The sheer technical cumbrousness of early sound-recording technique severely limited what the sound film could do with anything more complicated than one singer singing one song at a time. So no clear commitment of any sort was made. Jolson, of course, was from vaudeville rather than operetta, but basically all he did was sing. Most of the silent stars, having made their names in lush romantic dramas, saw themselves more at home putting the romance to music, as Gloria Swanson did in *The Trespasser* and Ramon Novarro in *The Pagan*, by singing the odd romantic ditty in a more or less straight dramatic context. Musical stars of all sorts were tried out, often disastrously: Gertrude Lawrence from the stage in *The Battle of*

Paris, Rudy Vallee from radio in *The Vagabond Lover*, Sophie Tucker from night-clubs in *Honky Tonk*, Fanny Brice from revue in *My Man*. Stage hits of both sorts were hurried on to the screen, sometimes stripped of all their original songs, as with the 1930 *No, No, Nanette*, or most, as with the same year's *Sunny*, sometimes with more care and attention, as when Grace Moore and Lawrence Tibbett were teamed (though none too successfully) in *The New Moon*, or with Jeanette MacDonald and Dennis King together in *The Vagabond King* (both also in 1930). Elsewhere, each studio turned out its own spectacular revue, with a bit of everything to test audience response: MGM's *Hollywood Revue*, Warner's *Show of Shows*, *Paramount on Parade*, Universal's *King of Jazz*.

And in all this confusion, there were only two people, in Hollywood at least, who showed any real sign of knowing what they wanted to do with the new monster medium. They were Ernst Lubitsch and Rouben Mamoulian. With Jeanette MacDonald or Maurice Chevalier, or both, Lubitsch created a new form of specifically screen musical in *The Love Parade* (1929), *Monte Carlo* (1930), *The Smiling Lieutenant* (1931) and *One Hour With You* (1932), and Mamoulian, with the same two stars, carried it a few steps further in *Love Me Tonight* (1932). The style was light, fast and sophisticated, using to the full the cinema's ability to cut corners in plotting and reduce everything to essentials. Though there was little or no dancing as such in these films, they were far nearer to musical comedy than to operetta; Jeanette MacDonald had had some operatic training, but was here required to do little more than trill prettily and exhibit a fresh, fun-loving personality, while Maurice Chevalier was of course personality all the way, with whatever other talents he might have as quite unimportant

Maurice Chevalier and Jeanette MacDonald / The Love Parade

Production still | Lady in the Dark (above);
Vera-Ellen | On the Town (below)

extensions. Their films were quite on their own, unlike anything else being made in Hollywood.

Clearly the mood was swinging more and more in the direction of the musical comedy, as distinct from the operetta, and this was unarguably confirmed in 1933 with two significant events: *Forty-Second Street* and *Flying Down to Rio*. Each introduced a talent (or in the case of *Flying Down to Rio* a pair of talents) which was to mark the musical of the next decade or so decisively. *Forty-Second Street* had catchy songs, appealing stars in the shape of smooth, baby-faced Dick Powell and pert, pint-sized Ruby Keeler; but above all it had dance director Busby Berkeley. He had already shown something of what he could do in the margins of such Eddie Cantor vehicles as *Whoopee*, *The Kid from Spain* and *Roman Scandals*, but it was *Forty-Second Street*, with its extravagant camera patterns of gorgeous girls by the dozen (in 'Young and healthy'), its extended 'dramatized' versions of the title song and 'Shuffle off to Buffalo', which really established him, and set him off on the long succession of ever more spectacular and bizarre pieces of musical staging with which he decorated musicals of the thirties and sometimes after. *Flying Down to Rio*, more simply, introduced as foils to the forgotten leads, Dolores del Rio and Gene Raymond, a couple of singer-dancers, Fred Astaire and Ginger Rogers. Fred Astaire had been a big star on the stage in the 1920s, with his sister Adele, but had yet to make any major dent on films. Ginger Rogers had played tough young chorus girls in half a dozen films, including *Forty-Second Street* itself. Together, they at once became a hit and an institution, singing and dancing to the tunes of Vincent Youmans, Cole Porter, Irving Berlin, Jerome Kern, the Gershwins and other great figures of the musical comedy stage. It looked as though the reign of the operetta in films had ended before it had even begun.

But then came a surprise. Mainly on the strength of Lubitsch's name and reputation as a maker of hit films, MGM in 1934 agreed to have another go at the despised romantic operetta with a remake of the most famous of them all, Lehar's *The Merry Widow*, starring (on balance, rather surprisingly) Jeanette MacDonald and Maurice Chevalier. To everyone's amazement, it was an enormous success. Nothing loth to follow it up, MGM then looked back among the old standbys for another possible vehicle for their leading lady, and came up with Victor Herbert's *Naughty Marietta* of 1910. This time they needed a real singer for a leading man (one could hardly imagine Chevalier tackling 'Ah, sweet mystery of life' or 'Tramp, tramp, tramp along the highway'), and to fill the place they picked a near unknown, Nelson Eddy. Again, unexpectedly, the die was cast: another institution had been created, and for the next five years, together or separately (but best when together) they worked their way through films of just about every classic stage operetta – *Rose Marie* (1935), *Maytime* (1937), *The Firefly* (1937), *Balalaika* (1939), *The New Moon* (1940), *Bitter Sweet* (1940), *The Chocolate Soldier* (1941) – not to mention an original score by Romberg for *The Girl of the Golden West* (1938). Fantastic throwback though it might seem, operetta became big business in the cinema just at the very moment that it had finally succumbed in the theatre, and operetta and musical comedy continued to survive on the screen, side by side, right up to the advent of war.

During all this time, though, since *Love Me Tonight*, in 1932, there had been hardly anything that could be reasonably called a director's musical. The good things in the films Busby Berkeley worked on, or in the Astaire-Rogers musicals, were like cherries in a cherry-cake: self-contained, very little affected by their environment, easily removable. Apart from Mamoulian's Jerome Kern musical *High, Wide and Handsome* (1937) little came along to suggest a different arrangement until the end of the decade and *The Wizard of Oz* (1939). Even that could hardly be called a director's film, since at least two other directors worked on it besides Victor Fleming, the credited director, but it was conceived as a whole, and, moreover, as an original film creation, with a new book, a new score by Harold Arlen and a new star, Judy Garland, to play Dorothy, the girl who gets carried over the rainbow. Also associated with it was a new producer, Arthur Freed, who rapidly went on to make his own niche for himself in film history as the man who, virtually alone, master-minded the whole great MGM renaissance of the musical in the 1940s. He it was who first gave their chances to directors such as Vincente Minnelli, Stanley Donen, Gene Kelly, George Sidney and others, and produced practically every outstanding musical of the next fifteen years.

Before we get very far into the Metro era of the musical, however, a new ideal has appeared. Not necessarily new in the sense that no one had ever before tried to approximate to it, but new rather in the sense that never before had it been produced as a formulated theory of what the musical ought to be. This ideal we might for convenience call that of the integral musical, and the decisive moment in its formulation came with the triumphant opening of the first collaboration between Richard Rodgers and his new lyric-writer, Oscar Hammerstein II, *Oklahoma!* in 1943. The idea, briefly, is that the musical should aspire towards the condition of opera by integrating dialogue, song, dance and incidental music in such a way that the whole work 'flows', each constituent element serving to forward the story in its own fashion, with no obvious breaks, cues for song or excuses to drag in a musical number by its hair. Needless to say, this had been done, on and off, for years. In films Lubitsch and Mamoulian had done it, on stage the Gershwins did it in *Of Thee I Sing* (and won a Pulitzer Prize for it), Cole Porter did it in *Nymph Errant*, Rodgers himself had done it with Hart not so long before in *Pal Joey*. But no one then seemed to feel the need to theorize about it, to argue that this alone was the proper, the only admissible way of putting together a musical. The motive for this hardening of attitudes in, and especially after, *Oklahoma!* seemed to be the somewhat dubious one of a desire to lend dignity to a form which up to then had not always enjoyed it. Oscar Hammerstein II had in fact nurtured for many years the desire to use the form of the musical in such a way as to turn it into a sort of American folk opera. Having come near to doing so once, with Jerome Kern in *Show Boat* back in 1927, he had been seeking ever since an occasion to do the job properly, and in *Oklahoma!* he seemingly found it. But not, some might say, without falling into some falsity of sentiment and phony poetizing which any self-respecting old-fashioned, unintegrated musical would have scorned.

With the further history of Rodgers and Hammerstein in the theatre we need not for the moment concern ourselves, except to observe that the ambitions to turn the musical into folk opera before long, in *The King and I* and *The Sound of Music* especially, brought us back very close to square one, your basic romantic operetta *à la* Victor Herbert. In the cinema, meanwhile, the notion was more

new

fruitful. The idea of a film musical as a unified creation, essentially the work of one man, gave birth to a considerable number of films in the great days of the MGM musical, between, say, *Cabin in the Sky* (1942) and *It's Always Fair Weather* (1955), which stand at the very peak of film achievement, any time, anywhere. What finally defeated this movement in films was apparently nothing inherent, but one of those unaccountable changes of public taste, which brought about a turning-away from musicals in the cinema, unless they were pre-sold re-creations of some theatre show that had been enjoyed. Such a re-creation need not necessarily lack life of its own, as witness the Stanley Donen versions of *The Pajama Game* (1957) and *Damn Yankees* (1958). But more usually they turn out like Fred Zinnemann's *Oklahoma!* (1955), Henry King's *Carousel* (1956), Joshua Logan's *South Pacific* (1958) and *Camelot* (1967), George Cukor's *My Fair Lady* (1964), of which the less said the better.

Then, suddenly, a new burst of interest in the musical, so that from musicals being box-office poison (alleged) we find that they suddenly turned into an unfailing box-office bonanza (equally alleged). The first allegation is difficult to prove or disprove if no films are made to try it, but the second can hardly help being tried and found wanting. Though the film of *West Side Story* in 1961 perhaps remotely got the musical going again, the real jackpot was hit only with *Mary Poppins* (1964) and *The Sound of Music* (1965), both starring Julie Andrews. Other musicals, good, bad and indifferent, have come and gone since then, the only thing they have in common being their sheer magnitude. But apart from the perennial Elvis Presley with his low-budget vehicles the only reliable thing seemed to be Julie Andrews herself. Even with her, the large success of *Thoroughly Modern Millie* (1967) was followed by the relative failure of *Star!* (1968), so who is to know what will happen next? Another star-panacea, no doubt, in the formidably talented shape of Barbra Streisand: long before her first film, *Funny Girl* (1968), had even been shown millions of dollars had already been invested on her in *Hello Dolly!* and *On a Clear Day You Can See Forever*. And if that gamble fails, well, there is always something else, someone else, to pin one's hopes on.

So, there is a rough-and-ready framework for our study of the Hollywood musical. Or rather, all the many, various and contradictory phenomena which sneak in under that one blanket heading. To make some sort of sense of it all, let us approach the subject first of all like a mosaic, adding piece to piece, little by little, in the hope that some coherent and convincing pattern will finally appear. And what better piece to start with than that essential of essentials, the music?

Irene Dunne | High, Wide and Handsome

Say It With Music

Of all the vital contributors to the musical film, it is probably the composer who has been least fairly treated. Generally he is the last to be mentioned in reviews, if he is mentioned at all, and then often with an off-handedness, if not downright ignorance, which is positively astounding. When Charles Walters's *Easy to Love* opened in London, for instance, one critic noted that it had a new Cole Porter song, but you would never know, as it was substandard and totally unmemorable. The 'new', 'totally unmemorable' song was, of course, the title number 'Easy to love'. Nor, on the whole, has Hollywood done much better by composers in the screen tributes to them: the fictionalized biographies of Gershwin (*Rhapsody in Blue*), Rodgers and Hart (*Words and Music*), Cole Porter (*Night and Day*), Jerome Kern (*Till the Clouds Roll By*), Romberg (*Deep in My Heart*), Kalmar and Ruby (*Three Little Words*), De Sylva, Brown and Henderson (*The Best Things in Life are Free*), and many more, have made little or no sense as stories and have usually done hardly more than wheel on a host of more or less suitable guest stars to sing the better-known numbers of their nominal subjects.

Despite which, on a more practical level, it can hardly be said that Hollywood has failed, at least until the recent past, to make proper use of the great popular composers' talents. Their talents have been used in two ways, direct and indirect. With composers as with other artists, Hollywood tends to have most respect for those who have already proved themselves elsewhere, and that elsewhere for composers is generally Broadway. To begin with, any hit Broadway musical at once becomes a desirable property because pre-sold: millions who have never seen it are presumed at least to have heard the name, and possibly to know the principal tunes. So it is virtually certain that the film rights of a hit stage musical will at once be bought, and quite possible, though far from certain, that once the rights have been bought the film will eventually be made. So important is the selling power of a Broadway success, indeed, that cases are known of a film musical being planned and a Broadway version of it presented first in the hope of providing the supposed guarantee: the film-makers Norman Panama and Melvin Frank, for instance, did precisely this with their musical version of *Li'l Abner* (1959) before filming it.

Thus, indirectly through their stage shows, nearly all the leading popular American composers have had some showing on film, though in the 1930s Hollywood's treatment of the dearly bought successes was sometimes perverse. The film version of the big Rodgers and Hart hit *On Your Toes* (1939), for example, retained the two big dance sequences from the show ('Slaughter on Tenth Avenue' being the better known of them), but dropped all the songs, except as so many spectral shapes on the soundtrack, so that one is faced with the absurd spectacle of hero and heroine going through a succession of cues for song ('Wouldn't it be marvellous if we could go away somewhere together? Somewhere in the country? Far away from other people . . . ') while in the background the band plays the tune of 'There's a small hotel', insistently but to no avail. A lot of other notable musicals were filmed shorn of most of their numbers, or with substitute songs by other composers, so that the first film of Cole Porter's *Anything Goes* (1936) has six of the original songs plus six new numbers by various composers (Hoagy Carmichael, Frederick Hollander, Richard Whiting) instead of Porter himself. Not that the second film version (1956) did much better: that kept six of the original songs, but inserted them in an entirely new story and added three new songs by Cahn and Van Heusen. And sometimes film producers' oddities bring a bonus. Vernon Duke was not available to write additional songs for the 1942 film version of his stage hit *Cabin in the Sky*, so the equally distinguished Harold Arlen obliged with three new songs, including one of his best, 'Happiness is just a thing called Joe'. When Columbia, who owned the film rights of the novel *My Sister Eileen*, wanted to film the hit stage musical version of it, *Wonderful Town* (music by Leonard Bernstein), they considered the asking price for the score too high, so went ahead with a completely new score, by Jule Styne, which (in my opinion at least) is far superior to the rather more pretentious Bernstein original.

Nevertheless, the scores of successful stage musicals do from time to time reach the screen in fairly complete form, and with additional numbers, if required, usually written these days by the original composers, as Cole Porter did for *Silk Stockings* ('Fated to be mated' and 'The Ritz roll and rock') and Richard Rodgers for *The Sound of Music* ('Confidence' and 'Something good'). It is, no doubt, some little compensation for their seldom being asked to write original scores for films. The last complete original score for a musical by a major composer was Harold Arlen's for the UPA cartoon feature *Gay Purr-ee* in 1961. Before that, we must go back to Cole Porter's *Les Girls* (1957) and *High Society* (1956), and *Gigi* (1958) by Lerner and Loewe. And there were the three songs Ira Gershwin put together from the Gershwin archive for Billy Wilder's *Kiss Me Stupid* (1965), which was a nice idea, but only a tiny, marginal exception to the rule. For the heyday of the original film score we have to go back to the late 1930s and the early 1940s. It was during this time that most of the great American song-writers were lured, briefly or long-term, to Hollywood, and that some of them did some of their best work there.

That they did so has not always been recognized. For one thing, there has always been a tendency, as much in Hollywood as anywhere else, to look down on songs written for Hollywood as inferior to those written for Broadway. Broadway success has always been 'legitimate' in a way that Hollywood has not. And sometimes chance has served to confirm this prejudice. Vincent Youmans's first full score for a Hollywood film, *Flying Down to Rio* (1933) came after a couple of Broadway flops and proved to be his last work, since the rest of his life was spent fighting the illness which finally killed him in 1946. Rodgers and Hart, early arrivals in the sound cinema and triumphant collaborators on Mamoulian's early masterpiece *Love Me Tonight* (1932) soon found themselves shackled to unsuitable assignments (which nonetheless produced a number of memorable songs) and in sore need of re-establishing their position on Broadway, which they did with *Jumbo* (1935), after which they looked back to Hollywood only briefly and incidentally. Jerome Kern settled in Hollywood definitively only when discouraged by his failure to come up with a Broadway hit after *Roberta* in 1933, and continued throughout the rest of his life to look longingly towards the stage, though in vain. Gershwin preserved his position by making only brief forays to Hollywood between his stage shows; Cole Porter steered clear until 1936, when he had an experience on *Born to Dance* which decided him to avoid any too deep involvement with the cinema and judge each suggested assignment strictly on its own merits (but all the same, *Born to Dance* produced, among other songs, 'Easy to love', 'I've got you under my skin' and 'It's de-lovely' which, dropped from the film, found a happy home in his next stage show, *Red, Hot and Blue*). The next year he was back in Hollywood to write, of all things, a new score for the screen version of *Rosalie*, a 1928 stage success composed half by Gershwin and half by Romberg!

Despite these often muddled and inauspicious circumstances nearly all the major American popular composers have been able to make an important contribution to the Hollywood musical – or, to look at it another way, Hollywood has been able to make an important contribution to their overall careers, from all points of view, financial and artistic. At least one of the great names, Irving Berlin, has indeed done his very best work for the cinema, to which he devoted his energies exclusively between his stage success *As Thousands Cheer* (1933) and his next Broadway show *Louisiana Purchase* (1940). Berlin, born in 1888, was nearly the oldest of a group of great song-writers born in the last few years of the nineteenth century, and musically he was one of the most conservative. Though his first all-time success, 'Alexander's ragtime band' had revolutionized the popular music scene by really selling as nothing else had done the idea of ragtime, he was fundamentally most at home with very straightforward romantic ballads or comedy songs which would have offered little which was musically unfamiliar to the average Victorian drawing-room. And as Berlin always wrote his own lyrics (indeed, he was a lyric-writer before he was a composer), he never found himself being pushed slightly out of his natural bent by the challenge of having to match some other lyric-writer's verbal intricacies with similarly complex music.

And, of course, the almost infallible common touch of Berlin's melodies, which could produce songs like 'White Christmas', 'Easter parade', 'How deep is the ocean', 'All alone', 'Remember' and 'God bless America' (not to mention 'Always', his wedding present to his second wife, estimated on royalty earnings to date to be the most valuable wedding present on record), is not to be sneered at. On the other hand, it is not finally at all as exciting as the special gifts of, say, Cole Porter, Jerome Kern or George Gershwin: the melodies are unforgettable, certainly, but somehow ordinary, and the words are adequate but little more. Moments of real musical ingenuity, like the contrapuntal combination of two distinct themes in the 'You're just in love' number of *Call Me Madam*, are infrequent, and Berlin really has no personal style one can put one's finger on. He takes colour so completely from his surroundings – witness the folksy homespun of *Annie Get Your Gun*, much closer to Rodgers and Hammerstein than to his other work – or from the performers he has in mind – surely it was the presence of Ethel Waters in the cast of *As Thousands Cheer* that inspired the memorable but totally uncharacteristic 'Harlem on my mind' and 'Suppertime'.

Similarly, three of his most compelling scores were written for Fred Astaire films: *Top Hat* (1935), *Follow the Fleet* (1935) and *Carefree* (1938). For Fred Astaire is one of those performers with a style so personal to himself that composers seem to find themselves automatically adopting or exploiting its salient features as a positive source of inspiration. The essence of Astaire's singing style is that he sings as he dances – lightly, swiftly, with a buoyancy which seems to defy gravity and chafe at confinement to the ground. Think, for example, of the melodic line of 'Cheek to cheek': the way it plunges *in medias res* with 'Heaven, I'm in heaven', the voice being forced to cherish the words lovingly,

26

as perhaps only Fred Astaire's can really do, before tripping lightly, almost breathlessly up to the high point of 'and my heart beats so that I can hardly *speak*'. Equally, Astaire the dancer, and the necessity for the songs to be also music he can dance to, encourage rhythmic subtlety of a kind not elsewhere to be much observed in Berlin's songs. The overall structure of songs like 'Isn't this a lovely day to be caught in the rain' from *Top Hat*, or 'I'm putting all my eggs in one basket' from *Follow the Fleet*, or 'Change partners' from *Carefree*, with their collocation of sections in different rhythms, their restless changes of key, reflects the demands of Astaire the dancer as much as it exploits the particular qualities of Astaire the singer.

Not that Berlin's golden period in the 1930s was entirely confined to his contribution to Fred Astaire films. 'Puttin' on the Ritz', which we nowadays think of as an Astaire number par excellence, was originally written for Harry Richman in the 1930 film of the same name. Dick Powell's bland, even tenor inspires two first-rate songs in *On The Avenue* (1937): 'I've got my love to keep me warm' and 'This year's kisses'. *Second Fiddle* in 1939 saw Berlin agreeably spoofing himself and 'Cheek to cheek' in 'Back to back', and matching Sonja Henie's grace on skates with a classic tango-bolero 'When winter comes'. But, perhaps unfortunately, this period of dedication to the cinema was only temporary. In 1940 Berlin was back on Broadway with *Louisiana Purchase*, after which came his long involvement with the American war effort and the big patriotic army show *This is the Army*. His only film score during this time was for *Holiday Inn* (1942), which provides a perfect demonstration of his sensitivity, over-sensitivity even, to the performer he was writing any particular song for. The two stars were Bing Crosby and Fred Astaire, and while Astaire inspired Berlin to a characteristically spry and limber song, 'You're easy to dance with', Crosby, the 'old groaner' whose slack, easy-going delivery nearly always seems to bring out the worst in song-writers, produced only 'White Christmas' and 'Be careful, it's my heart'. 'Only' may seem an unlikely word to apply to the best-selling, biggest money-making popular song of all time, but whatever the sentimental and other merits of 'White Christmas', I would gladly swap a hundred of them for one more classic Berlin-Astaire number.

For the most part thereafter Berlin's involvement with the cinema has been in a poorish screen version of *Annie Get Your Gun* (1949), a respectable but unexciting screen version of *Call Me Madam* (1953) and a succession of old-and-new mixtures like *Easter Parade* (1948), where new predominated and one more great Astaire number, 'It only happens when I dance with you', made its appearance, and *White Christmas* (1954) and *There's No Business Like Show Business* (1954), where old predominated. Plus, for the record, the title song to the Marlon Brando vehicle *Sayonara* (1957), which at least informed the world decisively what the Japanese for 'goodbye' was. Looking back over Berlin's whole career, there is one thing, I think, that no critic could tax him with, and that is lack of variety. What one might complain of, though, is lack of that underlying consistency beneath apparent diversity which marks the really great composer, in the popular song quite as much as in any other field of musical endeavour. It is hard to like all the numerous Irving Berlins which have been presented to us through the years, impossible to dislike them all: one can only take one's pick. For me the best Irving Berlin is Fred Astaire's Irving Berlin, and it is in the Fred Astaire films he has written for that Berlin has made his distinctive, indeed irreplaceable contribution to the film musical. Perhaps the elegance and sophistication of Astaire's screen personality has proved a unique challenge to Berlin's native directness and simplicity: perhaps it is as native to the chameleon Berlin as 'God bless America' or 'White Christmas'. But in the five films which allowed Berlin to write for Astaire, they somehow between them produced something individual and in its way definitive: a perfect amalgam of the singer and the song, the song and the dance, the dance and the film.

Jerome Kern, three years Berlin's senior, also devoted himself largely, and at last completely, to the cinema for a significant period of his life. But unlike Berlin, he really settled to the cinema only when he was considered (perhaps prematurely) washed up elsewhere, particularly on Broadway, which remained for him, as for most American popular composers, the Mecca and enduring dream. Berlin always had his Broadway success, and his proven ability to repeat it, as a threat to hold over Hollywood, something to give him bargaining power to pick and choose what films he would do. Kern, despite his tremendous reputation on the strength of his many big Broadway hits of the 1920s – *Sally* (1920), *Sunny* (1925), *Show Boat* (1927), *Sweet Adeline* (1929) – and even more recently with *The Cat and the Fiddle* (1931), *Music in the Air* (1932) and *Roberta* (1933), did not have this reassurance, and as three successive attempts at a major stage comeback failed – *Three Sisters* flopped in London in 1934, *Gentlemen Unafraid* closed on tour in 1938, and *Very Warm for May* flopped on Broadway in 1939 – his bargaining power in Hollywood grew less and less.

Hence, no doubt, the very curious and sometimes very undistinguished list of films to which he contributed scores as the 1930s turned into the 1940s. The nadir, I suppose, was an Abbott and Costello vehicle, *One Night in the Tropics*, in 1940 – for which, nevertheless, he wrote one of his most attractive love-songs, 'Remind me (not to find you so attractive)', sung by Allan Jones. And yet, throughout the last twelve years of his life, Kern continued to develop musically, and to write great songs even in the most unpromising circumstances. Of course, when he came to Hollywood, he had, like Berlin, a formidable string of successes to his credit and a well-established reputation. But also, he had something Berlin hadn't: a formed and personal musical style. Thus while Berlin always tends to take colour from his surroundings, hitting instinctively on a style for each assignment, each particular performer he has to deal with, Kern tends always to be, if not the dominant partner, at least a fixed point, an important individual voice to be taken into account in the sum total of any film to which he contributed.

It is instructive, for example, to compare his collaborations with Fred Astaire and Irving Berlin's collaborations. Astaire works a decisive influence on Berlin; but when Kern writes for Astaire in *Swing Time* (1936) and in *You Were Never Lovelier* (1942), no doubt Astaire's particular qualities play some part in the precise shape the songs take, but it is not radical: the songs remain decidedly Kern songs, using Astaire's talents and his personal quirks of style rather than being used by them. The aspect of Astaire's screen personality Kern concentrates on to the greatest effect is his wry lyricism, the fundamental romanticism always held in check and moderated by the cool, disenchanted surface. The tone is set at once in 'I won't dance', a song added to the screen version of *Roberta* in 1934. And yet curiously enough, that song, so indissolubly linked with Fred Astaire, was not originally written for him at all, but composed the previous year for the London show *Three Sisters*, which never reached America. But, while a typical Kern tune, it proved also to be an ideal Astaire tune, and Kern was not slow to take the hint when writing specifically for Astaire later on. The scores of both the later films are remembered primarily for three pieces of elegant lyricism each from Fred Astaire: in *Swing Time* the somewhat unwilling, backhanded 'A Fine Romance' and 'Pick yourself up' as well as the surrender of 'The way you look tonight'; in *You Were Never Lovelier* the more relaxed title song, 'Dearly beloved' and 'I'm old-fashioned', as befitted the more un-ashamedly romantic relationship in the film between Astaire and his leading lady, Rita Hayworth, as opposed to the rather spiky, competitive relationship he always had with Ginger Rogers.

Musically these songs nearly all exploited Astaire's remarkable control of phrasing, his ability to sing, with what appears to be very little voice, long and difficult lines which require a smooth, even, legato delivery and perfect breath control: the smooth sweep of 'Lovely – Never ever change Keep that breathless charm, Won't you please arrange it 'cos I love you, Just the way you look tonight', the unexpected twists and turns of 'Make a note And you can quote me, Honour bright: You were never lovelier Than you are tonight'. These songs, like the Berlin songs, seem to belong to no one but Astaire, and yet they are a different Astaire: they are as much Astaire's tribute to Kern as Kern's tribute to him. *— most radued scarce written for*

Apart from the Astaire scores much of what Kern wrote for the cinema belonged to his slightly folksy side – the aspect of his talents represented most famously by *Show Boat* and which would no doubt have been brought to the fore again had he lived to write the score for *Annie Get Your Gun*, as planned. For many this is the least appealing part of Kern's œuvre, though I suppose, given its persistence, one can hardly say the least characteristic. Certainly it is true that the Deanna Durbin vehicle *Can't Help Singing* (1945) did not bring out the best in Kern, and the score contains no out-and-out classic song, though 'More and more' is a typically subtle and intricate romantic song, and the title number is energetic and catchy. *High, Wide and Handsome* (1937) is a score in much the same vein, though distinguished by two of Kern's most memorable romantic songs, 'Can I forget you' and 'The folks who live on the hill', and in general rendered more striking by the outstanding qualities of the film itself, beautifully directed by Rouben Mamoulian. Kern's last two scores for musicals, *Cover Girl* (1944) and *Centennial Summer* (1946) are well up to standard: *Cover Girl* indeed shows almost the full range of latter-day Kern, with the bouncy 'Make way for tomorrow', the classic love-song 'Long ago and far away', the deceptively intricate 'Sure thing', 'Put me to the test', which provided Gene Kelly with a perfect pretext for a big dance number (interesting, incidentally, to compare Kern's musical response to Kelly's brashly extrovert personality with the songs he wrote for Astaire in similar situations), and some skilful pastiche of early Kern for the flashback scenes. *Centennial Summer* was composed as a score just before Kern's death, the lyrics by Oscar Hammerstein II and Leo Robin being written later: it also contains treasures like 'In love in vain' and 'All through the day', though the film itself is only another not very distinguished addition to a line of

turn-of-the-century Americana which was at that time fashionable in Hollywood.

Compared with Berlin and Kern, Cole Porter (1892–1964) made, over his whole career, a far more marginal contribution to the cinema, at any rate in terms of work directly inspired by the medium. Again, he composed for Fred Astaire (as did all the 'big five' except Richard Rodgers, as well as several slightly lesser composers such as Youmans and Arlen). One of the first Astaire-Rogers films, *The Gay Divorcee* (1934), was based on the Porter stage show written for Fred Astaire (as *The Gay Divorce* – the original title was considered too *risqué* for the cinema), but in fact only one song by Porter, 'Night and day', survived into the film. However, in 1941 Porter was able to repair the omission with *You'll Never Get Rich*, one of the two films in which Astaire found a congenial dancing partner in Rita Hayworth. This is not exactly one of Porter's more memorable scores, despite the topicalities of 'Shooting the works for Uncle Sam' and 'Boogie woogie barcarolle', but it does contain one excellent Astaire song, 'So near and yet so far', and a striking but very uncharacteristic Porter piece, 'Since I kissed my baby goodbye', which with its flowing melismatic lines might almost be mistaken for the work of Harold Arlen.

Oddly enough, in films generally Cole Porter has been represented not so much by his famous 'point'-lyric songs, the witty lists like 'You're the top', 'Anything goes', 'Just one of those things', 'Let's do it' and so on, as by his more deeply felt romantic numbers. Perhaps this is because on the whole film audiences have been felt to be less sophisticated than theatre audiences, but whatever the reason only one notable list-song by Porter has originated in a film, 'They all fall in love', which was his very first film work, written in 1929

Ginger Rogers and Fred Astaire/Follow the Fleet

for a Gertrude Lawrence film, a flop called *The Battle for Paris*. Otherwise, the wittier Porter songs have all been tested on the stage first before being let on to the screen. The more romantic streak in Porter has been given full scope, however, to produce original film songs like 'I've got you under my skin' and 'Easy to love', both written for the Eleanor Powell film *Born to Dance* in 1936, 'In the still of the night' (*Rosalie*, 1937), 'I concentrate on you' (*Broadway Melody of 1940*) and 'You'd be so nice to come home to' (*Something to Shout About*, 1943).

The scores of Porter's later films, *The Pirate* (1947), *High Society* (1956) and *Les Girls* (1957) are much more coherent as total works than those for the earlier films, which were rather arbitrary collections of separate songs, but honesty forces one to admit that the last two are decidedly inferior, the only song of any interest from either being 'Mind if I make love to you?' from *High Society* ('True love', from the same film, is, despite its enormous popularity, one of the least distinguished songs Porter ever wrote). *The Pirate*, on the other hand, has never really had its due – not even from Porter himself, who was apparently unhappy with the whole project and consequently disliked the result. But 'Love of my life', with its insidious Latin lilt, is one of the best love-songs he ever wrote, 'You can do no wrong' is almost equally memorable, and 'Niña', as well as providing the material for a virtuoso dance number from Gene Kelly, is full of typical Cole Porter pieces of outrageously ingenious rhyming (the title word is rhymed with 'seen ya', 'schizophrenia', 'neurasthenia', 'mean ya', among others). Moreover, the whole score is an excellent example of a composer finding a uniform style, still recognizably his own yet coloured by the particular locale of the story (a Caribbean island) and the personalities of the stars (Judy Garland and Gene Kelly), to give his score an overall unity within the larger unity of the film as a whole.

Another major composer in whose career the cinema played a fairly minor role was George Gershwin (1898–1937). And in his fairly brief and peripheral entanglement with the cinema the ubiquitous Fred Astaire again figures prominently: he starred in the two most important of the four films for which Gershwin wrote original scores, *Damsel in Distress* and *Shall We Dance?* (both 1937). Of the other two, *Delicious* (1931), a Janet Gaynor vehicle, inspired no lastingly important song, and Gershwin's work on *Goldwyn Follies* (1938) was interrupted by his death, though two of his songs for it, among the very last he wrote, 'Love walked in' and 'Love is here to stay', are by any standards among his best, full of melodic inventiveness and delicate but unsentimental feeling. Gershwin's gift *par excellence* was as a melodist: in the writing of tunes which stay in the memory – and which stay as a whole, not merely the striking turn or phrase – Gershwin has few rivals in music, popular or 'serious'. Many popular song tunes stay in the mind because of their beginning or their end or both, without any very convincing or functional middle section: Gershwin's great triumph is his gift for through-composing his tunes, so that there is a constant, cumulative progression from beginning to end (his only rival in this area is Harold Arlen). This gift is particularly in evidence in the scores he wrote for the two Astaire films: the very titles of the songs are enough – *Damsel in Distress* had 'A foggy day', 'Nice work if you can get it', 'Things are looking up' and 'I can't be bothered now'; *Shall We Dance?*, the title number, 'Let's call the whole thing off', 'They can't take that away from me', 'They all laughed', 'Slap that bass' and 'I've got beginner's luck'. And of course Gershwin's stage scores for many shows have been transferred in whole or in part to the screen, most notably perhaps in *Girl Crazy* (1943) and *Funny Face* (1956). Not to mention two scores assembled from his sketches after his death, *The Shocking Miss Pilgrim* (1947) and *Kiss Me Stupid* (1965), which are still clearly far from scraping the bottom of the barrel.

Richard Rodgers (b. 1902) is unique among the five major popular composers in that he never wrote a score for Fred Astaire. Indeed, Fred Astaire has never even sung a song by Rodgers and Hart – let alone Rodgers and Hammerstein – on screen. No doubt this is absolutely by chance: at the time of the early Astaire-Rogers musicals for RKO, Rodgers and Hart were working mostly at Paramount, and later on, when Rodgers and Hart did two original musicals at RKO, *Too Many Girls* (1940) and *They Met in Argentina* (1941), Fred Astaire had moved elsewhere; while after the death of Lorenz Hart, Rodgers wrote only the original film score, *State Fair* (1945, extended for the remake of 1962), and it is hard anyway to imagine Fred Astaire fitting into the richly sentimental world of Rodgers and Hammerstein. Not so much so that of Rodgers and Hart, for surely his wry, offhanded, backhandedly romantic delivery would have perfectly matched Hart's brittle, witty and often, underneath, deeply felt lyrics. As it was, Rodgers and Hart scores often decorated rather unlikely films, to be sung by rather unlikely singers – notably *Mississippi* (1935), a period comedy set on a river paddle-steamer, in which Bing Crosby introduced, among others, 'Easy to remember', a classic example of a song's first appearance providing no reliable guide to how it is meant to be sung, since there it turns up as a fast waltz!

However, among the early films for which Rodgers and Hart wrote scores are two of unusual interest, Rouben Mamoulian's *Love Me Tonight* (1932) and Lewis Milestone's *Hallelujah, I'm a Bum* (1933). Both of them made extensive use of rhymed dialogue (written by Lorenz Hart) outside the strictly musical numbers, suggesting a new unified convention in the musical somewhat similar to an opera's alternation of recitation and aria. *Hallelujah, I'm a Bum*, written for Al Jolson when his film popularity was already in decline, was not a success, and the team's later assignments were more conventional, but *Love Me Tonight* remains a classic example of the singer's musical as opposed to the dancer's musical – a film in which, with virtually no actual dance, all the action, musical and non-musical, seems to be choreographed. The most famous instance of this is no doubt the long introduction in which the sights and sounds of Paris at dawn – brushes sweeping the pavement, carpets being beaten, workmen going about their various jobs – are all 'orchestrated' as a prelude to the first number, the 'Song of Paree'. That, in its turn, leads almost immediately into 'Isn't it romantic?' by way of a short dialogue exchange which slips into rhymed doggerel and then into the song, sung by Maurice Chevalier as he fits a customer with a wedding suit (he is a tailor), then on through a series of rapid transitions via a taxi-driver, a composer who takes down the melody, a group of soldiers who turn it into a march, a gypsy fiddler who plays a soulful variation of it in a forest encampment, and so to Jeanette MacDonald, our heroine, singing the song on a balcony of the chateau where she lives, and thus into the next scene and the next phase of the plot. All achieved by, for once, an ideal combination of director, sensitive to every slightest nuance of rhythm, every subtlety of word and music, and a composer and lyric-writer bringing together a matching mixture of wit, sophistication and unashamed romanticism.

Rodgers and Hart were destined never to hit on such a winning combination again in the cinema. Adaptations from their stage work like *Evergreen* (1935) – despite the enchanting presence of Jessie Matthews – *On Your Toes* (1939) – inexplicably made with none of the songs actually sung – *Babes in Arms* (1939) – with only three numbers surviving – and the more agreeable, and rather more respectful *The Boys from Syracuse* (1940), *Too Many Girls* (1940) and *I Married an Angel* (1942), adapted as an unlikely final vehicle for Jeanette MacDonald and Nelson Eddy, were at best routine, while their screen originals, such as *Fools for Scandal* (1938) and *They Met in Argentina* (1941) were quite unremarkable. It cannot even be said that very distinguished songs have emerged from these undistinguished films: the classic score of *Love Me Tonight* ('Isn't it romantic?', ever since a virtual trademark of Paramount films, 'Lover', 'Mimi' and the title number) apart, I suppose the only really important Rodgers and Hart songs to make their first appearance in films have been 'Easy to remember' in *Mississippi* and 'You're nearer', written for the screen version of *Too Many Girls*, while ironically that great Rodgers and Hart standard 'Blue Moon' was dropped, in various forms, from two films before it emerged as we know it, unconnected with any film, as a piece of independent sheet music.

While the cinema could have exploited the combined talents of Richard Rodgers and Lorenz Hart but on the whole did not, it had very little chance to exploit directly the combination of Richard Rodgers and Oscar Hammerstein II. From their very first collaboration on *Oklahoma!* (1943), the succession of

Broadway successes continued almost unbroken up to Hammerstein's death in 1961. One or two of the shows admittedly fell slightly below smash-hit rating, and *Allegro* (1947) and *Pipe Dream* (1955) actually lost money. But against that must be set the triumphant series of *Carousel* (1945), *South Pacific* (1949), *The King and I* (1951), *Flower Drum Song* (1958) and *The Sound of Music* (1959). Rodgers and Hammerstein's six big hits have all been filmed, seldom with any special distinction. *The King and I* (1956) at least exploited the charm of the original rather well, if without any special cinematic flair; *Oklahoma!* (1955) was assigned experimentally to a non-musical director, Fred Zinnemann, who made rather a hash of it; *The Sound of Music* (1965), also directed by a director not normally associated with musicals, Robert Wise, unexpectedly turned out to be a great improvement on the stage original, really making the most of the cinema's potential for exciting locations and finding the ideal star in Julie Andrews.

Amid all this stage activity it is hardly surprising that Rodgers and Hammerstein found little time to write directly for the screen. The only full score they wrote was for the 1945 version of *State Fair*, which is right in the centre of their *Oklahoma!* range, full of folksy poetry and pawky rustic humour. As a matter of fact, 'It might as well be spring' is one of Oscar Hammerstein's most telling 'poetic' lyrics, and the score in general has a good deal of charm, though the film itself is not very distinguished. Nor, for that matter, is the 1962 remake, for which Richard Rodgers alone wrote several new songs, showing an occasional sprightly turn of wit on his own account as a lyricist – notably in 'Never say no to a man', a surprising reversion to the style of his earlier collaborations with Lorenz Hart.

Apart from the 'big five' of American popular composers, Gershwin, Porter, Kern, Rodgers and Berlin, there are of course many others whose contributions to stage and film have been valuable and individual. Only one of them, Harry Warren (b. 1893) has devoted his talents almost entirely to the screen, and has continued writing songs for film musicals right through from the early days of *Forty-Second Street*, *Footlight Parade* and *The Golddiggers of 1933* to comparatively recent times. Many of his early film songs are unforgettable, in themselves, and for the sequences with which they are associated: one has only to mention 'We're in the money', 'My forgotten man', 'Petting in the park', 'Honeymoon Hotel', 'Forty-Second Street', 'Shuffle off to Buffalo', 'I'll string along with you' or 'I only have eyes for you' to summon up visions of hundreds of chorines arranged in Busby Berkeley's inimitable style, vast and extravagant transformation scenes, and pre-Hayes Code details of sexual fantasy.

But for me his most interesting period is the 1940s, when he wrote, it seems, nearly all the great songs everyone remembers: the Glenn Miller numbers like 'Chattanooga choo-choo', 'I know why and so do you', 'At last', 'Kalamazoo', and 'Serenade in blue' from *Sun Valley Serenade* (1941) and *Orchestra Wives* (1942), the famous Carmen Miranda pieces from *That Night in Rio* (1941) and *Weekend in Havana* (1941), like 'I-yi-yi-yi-yi-yi like you very much' and 'When I love I love', and others equally redolent of their period, such as 'You'll never know' (*Hello, Frisco, Hello*, 1943), 'No love, no nothing' (*The Gang's All Here*, 1943), 'The more I see you' (*Diamond Horseshoe*, 1945) and 'This heart of mine' (*Ziegfeld Follies*, 1944), as well as a couple of first-rate integrated dramatic scores inspired

Fred Astaire and Lucille Bremer/The Ziegfeld Follies

Arlene Dahl/Three Little Words (above); Sheree North/The Best Things in Life Are Free (below)

by the small town and western Americana of *The Harvey Girls* (which produced another of his classic train songs, 'The Atcheson, Topeka and the Sante Fe', 1945) and *Summer Holiday* (1946). It seems, indeed, that Harry Warren's particular gift throughout his career has been to capture exactly the tone of the time: the brash materialism of the early thirties, the era of 'We're in the money', which yet held just beneath the surface the harsh observation of 'My forgotten man'; the wry romanticism of the forties, allied with the bounce of Glenn Miller's solid beat and the quest for escape via the exoticism of Carmen Miranda, or the appeal to the supposedly settled, reliable values of an earlier time in *Summer Holiday*.

In contrast, Harold Arlen (b. 1905) is remarkably timeless and consistent in his style. His songs are immediately recognizable, by the length and intricacy of their melodic lines. Whereas most Tin Pan Alley songs, even the best, are fairly strictly constructed on a symmetrical scheme of 32 bars in all, Arlen's are as a rule long and asymmetrical, like 'The Man that got away' from *A Star is Born* (1954), which runs to 62 bars, or 'One for my baby' from *The Sky's the Limit* (1943), a 48-bar melody which Fred Astaire considers 'one of the best pieces of material that was written specially for me', or 'That old black magic' from *Star Spangled Rhythm* (1942). Undoubtedly an important element in Arlen's style is the synagogue music of his childhood – his father was a cantor and he sang in a synagogue choir – with its oriental slides and melismata. Probably this has contributed more than the Negro blues, though some of Arlen's most famous songs, like 'Blues in the night' (*Blues in the Night*, 1941), observe at least the outer forms of the traditional blues, and some of his best scores are conceived as evocations of Negro folk-lore, notably the stage shows *St Louis Woman* (1946) and *House of Flowers* (1954). But whatever the constituents of his style, Arlen remains extraordinarily consistent in it throughout his career, which includes numerous film scores of distinction, among them, as well as those already mentioned, *The Wizard of Oz* (1939), *Cabin in the Sky* (additional numbers, 1942), *Casbah* (1948) and *The Farmer Takes a Wife* (1952). Writing for Fred Astaire he succeeds in using the positive qualities of a Fred Astaire performance without being any the less himself; writing for Bing Crosby, whose mannerisms have so often swamped even the strongest personalities among song-writers, he has been able to come up with songs as fresh and personal as 'Let's take the long way home' (*Here Come the Waves*, 1945) and 'Love and learn' (*The Country Girl*, 1954) while yet observing the requirement of accommodating his style to Crosby's.

Other composers have figured in Hollywood rather as birds of passage. The only Hollywood scores of Vernon Duke (1903–69) were for *April in Paris* (1952), the title of which came from one of his most famous songs, and the Virginia Mayo vehicle *She's Working Her Way Through College* (1951). Kurt Weill (1900–50) saw three of his stage musicals, *Knickerbocker Holiday* (1938), *Lady in the Dark* (1941) and *One Touch of Venus* (1943) filmed with more or less (usually less) fidelity in 1944, 1943 and 1948, contributing new songs to the first and third, and wrote an original score for *Where Do We Go From Here?* (1945), noted chiefly for its very extensive *scena*, almost a miniature opera, 'The Niña, the Pinta and the Santa Maria'. Vincent Youmans (1898–1946) wrote his first and last original film score for the first Astaire-Rogers film, *Flying Down to Rio* (1933) – this being, as it

37

Fred Astaire/Follow the Fleet

happened, the last thing he wrote before being attacked by the tuberculosis which made him an invalid for the rest of his life. Arthur Schwartz (b. 1900) has written scores for a number of films, the most successful *Thank Your Lucky Stars* (1943), in which Bette Davis, of all people, introduced 'They're either too young or too old', and *The Time, the Place and the Girl* (1946), which brought to our attention the urgent question 'What do they do on a rainy night in Rio?' Neither score shows Schwartz at his best: for that we must look to *The Band Wagon* (1953) which included such great songs as 'Dancing in the dark', 'You and the night and the music', 'Something to remember you by' and 'I guess I'll have to change my plan', none of them in fact originally written for films, plus one good new song, 'Entertainment'.

Evidently, the contribution which all these composers, and a host of lesser musical figures, make to the musical film is of enormous importance. A first-rate score does not, admittedly, guarantee a first-rate film, but a first-rate film can hardly hope to emerge without a first-rate score. And there are films which deserve notice largely on the strength of their scores – films like *Centennial Summer*, or *The Farmer Takes a Wife*, or *Where Do We Go From Here?* which are otherwise fairly undistinguished, yet find their due place in a study of the musical because of the contribution made to them by Kern, Arlen and Weill respectively. Nor, of course, does the importance of music in the musical stop with the bare contribution of the composer: so much of the atmosphere of a characteristic musical from MGM's heyday comes from the art of the arrangers, the orchestrators – like Conrad Salinger, Johnny Green, Lennie Hayton – who gave them the lush MGM sound, devised the magical transitions like the lead-in to the 'You were meant for me' number on an empty stage which is gradually filled with lights and colour in *Singin' in the Rain*, or elaborated the ballet sequence in *An American in Paris* from Gershwin's tone-poem. But, like everything else in the musical, the strictly musical element is only one constituent, interdependent with many others in the elaboration of a complex whole. Not least important among these others is the performer: already such names as Fred Astaire, Dick Powell and Bing Crosby have forced their way into this section, and now perhaps it is time to turn the spotlight directly on to the musical star and his (or her) contribution to the finished product.

Ginger Rogers and Fred Astaire/Follow the Fleet

They Sing, They Dance

After the second great flowering of the film musical, at MGM in the late 1940s and early 1950s, had begun to fade a little, there was a curious vogue, not yet altogether past, for making musicals with stars who could not sing or dance. The point of this was always a trifle obscure. Obviously there is something to be said for providing a dancer who cannot sing – Rita Hayworth, Cyd Charisse and Vera-Ellen are famous examples – with someone else's voice on the soundtrack, and the marriage is often very successful. There may even be some sense in casting an actor who is right for the role in a dramatic musical, like Deborah Kerr in *The King and I*, and dubbing the singing voice (or in that case merely borrowing a few high notes). But films like *Guys and Dolls*, with the leading roles played by Marlon Brando and Jean Simmons doing their best to sing for themselves, or *West Side Story*, with Richard Beymer and Natalie Wood as non-dancing principals who cannot sing either and have to be dubbed, remain a complete mystery. At least they make one thing clear: the musical star, singer or dancer or preferably both, is a necessary part of the film musical, and ideally should be so from the very inception of the idea, so that his or her particular talents and qualities have a fertilizing effect on all the others involved – song-writers, script-writers, choreographers, directors. That, anyway, is how nearly all the best film musicals have been made, and to date no compelling evidence has emerged to make us change our ideas.

It would be possible, though much too one-sided, to write a history of the musical exclusively in terms of its stars and their shaping influence. The pattern was not, it is true, established right at the start. The early Lubitsch musicals remain primarily the creations of their director, though clearly stars such as Jeanette MacDonald and (especially) Maurice Chevalier have given a strong, specific colouring to the original conception; the early Warner Brothers musicals depend for their effect mainly on spectacle devised by Busby Berkeley, the charms of Dick Powell and Ruby Keeler remaining fairly incidental. But with the first appearance of Fred Astaire and Ginger Rogers together on film the pattern of the musical as very importantly, if not exclusively, a star vehicle was established. In the Astaire-Rogers musicals the stars are the conditioning factor in the overall conception and the most important single factor in the film's success. The films are nearly all directed by competent journeyman directors like Mark Sandrich and H. C. Potter, the only arguable exception being *Swing Time*, directed by George Stevens, who could hardly at that time count as a major talent. The choreography is by Hermes Pan, whose main claim to fame is as Astaire's constant aide through the years. The songs, admittedly, are by really topline composers, but almost all of them, as we have seen, were written as material for the exercise of the couple's special talents, Fred Astaire's especially. In consequence, everyone remembers the films as Astaire-Rogers films rather than as the work of their directors or writers. And everyone, for once, is right: that is exactly what they are.

It is hard to define exactly what is the special magic of the Fred Astaire-Ginger Rogers combination, that quality which keeps their names linked in the public mind thirty years after the partnership broke up and despite everything they have both done separately since. Evidently, they danced well individually and together, but Fred Astaire has had other good, probably in some cases better, dancers as partners since, without any of them supplanting Ginger Rogers in the public memory as the perfect partner. Evidently, there was some sort of chemistry between them on screen, but one would be hard put to it to say what, especially considering that in real life they had little in common, hardly knew each other away from the set, and did not even, if one may read between the gentlemanly lines of Fred Astaire's autobiography *Steps in Time*, like each other very much. Perhaps best is the old formulation of their complementary qualities: she brought him down to earth, he made her look like a lady. But whatever it was, it worked.

Significantly, nearly all the plots (such as they are) of Astaire-Rogers musicals are based on antagonism rather than passion between the two principals. As a rule, Astaire falls for Rogers right away, but she for some reason doesn't like him or determines to play hard to get. She devises various ways of evading, needling and generally irritating him until towards the end she succumbs and permits a happy-ever-after ending. The plots, anyway, are only slender excuses for a string of musical numbers demonstrating the talents of the central duo – and as a rule only them: the spectacular concerted pieces are nearly always saved for the grand finale. The dialogue scenes between the numbers are sometimes rather perfunctory, although they can also be amusing in their own right when decorated by such pleasing comic character actors as Edward Everett Horton, Edna May Oliver or Eric Blore. But at least everything in the films – plot, song, dance – has the advantage of embodying a single informing idea: that of the love-hate, romantic-antagonistic relationship between the principals.

This is true even in detail of their dance routines together. Again and again the movements play out a small drama of difficult courtship: Astaire beckons and enchants with mesmeric hand-movements, Rogers almost against her will drifts into concerted movement with him, a close rapport is established and grows in romantic abandon to a point where everything stops, a pause, a decisive moment after which Rogers breaks away, or tries to, and has to be summoned back. A constant hallmark of Astaire-Rogers routines is this magical pause, a hesitation seemingly on the brink of something attractive yet frightening – of really deep emotion perhaps. Of course, all this is by no means without precedent: I suppose it goes back to the grander routines of exhibition dancers like Vernon and Irene Castle, whose fictionalized life Astaire and Rogers portrayed in the last of their RKO films. In these the accepted convention was that the woman is slightly haughty, imperious, and the man ardently wooing; he more or less has to make her dance with him. But in Astaire-Rogers routines the convention is revivified and given some real dramatic *raison d'être* in terms of the established personalities of the dancers and their established relationship with each other.

Also, from film to film and even within one film the variations on the formula are infinite. Nor does it cover quite all the routines shared by the couple. Most of the films have, as well as the obviously romantic numbers like 'Night and day' in *The Gay Divorcee*, 'Cheek to cheek' in *Top Hat*, 'Let's face the music and dance' in *Follow the Fleet*, 'Never gonna dance' in *Swing Time*, 'Change partners' in *Carefree* and so on, one or two jolly, asexual, almost brother-and-sister routines harking back to the days when Fred Astaire danced with his sister Adele. Most of the specialty dances introduced in the films come into this category: the Carioca in *Flying Down to Rio*, the Continental in *The Gay Divorcee*, the Piccolino

43

in *Top Hat*, the Yam in *Carefree*. But there are others, like 'I'm putting all my eggs in one basket' from *Follow the Fleet*, a brightly competitive piece of mad showmanship, or 'Let's call the whole thing off' from *Shall We Dance?* which climaxes with the two of them on roller-skates in the park, energetically reflecting on their evident unsuitability for each other.

In all these numbers the contribution of Ginger Rogers is very important, and has tended in recent years to be underestimated, perhaps because in her later years as an actress she lost much of the real freshness and charm she then had. But the Astaire-Rogers films are very much duets, founded securely on the two of them and their relationship with each other. True, in nearly all of them Fred Astaire does have his solo spots, those special exhibition pieces in which he dresses while singing and dancing ('Looking for a needle in a haystack', *The Gay Divorcee*), drills a group of young sailors to the rhythm of his taps ('I'd rather lead a band', *Follow the Fleet*), plays golf to music (*Carefree*), dances with three silhouettes of himself ('Bojangles of Harlem', *Swing Time*) or shoots down a row of doubles with his cane (the title number in *Top Hat*), but these obvious cadenzas do not much detract from the coherence of the films' overall concept as dual creations, built round two contrasted personalities.

Nor do the interludes for other performers, such as that in *The Gay Divorcee* which allows Edward Everett Horton and Betty Grable to introduce 'Let's k-nock k-nees' or Fred Astaire to dance a rather ill-advised balletic version of 'They can't take that away from me' in *Shall We Dance?* with a ballerina called Harriet Hoctor. The only exception to the rule, I suppose, is the long-unseeable *Roberta* (out of circulation following a 1952 remake, *Lovely to Look At*), since there Irene Dunne plays a leading role and gets to sing important songs, including 'Smoke gets in your eyes', a more serious and obviously 'singer's' number than the Astaire-Rogers usual. Fred Astaire and Ginger Rogers are both, of course, singers to some extent (and in his case a very individual vocal stylist), but they both remain primarily dancers. The Astaire-Rogers films are above all dancers' musicals, musicals founded on the dance. Every musical number drifts sooner or later into dance, and dance somehow seeps into sections which are not primarily musical set-pieces, like the dog-walking in *Shall We Dance?* (to special music by Gershwin) or the sequence in the very first film, *Flying Down to Rio*, in which Astaire is constantly distracted from his instruction of a dancing class by the sound of a band seeping through from a nearby rehearsal room. As far as the camera is concerned, the directors of the films are mostly content to record the dances with a minimum of fuss, keeping everything as mobile as possible and the joining of shot to shot as invisible as they can, by cutting on a turn, moving from close-up to medium shot at the precise moment that song goes over into dance and so on. Again, everything is subordinated to the stars, and to showing them off as clearly and directly as possible.

Only very rarely, in the big concerted numbers which usually come at the end of the film, do the directors have a chance to do anything independently impressive. In the Carioca number of *Flying Down to Rio* Thornton Freeland seems to have learnt a thing or two from Busby Berkeley in his use of striking single shots like that, very Berkeleyish, which shows a series of couples whisking past a camera practically at floor level, or the spectacular crane shot which gradually reveals to us a whole large set packed with dancers. The finale, too, with its aerial choreography, has more than a hint of Berkeley fireworks. Mark Sandrich does a few similar tricks in the Continental finale of *The Gay Divorcee*, moving his camera in among the dancers at crucial moments to involve us in a whirl of movement. But the later films have less and less of this conscious directorial display. The little story-ballet which climaxes *Follow the Fleet* with Astaire as an unsuccessful gambler (somewhat suggestive of the 'This heart of mine' number in *Ziegfeld Follies*) and Rogers as his love-at-first-sight, is effectively reduced to an intimate romantic *pas de deux*; the 'Bojangles of Harlem' number in *Swing Time*, for all its lavish set and large chorus, makes its main point as an Astaire solo.

This intimate approach to the musical, as first and foremost a star vehicle, gradually came to predominate in the later 1930s. Many Warner Brothers First National films were conceived on a more intimate scale than the great Busby Berkeley machines; and something like *Twenty Million Sweethearts* (1934), bringing together Dick Powell, star of most of the big Berkeley films, and Ginger Rogers, was much closer in conception to the Astaire-Rogers films, exploiting its stars' personalities in much the same way, than it was to the previous Dick Powell spectaculars. True, Busby Berkeley went right on, transferring his centre of activities around 1939 from Warners to MGM, where already his lessons had been learnt by directors eager to show off the spectacular tap-dancing of Eleanor Powell to best advantage in films like *Broadway Melody of 1936*, *George White's Scandals* and *Broadway Melody of 1940*. But the most important and memorable series of musicals in the 1930s, the only real rivals to the Astaire-Rogers films in popularity, also featured a romantic twosome, and were also based very closely and deliberately on the exploitation of their stars' talents: the Jeanette Mac-Donald-Nelson Eddy films being made at MGM between 1935 and 1942.

These, no doubt, appealed to a very different public from the Astaire-Rogers films, or at least to the same public for very different reasons. To begin with, they are very decidedly singers' musicals, with little or no dancing and what there is very incidental. Also, to revert to the distinction between the musical comedy tradition and the operetta tradition in American show-business, these belong entirely and unashamedly to the operetta side, while the Astaire-Rogers films are just as unmistakably modern musical comedies (significantly, the single foray of Jeanette MacDonald and Nelson Eddy into modern musical comedy, *I Married an Angel*, based on the Rodgers and Hart stage show, was not a success and proved to be their last film together). Nor was there much play on a contrast or conflict of personalities between the two, as in the Astaire-Rogers films. Nelson Eddy was called on mainly to look solid and handsome, sing lustily, and provide a succession of colourless hero-figures. Jeanette MacDonald was required by the plots every now and then to be playful and coquettish: 'For I'm falling in love with . . . someone' she might avow with a coy shake of her head, withholding a full admission with an admonitory wag of her finger, but no one was ever in any real doubt where her heart lay. But essentially they were a twosome, combined against fate, which might finally bring them cheerfully together or occasionally, as in *Maytime*, hold them tearfully apart.

In keeping with this relative lack of interest in the personalities of the performers, the films had far more elaborate, well-upholstered plots than those of the Astaire-Rogers films. Not, naturally, very sensible plots – that would be asking too much of any operetta – but reasonably complicated plots involving pirates and princesses in disguise (*Naughty Marietta*), a mountie who always gets his man, even if it involves misunderstandings with the prisoner's sister (*Rose Marie*), the sufferings of a singing star torn between two men, one of whom she loves and the other to whom she owes everything (*Maytime*), a bandit who poses as a soldier to win the heroine's love and has it returned (*The Girl of the Golden West*), or the agonies and ecstasies of artistic life in old Vienna (*Bitter Sweet*). And all the films are packed with songs, on the reasonable assumption that audiences have come in principally to hear Jeanette MacDonald and Nelson Eddy singing famous romantic duets full in each other's faces or soulful solo ballads gazing into the far blue yonder. This may not exactly make for lively film-making, and (again like the Astaire-Rogers films), the MacDonald-Eddy vehicles were never given to first-string directors, but left to the competent craftsmen of the studio like W. S. Van Dyke and Robert Z. Leonard.

Finally the films stand or fall on the appeal of the scores, the singing of the two principals, and the personality of Jeanette MacDonald (Nelson Eddy either never had any or was called on to be fantastically self-effacing). As far as the first two are concerned, there is little room for argument. The singing of the stars, both with some operatic training, may well fall short by the highest operatic standards, but is more than adequate for the songs they are here called on to sing. The scores too, drawn from the works of old reliables like Victor Herbert (*Naughty Marietta*), Romberg (*Maytime*, *The New Moon*), Friml (*Rose Marie*) and Noël Coward at his most romantic (*Bitter Sweet*) – and, for the one complete original score, *The Girl of the Golden West*, turning again to the veteran Romberg – may not be to everyone's taste, but they are undoubtedly among the best examples of their genre. And who, after all, is musically so strait-laced that he does not leave the cinema after *Naughty Marietta* humming, even if shamefacedly, 'Ah sweet mystery of life', or 'I'm falling in love with someone', or find 'Wanting you', 'Lover come back to me' or 'Stout-hearted men' hanging round uncomfortably after *The New Moon*?

The real problem, though, remains the personality of Jeanette MacDonald. Certainly the films are not built round it, as the Astaire-Rogers musicals are built round their personalities, and yet it is an important, even a crucial, constituent of their charm. I should, perhaps, confess at once to my own weakness for Jeanette MacDonald: she always seems to me a much funnier lady (quite deliberately) than most people are willing to suppose. That she could be a charming comedienne is at once evident if we look back from her romantic heyday at MGM to such effervescent delights as the Lubitsch/Cukor *One Hour With You* or Mamoulian's *Love Me Tonight*. Even in the silliest of her later vehicles an impish touch of fun peeps out from time to time, forbidding us to think that she took them quite so seriously, or expects us to take them quite so seriously, as one might at first glance suppose. Admittedly, some of the later films require her to behave kittenishly at an age when such goings-on are beginning to seem a trifle incongruous. But that seems to be much more the fault of the scripts than of the star. And when all is said and done, the films do have a continuing vitality of their own: idiotic maybe, acceptable these days primarily as camp. But their sincerity and their air of conviction even during the worst absurdities, allow them still to be not only funny, but also at times curiously, unaccountably touching.

Apart from the two great series, there were of course many more or less interesting individual musicals in the 1930s, and many stars of temporary or lasting
importance for whom they were devised. Sometimes rather unexpected people.
We tend to forget, for example, that all the films of Mae West and the Marx
Brothers were musicals to the extent of at least including a few songs, and often
much more importantly than that. Mae West hymning the advantages of 'A
guy what takes his time' in *She Done Him Wrong* or explaining that 'They call me
Sister Honky-Tonk' in *I'm No Angel* is, of course, *sui generis*, coolly combining
outrageous sexual innuendo with a sharp sense of how funny the whole business
is. The Marx Brothers' films always featured solo spots in which Chico could
play the piano and Harpo the harp, both to rather embarrassing effect, I always
thought. But as well the earlier films such as *Animal Crackers* (1930), *Horse
Feathers* (1932), and *Duck Soup* (1933), the first of them based on a stage show in
which the brothers had starred, were equipped with full-scale concerted musical
numbers integral to the story, the most ambitious of them being composed by
Kalmar and Ruby in a surprisingly close imitation of the Gilbert and Sullivan
manner. The later films, from *A Night at the Opera* (1935) on, were cursed with
irrelevant production values, generally including colourless singing romantic
leads like Allan Jones, Kitty Carlisle, Kenny Baker and Florence Rice, but at
least *At the Circus* contained an excellent score by Harold Arlen and offered

Groucho one of his best musical outings in the saga of 'Lydia, the tattooed lady'.

To this gallery of oddities should perhaps be added the films of Shirley Temple, who was already a star introducing one of her most famous numbers at the ripe old age of six, when she sang 'On the good ship Lollipop' in *Bright Eyes* (1935). Most of Shirley Temple's films as a child star were musicals or semi-musicals, and there are at least a few numbers in them that everyone remembers, like 'Animal crackers in my soup' from *Curly Top*. Insufferably cute she may have been with her piping voice, her dimples, her golden curls and her inexhaustible determination to do everybody good, but seen again at this distance of time she proves also to have surprising, and quite genuine, star quality. You may loathe her, but you stay watching. Also, it must be admitted that she did have some real talents. There is no hint in her films of a performance put together in the cutting room: she really acted long and sometimes quite difficult scenes herself, in continuous takes, and she could clearly master quite long and complicated musical routines, like the 'Catfish ball' number she sings and dances with Buddy Ebsen in *Captain January*. Sometimes, too, she was teamed with real grown-up stars like Alice Faye in *Poor Little Rich Girl* and *Stowaway*, and managed to hold her own – though admittedly children are notorious scene-stealers anyway. She didn't quite manage to steal the scene from Bill Robinson, the great Negro tap-dancer in *The Little Colonel* and *Rebecca of Sunnybrook Farm*; in the first he even managed to do a bit of his famous staircase dance, in company with the little colonel herself.

But much more in the mainstream of the musical than any of these were the singing stars like Bing Crosby and Alice Faye, joined as the decade wore on by such slightly newer faces as Betty Grable and real youngsters like Judy Garland, Deanna Durbin and Mickey Rooney. The earliest of these to assume front-rank importance, and by far the longest to keep it, was Bing Crosby. There is no doubt that Crosby has been a major musical star. He has appeared in many films, including quite a few of some real distinction, one way or another. He has a very distinctive singing style and screen personality which have helped to form many who have come after him, and not only obvious imitators. He has had songs written for him by a few really important composers and a lot of lesser figures. In fact, I have great difficulty in deciding why I do not really like him and tend to find films devised to accommodate his personality tiresome. I suppose it is a largely subjective reaction to that personality; but also it is a feeling that 'the old groaner' is much too slow and expansive in style not to bring out the worst in most of the composers who have written for him and weigh down his vehicles to a steady but often wearisome jog-trot.

However, there is no arguing with his real and consistent popularity from the very early thirties (his actual film début was in *King of Jazz* in 1930 as one of a trio called the Rhythm Boys, and two years later he was already a top musical star) through at least to *High Society* in 1956. He has rarely made a film with a really notable director, and when he has, as in *Anything Goes* (1936 version, Lewis Milestone), *Emperor Waltz* (Billy Wilder), *Riding High* and *Here Comes the Groom* (Frank Capra), it has been with directors who, whatever their achievements elsewhere, are hardly associated with musicals or at their best with them. Probably the best combinations from this point of view have been with Charles Walters, who is at least an established musical director, though *High*

Society is scarcely one of his best, and Leo McCarey, who managed *Going My Way* and *The Bells of St Mary's* as well as could be expected, though neither is among his more appealing works (and neither, incidentally, is really a musical). Nor has Crosby been very often associated with first-rate composers. Apart from the two versions of *Anything Goes*, there are to his credit one Cole Porter score (*High Society*), one Rodgers and Hart (*Mississippi*), two Harold Arlen (*Here Come the Waves* and *The Country Girl*, plus a guest appearance in *Star Spangled Rhythm*), and three Irving Berlin (*Holiday Inn*, *Blue Skies*, *White Christmas*). Considering the number of films he has made, the total is not impressive, and it is probably no accident that the most satisfactory scores overall are those by Irving Berlin, the most straightforward and traditional of all the important American song-writers. It cannot even be said that Crosby brings out the best in his more usual song-writers, Burke and Van Heusen – though of course the jolly, easy-going cracker-barrel philosophy songs for which their Crosby film scores are best remembered, like 'Swinging on a star', 'Aren't you glad you're you?', 'Busy doing nothing', 'Sunshine cake' and 'Life is so peculiar', do certainly reflect very accurately a familiar, and seemingly much-loved, aspect of the Crosby screen character.

What is that screen character? From the very first the hallmarks of Crosby's singing style and his film characterizations have been casualness, effortlessness, even laziness: he gives the impression of being perfectly relaxed, walking through his roles, creeping up to his notes belatedly if at all. It is hardly conceivable that anyone who has worked so hard and constantly for so many years can in fact be like this; but that is another matter. Crosby on the screen is all sleepy charm, a personal embodiment of *laissez-faire*. He can be rather funny when galvanized into it by the saving proximity of Bob Hope, as in the series of 'Road' films in which they vied in various sneaky and dubious fashions for the attentions of Dorothy Lamour. But normally you go to a Bing Crosby film for a nice rest, in the confidence that everything will be smooth, pleasant, undemanding and unchallenging. In this, at least, his films are quite unlike anyone else's, especially in the musical, where normally some bouts of energetic singing or dancing, some moments of spectacle or virtuoso camerawork can be expected. For those that like them, of course, that is what they like. But the long-lasting appeal leaves me totally mystified.

Alice Faye, now, is a very different matter. It would be difficult, and pointless, to maintain that her films were in general any better than Bing Crosby's; indeed, from most points of view they were a good deal worse. But she really had something. Perhaps it was just a quality of personality, perhaps it was that inexplicable incandescence which lights up the screen when a real film star crosses it. The classic demonstration of this is *Alexander's Ragtime Band*, in which the other female star is Ethel Merman. Now no one doubts that Merman is a better singer, a better actress, a more forceful and individual presence, a more dashing technician than Alice Faye. But in this film she can be seen working her heart out to precious little effect (on the stage, no doubt, it would be electrifying, but . . .). And all Alice Faye has to do is come on, open her mouth, and the effect is made, simply, effortlessly; the competition from Ethel Merman is annihilated. A pity, of course, that this gift was not better used in better films. But *In Old Chicago*, *Rose of Washington Square* and *Hello, Frisco, Hello*

vibrate nostalgically in the memory, bringing back a vision of generous curves, trembling bee-stung lips, platinum blonde hair and a throaty, throbbing voice that gave vintage Irving Berlin and Harry Warren a special, distinctive flavour. Alice Faye vanished from films after giving one of her best performances in a completely non-musical role, in Otto Preminger's black thriller *Fallen Angel* (1945), and has reappeared only once more, in the second *State Fair* (1962). But at least for once the new encounter was not a disappointment: one of the most satisfying moments of a not too satisfactory film was Alice Faye's delivery of that sensible piece of parental advice from Richard Rodgers, 'Never say no to a man'.

Already by the time Alice Faye reached the peak of her career the new faces were crowding in, the stars who would dominate the musicals of the 1940s. In 1935 a short called *Every Sunday* (or *Every Sunday Afternoon*, say some) introduced two new little girls, Judy Garland and Deanna Durbin, respectively thirteen and fourteen years old. Betty Grable had been around in films since 1930, but did not achieve top-billing until about 1940 and *Down Argentine Way*. Mickey Rooney had also been in films on and off for some time (he began in silent shorts at the age of four, in 1924), but in 1938 made *Love Finds Andy Hardy*, which began his musical partnership with Judy Garland. In 1941 a newly popular singer named Frank Sinatra made a guest appearance in something called *Las Vegas Nights*. The same year a singer and dancer named Gene Kelly got his big stage break in *Pal Joey*, and the next year made his first film starring with Judy Garland in *For Me and My Gal*. And of course, still and always Fred Astaire, a Fred Astaire separated from Ginger Rogers after *The Story of Vernon and Irene Castle* in 1939, ready to find a new partner and perhaps a new image, to keep us happily entertained right through the forties, the fifties and the sixties.

The very list of names gives some idea of the variety of the period. It also at

*El Brendel, Shirley Temple and Jimmy Durante/
Little Miss Broadway*

once suggests something of the melting-pot into which the musical as a genre was about to be flung. We have been talking up to now in this chapter of musicals which were first and foremost star vehicles, to such an extent that they could hardly be fruitfully considered as anything else. During the 1940s a new kind of musical is to emerge: one in which the contribution of a star, the importance of having the *right* star, is hardly any less, but in which it remains just that, a contribution, only one part of a far more complex and varied whole. The revolution, of course, was not complete: revolutions never are. While some studios – MGM especially – forged ahead to develop the new kind of musical to its maximum potential, others, like Universal and Twentieth Century-Fox, remained true to the older model. And so, willy-nilly, did their stars.

At Universal Deanna Durbin was musical queen, and though her reign extends as far as 1947 her films always remained vehicles as that term was understood in the 1930s. At Fox Betty Grable was the big attraction, and her films remained simple, energetic, totally unpretentious (and therefore often rather appealing) star vehicles well into the 1950s – up to *The Farmer Takes a Wife* in 1952 to be precise, after which CinemaScope put an end to that kind of film for ever. Even as late as 1948 Warner Brothers may be observed *starting* a cycle of this sort by building up a new star, an erstwhile band-singer called Doris Day, in exactly the same way that Alice Faye and Betty Grable were built up in their day, with a string of low-budget musicals built to specification for the current discovery. Other stars belong partly to the old and partly to the new scheme of things. Judy Garland and Mickey Rooney began in carefully tailored vehicles, but were able to make the transition; so, triumphantly, was Fred Astaire. Gene Kelly belongs entirely to the new order, being, indeed, one of the people most importantly responsible for bringing it about. Frank Sinatra never caught on in attempts to provide him with an old-style vehicle; his first screen successes are in films which require him to contribute as a member of a team rather than be shown off on his own.

Before we come on to this new order, and to consider exactly how it used stars and what they contributed to it, it will be easiest to dispose first of those stars who had nothing to do with it but went on in their own sweet way. There were those who were just wheeled on to do their particular specialities: Sonja Henie to skate to music, which she did pleasingly in *Second Fiddle*, *Sun Valley Serenade* and many more; Esther Williams to swim to music in a succession of simple musical spectacles like *Bathing Beauty*, *Neptune's Daughter* and *Easy to Love* (as well as appearing in one film of more significance, *Take Me Out to the Ball Game*); Carmen Miranda to cavort in a tutti-frutti hat and deliver a variety of incomprehensible but vivacious Latin American numbers in films with titles like *That Night in Rio*, *Weekend in Havana* and *Copacabana*.

Much more important, both in her standing and in the number and variety of the films she made, was Deanna Durbin, a very big star in her time and now, rather unfairly, almost forgotten. She started as a teenage star, not exactly a child but not yet, in those innocent far-off days, considered ready at fifteen or so for her first screen kiss (she got that at the age of eighteen, with much publicity, in *First Love*). What she had going for her were a pretty face and a ringing coloratura soprano voice (or at least what passed for such in Hollywood). She also had a pert sense of humour which enabled her to make her way painlessly through such edgy *rapprochements* of pop music and the classics as *One Hundred Men and a Girl* (the 'hundred men' were a symphony orchestra of unemployed musicians conducted by Stokowski) and *Mad about Music*, in which she brings her film-star mother together with a conveniently available composer. Once grown up she branched out into light comedy with music (*His Butler's Sister*), comedy thriller with music (*Lady on a Train*) and even classy psychological drama (*Christmas Holiday*, which had her married to unbalanced murderer Gene Kelly, mournfully intoning 'Always' in a sleazy club).

From most points of view her most interesting later film was *Can't Help Singing* (1945), which had a Jerome Kern score of lively Americana (in the wake of *Oklahoma!*) and some agreeable photography of Western locations; *Up in Central Park* (1948), her very last, also had much to recommend it, including a handful of songs from Sigmund Romberg's last big stage hit. But probably she is best remembered, when remembered at all, by the early films in which she was directed by Henry Koster, who seemed to have the knack of bringing out her infectious gaiety and not letting us take her too seriously. At that time, in *Three Smart Girls*, *Mad about Music* and others, she was really delightful, not quite like anyone else. It was a change of fashion in screen musicals, and more specifically in screen music, which hastened her retirement from films; if her studio had given her some chance to become a contributor to a team musical rather than a star enshrined it might have been less premature, or less final.

Betty Grable is a less complicated case. She rose slowly from the ranks (she can be recognized, just, as a chorine in some early Goldwyn musicals) to a featured role in *The Gay Divorcee*, to top feminine billing in *Man About Town* (1939), and top stardom in *Down Argentine Way* the following year. She was pleasant looking, in an athletic, clean-cut, all-American way; she could sing a bit, and dance rather more than a bit; and she had famous legs, which in the leg-conscious 1940s were the real key to millions of sex-starved servicemen's hearts. In consequence she was the forces' pin-up *par excellence*, and quite a sex star, though only in the nicest possible, no-nonsense way. There is not much more to say about her or her films, except perhaps that they both improved as time wore on: two with Dan Dailey, *Mother Wore Tights* (1947) and *Call Me Mister* (1951) were admirable examples of formula film-making that worked; *The Farmer Takes a Wife* has a fine Arlen score and a lot of imaginative open-air shooting in the musical numbers staged on a reconstruction of the Erie Canal in the 1850s (except for an unaccountable stylistic aberration in the finale, shot for no apparent reason on a formalized studio set); and *Three for the Show* (1955), after she had left Fox, is a very creditable attempt at a new-style team musical, with a couple of really inventive big numbers in which Betty Grable plays an effective part rather than being given an exclusive lease on the limelight.

Doris Day in her early films bade fair to fill the place about to be vacated by Betty Grable as the screen's lady you would most like to share a picnic hamper with. Blonde and freckled, healthy to a fault, brimming over with energy and eupeptic charm, she sang her way lustily through a succession of shaky vehicles like *Romance on the High Seas* (*It's Magic* in Britain), *Tea for Two*, *I'll See You in My Dreams* and *Lullaby of Broadway*, and rose to rather better things in *April in Paris* (by virtue mainly of its Vernon Duke score) and *Calamity Jane*, an original

Betty Hutton/Annie Get Your Gun

Gene Kelly/Summer Stock (above); Cyd Charisse and Gene Kelly/Singin' in the Rain

screen musical which makes sense in all departments, if achieving outright distinction in none. But by this time (1953) the simple star vehicle had at last really had its day, and Doris Day's hankering to be an actress took her further and further away from the musical: indeed since *Calamity Jane* she has made the odd drama with songs, like *Love Me or Leave Me*, but only two real musicals, *The Pajama Game* and *Jumbo*, both excellent examples of directors' musicals, concerted works in which Doris Day was used, and used well. But these films are exceptions in her career, and in general as a musical star she remains a weird anachronism, the last important passenger on a type of vehicle which had lost most of its momentum before she even boarded it.

Though Deanna Durbin and Judy Garland began their film careers in the same short, they seem to belong to different worlds, and their common point of departure is an unthinkable incongruity. Not only are their personalities and talents so utterly unlike, but the sorts of film they represent are totally different. Judy Garland's career began conventionally enough. After her discovery she was given small parts in which to show her ability and attract attention, such as the episode in *Broadway Melody of 1938* in which she sang a version of 'You made me love you' addressed as an expression of half-funny, half-touching calf-love to Clark Gable. Supporting roles built up her following, and in *Love Finds Andy Hardy* came the next stage, a romantic partnership with Mickey Rooney, the studio's principal young male star, which was then carried through a series of conventional enough vehicles in which they would be young and eager and bursting with vitality, and of course sing and dance to everyone's heart's content: *Babes in Arms*, *Strike Up the Band*, *Life Begins for Andy Hardy*, *Babes on Broadway*. At this stage it might look very much like a career along traditional lines, were it not for one thing, one film: *The Wizard of Oz* in 1939.

The Wizard of Oz had one significant thing about it, behind the scenes: the presence of Arthur Freed as associate producer. His name was to be associated with just about every musical of MGM's heyday and already *The Wizard of Oz* showed the shape of things to come. Though the role of Dorothy in L. Frank Baum's classic fantasy was an excellent one for Judy Garland (and permitted her to sing her most famous song of all, 'Over the rainbow'), the film was certainly not a conventional vehicle for her. It was written and composed as an original musical with a strong story and no stop-plot numbers, but every bit of music and dance closely integrated into the overall conception – as very few musicals had been since the early thirties and the first sound films of Lubitsch and Mamoulian.

If *The Wizard of Oz* pointed the way the screen musical was to go, *Meet Me in St Louis* (1944) confirmed it once and for all; and again Judy Garland was the star. Here the film as a whole came before the exploitation of any individual star. True, Garland did get to sing two more of her most famous songs, 'The boy next door' and 'The trolley song', but she was called on mainly to give a performance, to form part of the meticulously re-created turn-of-the-century world in which the story took place. Which she did immaculately: who can forget the scene of her comforting her little sister (Margaret O'Brien, the Shirley Temple of the forties, but here used as an actress in a way that Shirley Temple never was) about the prospect of their leaving their home town, St Louis, in which she sings with her special intensity of feeling and wide-eyed innocence

'Have yourself a merry little Christmas', aureoled in gentle, glowing light? From *Meet Me in St Louis* on, a succession of new-style musicals like *Ziegfeld Follies*, *The Harvey Girls*, *The Pirate*, *Easter Parade* and *Summer Stock*, not all of them equally successful, but all quite decidedly films with stars rather than films for stars. After *Summer Stock* (called *If You Feel Like Singing* in Britain) the over-publicized series of crack-ups and comebacks, the first comeback, in *A Star is Born* (1954), being her best performance and her best film of all, petering out into a series of unremarkable non-singing performances, television and night-club disasters, and an ever more complete retreat into the hysterical adulation of a small but passionate camp-following who took, it seemed, a morbid delight in suffering along with Judy. And in 1969 that death so many times anticipated.

Looking back over her life as a whole it is easier than it was to separate the true qualities from the hysteria. Pages of purple prose have been dedicated to her gifts and her personality, much of it over-keyed to the point where a reaction is bound to set in. But putting that aside and going back to the films, as the only things which really last and really matter, one can begin to distinguish certain clear lines of development. In her very earliest films what she had going for her above all was her freshness and unspoilt quality. The cine-camera, as has often been said, photographs people's insides much more vividly than their outsides; or as Orson Welles once put it, it is perfectly simple: the camera either loves you or it does not. Judy Garland it loved from the start. To watch her pouring out her devotion to 'Dear Mr Gable' and the great, big, wonderful world of the movies was a pleasure almost voyeuristic – one was seeing not the simulated emotions of an accomplished actress, but the real emotions of a real, vulnerable girl, passing across her face and colouring her urgent, husky, sometimes plangent voice with an extraordinary absence of self-consciousness, self-censorship. She gave everything, without reserve – which was to be her triumph and finally, perhaps, her tragedy.

The button nose, the limpid brown eyes, the generous mouth which seemed ever alive, betraying every slightest flurry of feeling, the twinkling toes – she was, in the early 1940s, a better dancer than she has ever been given credit for – and above all the inimitable Garland voice, were constant and familiar, fit to be exploited, if the studio had so chosen, if the logic of film history had permitted it, in a succession of routine vehicles. Most of her early films with Mickey Rooney are not quite that – but to my mind they are not very successful either. Busby Berkeley, who directed *Babes in Arms*, *Strike Up the Band* and *Babes on Broadway*, and staged the musical numbers for *Girl Crazy*, was really by temperament an abstractionist, much better at complicated logistics than at dealing simply with human beings. The result is that all the films seem now over-directed and over-elaborate, swamping their real assets – which are the personalities and talents of the two stars – in lavish but unappealing 'production values'. *Babes in Arms* is about the worst, with a really terrible supporting cast, a lot of heavy humour (very much of its period, but very tiresome) about jazzing the classics and showing what good Joes even the stuffiest classical musicians can be, and big numbers like the title song and the minstrel show which overwhelm the stars completely. It is a great relief when Judy Garland and we are permitted a brief respite while she sings 'I cried for you' quite straightforwardly into the camera, with throbbing sincerity and absolutely without tricks. *Strike Up the Band* and *Babes on Broadway* mark some improvement, mainly because they give more breathing-space to the stars, especially the second, which includes the famous cavalcade of show-business greats, imitated with varying success by Garland and Rooney, and the charming Harburg-Lane number 'How about you?' put over with the right kind of light-hearted relaxation by the two of them.

But all these films really point to the dilemma of the musical in this time of transition. They are not simple, unassuming vehicles for the stars, like the Astaire-Rogers films. But neither are they coherent dramatic musicals, using the stars as an ingredient only. They would probably be better if they were either, but as they are they stand as an uneasy compromise. Admittedly, the brash attractions of Mickey Rooney are not too badly served by them, but the more complex talents of Judy Garland need more time and more space to flower properly: they also need more to engage the actress in her. For looking at these early films one can see at once, quite clearly, that she is above all a musical actress. She is splendid in *The Wizard of Oz* and *Meet Me in St Louis*, as in the non-singing *Under the Clock* and finally in *A Star is Born*, because she has a part to play, a believable character which engages her whole being. In other words, she was ready for the new kind of integrated, dramatic musical even before it was ready for her.

As an actress, a star in terms the new musical could use and understand, there were many things she could do and certain things she could not. It seems to me, for example, that her attempts at being flip and sophisticated are always disastrous: 'A great lady has an interview' in *Ziegfeld Follies*, for example, where she is called on to parody a film-star interview with the press and to resume the plot of a film about a famous female inventor (shades of Greer Garson in *Madame Curie*) who at length presented the world with . . . the safety pin, is embarrassingly heavy-handed and lacking in edge or sparkle. It seems to me also that her talents just do not marry well with those of Fred Astaire in *Easter Parade* (or properly speaking one should perhaps rather say that his do not marry with hers, since his role in the film was originally meant for Gene Kelly): his light, glancing, indirect style seems curiously at odds with her head-on, all-out approach to her material: he is the model of reticence and suggestion, she the great exponent of direct emotion. Nor, to complete this catalogue of deficiencies, is she very well suited to the artificial comedy of *The Pirate*; neither for that matter is Gene Kelly, and the film's most important – indeed, virtually its only – defect is that a play originally devised for Alfred Lunt and Lynn Fontanne has not been sufficiently made over for stars whose talents are much broader and less sophisticated, whose weapon is the broadsword rather than the rapier.

As it turns out, though, this is a small mark against an otherwise almost perfect film, with a vintage score by Cole Porter, inventive choreography by Kelly and Robert Alton, stylish sets by Jack Martin Smith, and Minnelli directing at his virtuoso peak. Even if Garland is not, by the very highest standards, quite right, she is good enough to support and round out the film as a musical/dramatic entity, which is what is required of her; and she is given two great romantic numbers perfectly calculated to fit the tone and timbre of her voice, the almost tearful intensity of her sustained high notes, in 'Love of

my life' and 'You can do no wrong'. There is too, about this performance, the slightly hectic quality which is to be seen in all her films from now on. Having reached glowing maturity in *For Me and My Gal*, a straightforward vehicle by Busby Berkeley for herself and Gene Kelly, and radiating uncomplicated happiness in *Meet Me in St Louis*, *The Harvey Girls* and her numbers in *Words and Music* ('Johnny one-note' and a nostalgic return engagement with Rooney in 'I wish I were in love again'), she is now obviously under the pressure which will eventually send her right out of pictures and into a breakdown, but which meanwhile gives her performances, at best, a nervy electric quality as though she is walking an emotional tightrope. In fact, though not yet too obviously for comfort, there was creeping into her relationship with her audiences a sort of sado-masochistic element: they went along at least partly to share her suffering, vicariously, and in the fear (which was perhaps also the hope) of seeing her break down.

That was disturbing, and makes a film like *Summer Stock* a very uncomfortable experience, largely because these overtones are quite out of place in what was meant to be a simple, enjoyable putting-on-a-show musical with no pretensions to deeper meaning. But of course with a true artist – and Judy Garland was certainly that – everything can become grist to the mill. In her comeback film, *A Star is Born*, all these elements of her by now very complicated screen personality and presence are magisterially used. The role requires the actress to

Jack Haley, Ray Bolger, Judy Garland and Bert Lahr/The Wizard of Oz

mature rapidly from a complete innocent, emotionally unawakened, to a big star, a perfect show-biz professional, and a woman who has suffered, gone to the brink of breakdown and fought her way back. In the earlier, non-musical, version Janet Gaynor was fine as the innocent, less compelling as the high-powered pro. This time round Judy Garland is a trifle less likely as – in effect – her own younger self, but once, fairly early on, she makes the big time and starts right in suffering she is superb. Not only does the film include the last great song she introduced – Arlen's 'The man that got away' – designed as no other ever was to use her mature singing talents to their fullest extent, but the film's three big numbers, 'Born in a trunk', 'Someone at last' and 'Lose that long face' all bring every ounce of talent, every fragment of irrelevant association that every real star carries round wherever he or she may go, into full play.

'Born in a trunk' is the most conventional of the three (it was an after-thought, not directed by George Cukor, who directed the rest of the film): it reprises a long show-business career not so unlike Judy Garland's own in rough outline – like enough, at any rate, to permit an intensifying identification of performer and material – in a long episodic *scena*: beginning and ending with Garland in black-face singing 'Swanee'. More dramatically remarkable was 'Lose that long face', a number being staged for a film 'Vicki Lester' is appearing in while her husband 'Norman Maine' (James Mason), a fading star, has just been put in a sanatorium for alcoholics. In a long-held single shot, Vicki, in her dressing-room, struggles painfully to keep hold on her almost unbearable emotions as she talks about her husband, the hopelessness of their relationship, and her continuing love for him. Then, all at once, in a textbook example of the Pagliacci syndrome, she is called on stage and has to go straight into an energetic rendering of this cheery, optimistic number, so utterly at odds with her true feelings. Unfortunately, in the version of the film normally distributed in the States, and the fullest shown in Britain, the number itself, an essential constituent of this effect, was removed, though some idea of the intention can be gained by playing it on the soundtrack recording with the immediately preceding scene in mind.

And finally, there is the 'Someone at last' sequence, a brilliant invention in which Vicki/Garland tries to cheer up her husband by miming single-handed a whole involved production number, full of shifts in style, locale and mood, to a playback of its already recorded soundtrack, in their own living-room. Here the various elements – the sending-up of the idiotic routine, which we are left mind-bogglingly to imagine, the virtuosity of the whole thing as a solo performance, the dramatic significance of the number with the complicated motivation behind its performance in this way, at this point – are all held breathtakingly in suspension, so that it all becomes rich with unexpected resonances and ironies. Garland was never better, and never more fully, satisfyingly Garland than here: the ultimate in ideal confrontations of the singer with the song, the actress with the role, the person with her screen apotheosis.

Gene Kelly, somehow, seems in retrospect a sort of complementary figure to Garland in the new MGM musical of the 1940s. It was, no doubt, sheer accident that his first screen role should be in *For Me and My Gal*, which also gave Garland her first really grown-up part. But it was no accident, their personalities matching so well, that the teaming should be repeated effectively in *The Pirate* and *Summer Stock* (and would have been in *Easter Parade* if an accident had not kept Kelly out of it). Kelly's qualities on screen are a close male equivalent to Garland's, without of course that neurotic edge which latterly became an important Garland characteristic. Kelly is the open, confident, brash (but not insensitively so, like Mickey Rooney), straightforward American male, with a smile on his face for the whole human race, as one of his songs puts it. The personality is not altogether appealing. There is sometimes the feeling that the charm is laid on a little too thickly, that the smile is a trifle synthetic, that, to quote another of his songs, he may like himself just a fraction too much. The role of Pal Joey, cynical user of rich women, would exploit these hints and undertones perfectly, and it is a thousand pities Kelly did not have the chance to repeat his great stage success on screen. As it is, only Robert Siodmak in *Christmas Holiday* has had the wit to cast him in a slightly sinister role, as a man whose charm, if considerable, is all on the surface; and the result was one of Kelly's best performances.

However, reservations about his screen personality have little to do with the evaluation of his contribution to the Hollywood musical. Creatively, this is maybe more in the fields of choreography and direction than simply as a performer; nor is it easy, or desirable, to separate any one area of activity from the others. But for the moment it is the star personality that principally concerns us. This was established at once on screen in *For Me and My Gal*: somewhat in spite of the story, which required him to be torn by ruthless ambition which disrupts his emotional life and complicates his professional life in vaudeville, he came over as the simple, healthy good guy, as near as an accomplished song-and-dance man could be to the boy next door. In *Thousands Cheer* the following year he smiled his way through the trials of a circus star reduced to the humble role of a private by the advent of war, and got to do his first characteristic solo spot (choreographed, I suppose, by himself) in a send-up romantic duo with a mop to the strains of 'Let me call you sweetheart'. Here we get all the constituents of his style: the bouncy ebullience, the broad humour (always a nudge in the audience's ribs), the slightly over-emphatic masculinity, as well as some of his favourite steps and movements.

Cover Girl is a step further along the road: not, for once, an MGM musical, it is stylistically a curious hotch-potch, featuring a hefty flashback to Edwardian times decorated with old-fashioned spectacle as well as a series of lively, light-weight, thoroughly modern numbers bang up to (1944) date. Kelly is rather well teamed with Rita Hayworth (a star whose early training and considerable abilities as a dancer tend to be forgotten) and Phil Silvers as a trio of show-business hopefuls, striding out along the night streets – in gleeful anticipation of the title number of *On the Town* – exhorting passers-by and the world at large to 'Make way for tomorrow'. He also has another famous solo to himself – or rather a famous *pas de deux*, the 'Alter ego' dance in which he partners himself in double exposure. There are various opinions on this: no less an authority than Busby Berkeley considers it Kelly's greatest single number; others feel that it runs into the same trouble as most screen dance involving evident camera trickery, that, as Richard Griffith remarks in connection with another of Kelly's trick-camera numbers, 'When the pull of gravity is no longer felt by the audience, felt kinesthetically, the achievement of the dancer too is no longer felt, and the drama of the dance goes flat.'

The same could be said of Kelly's duet with a cartoon mouse in *Anchors Aweigh* (1944), but otherwise his extrovert, athletic dance style is seen here to full advantage: he is cast as a sailor on shore leave (he is the brash one, Frank Sinatra the sensitive, girl-shy) whose bold façade cannot long conceal that under it all he is decent, all-American, with a passion for home life, home cooking, and children. The Mexican girl in the 'Mexican hat dance' number is the first of a long line of children used as props in Kelly films, and it must be admitted that he manages them very well, whereas Fred Astaire avoids them and when actually trapped with one in *Easter Parade* shows every sign of deep distrust. *Anchors Aweigh* looks in many respects like a trial run for *On the Town*, which has Kelly and Sinatra, joined by Jules Munshin, doing their sailors ashore bit again. So, in other respects, does *Take Me Out to the Ball Game* (called in Britain *Everybody's Cheering*), in which a similar plot is given a baseball back-ground and staged and choreographed by Kelly and Stanley Donen, under Busby Berkeley's directorial supervision, as a preparation for their first solo outing in *On the Town*.

Meanwhile Kelly's personality had been explored and exploited in other ways. Briefly in *Ziegfeld Follies*, his only screen encounter with Fred Astaire, to do the Gershwin number 'The Babbitt and the bromide', in which the two great dancers regarded each other warily, like gods of dissimilar races. More exhaustively in *The Pirate*, in which he plays a travelling actor mistaken for the famous pirate Black Macoco, and is given ample scope for charming the ladies (in 'Niña', one of Kelly's most extravagantly virtuoso performances, as he shins up drainpipes, climbs over roofs and hops on to balconies in search of yet one more flower of Hollywood/Caribbean beauty), clowning and hamming – with more than one large wink to the audience. And also to good effect in *Living in a Big Way*, one of the most charming of the 'lost' films, the last to be directed by one of Hollywood's great odd-men-out, Gregory La Cava; and in *The Three Musketeers*, a straight version of Dumas which was directed quasi-musically by George Sidney and played almost choreographically by Kelly as D'Artagnan. There were also worrying glimpses of a more pretentious side to Kelly's nature, most noticeably in the ballet he danced and choreographed to 'Slaughter on Tenth Avenue' for the Rodgers and Hart screen biography *Words and Music*. This was a moderately effective variation on the traditional Apache routine, but vitiated by touches of self-conscious poetry which seemed to indicate unappealing areas of as yet unfulfilled ambition in Kelly's nature.

This danger raised its head again, though not I think very damagingly, in *An American in Paris*, which has as well as Kelly being his usual simple, good-hearted self, dancing with the kids in 'I got rhythm', carolling the joys of life in Paree with Georges Guétary to 'S'wonderful' and doing a bit of broad burlesque with the ordinary people in 'By Strauss', a long and elaborate dream ballet set to an expanded version of the Gershwin tone-poem which gave the film its title. This seems to me to work perfectly well in its own terms, with its brightly sophisticated décors by Preston Ames, evoking Paris through the eyes of various artists, Dufy and Lautrec in particular, and its moments of conscious virtuosity in Kelly's choreography and Minnelli's direction, such as the terrifying pursuit of Kelly across the set by a camera swooping down on him like some great bird of prey. But the cultural references, if here kept under control, are ominous: Kelly the ordinary guy shows signs of wanting to become Kelly the poet of screen dance – a role which fits him much less becomingly.

None of that, or not very much, in *Singin' in the Rain*, his next directorial collaboration with Stanley Donen: indeed, for most of the way, as a silent star who has made it from burlesque to classy period drama on screen, he is required effectively to deflate his own claims to be taken too seriously – the very first sequence juxtaposes his improbably la-di-da account of his upbringing and artistic training, designed for the doting millions of radio-listeners, with the considerably less dignified true story of his gradual and often unseemly climb to fame, which we see in counterpoint with his words. Even the big 'Broadway ballet' sequence is mostly sprightly and unpretentious, and its dream interlude, with Kelly and Cyd Charisse cavorting symbolically on a vaguely Daliesque astral plane, is brief and simple enough to be acceptable, despite Kelly's attempts to assume what he believes, or thinks we will believe, to be the proper soulful expression for 'serious dance'.

Brigadoon and *It's Always Fair Weather* add little to our knowledge of Kelly the screen star; though *Brigadoon* does contain, to the words 'there's a smile on my face for the whole human race' (in the song 'Almost like being in love'), perhaps the most terrifying big-screen close-up of his crocodile smile, and *It's Always Fair Weather*, his last collaboration with Donen, has an almost too characteristic solo on roller-skates to the tune of 'I like myself'. In *Invitation to the Dance* (1954), though, the first film Kelly not only starred in but choreographed and directed all by himself, his more loftily artistic ambitions were at last given free rein. The film consists of three episodes, entirely danced and all featuring Kelly. The final one, 'Sinbad the Sailor', is an elaboration of the idea Kelly first used in *Anchors Aweigh*, of combining live and cartoon characters in a dance (it must still fascinate him, since despite his expressed dissatisfaction with this episode he has repeated the experiment in a half-animated television spectacular, *Jack and the Beanstalk*). The middle episode, 'Ring Around Rosy', is a modern satirical story of amorous intrigue based on the same idea as *La Ronde*. And the first and worst is a stiflingly 'artistic' attempt at pseudo-classical ballet, 'Circus', with two classical dancers, Igor Youskevitch and Claire Sombert, and Kelly miming sadly as a lovelorn clown.

Perhaps 'Circus', like the comic's allegedly inevitable desire to play Hamlet, is something Kelly had to get out of his system, and is best regarded, and excused, in that light. Unfortunately the desire to do something of the sort has rather overshadowed his later career, and the failure, artistically and commercially, of *Invitation to the Dance* must have struck him more as a defeat in the light of which *Les Girls*, the last real Hollywood musical in which he has starred, no doubt seemed very like an enforced (though for us quite enjoyable) retreat into conventional routine. Since then he has turned almost entirely to direction, his best film (in which, surprisingly, he does not make even a tiny Hitchcock-type appearance) being the lavish film version of *Hello Dolly!* His monument in cinema history will always be *On the Town*, *An American in Paris* and *Singin' in the Rain*, which he starred in, choreographed and, the first and third, co-directed. His achievement in those was as much that of a choreographer and director as that of a star pure and simple – and those aspects of his work I shall return to in their place. But undoubtedly it was his prestige as a star, his popularity with the public, which was instrumental in getting the films made, and made as they were, in the first place. He was therefore, by extension, very important in bringing about the whole revolution in musical film-making in the 1940s and early 1950s. In the hearts of the public, it is as a performer that he is still best remembered – and usually as a solo performer. Dancing with his mop in *Thousands Cheer*, or with his alter ego in *Cover Girl*, celebrating 'The hat me dear old father wore' in the midst of the clambake in *Take Me Out to the Ball Game* or sublimely, ecstatically 'just singin', and dancin', in the rain' he was unique, like no one else in films.

The name of Frank Sinatra comes up almost inevitably as an annexe to any consideration of Gene Kelly as a star – far apart though their later careers have carried them. Sinatra's career has been weird even by Hollywood standards, presenting as it does a complete change of image, a total transformation indeed, in mid-stream. He started off as a crooner beloved of teenagers, who had only to open his mouth in song to have them screaming and swooning; he was certainly not expected to *act*. When something recognizable as a role came his way (though he was still nominally playing himself) in *Higher and Higher* (1943), he stepped straight into a stereotype: lean, hungry, vulnerable, in need of mothering, he was painfully shy with girls, and a natural prey to the voracious ladies of the 1940s, who shamelessly set about seducing him, like the maid in

Higher and Higher who announced musically to anyone interested 'I saw him first'. The character was fixed for the rest of his first film career. In *Anchors Aweigh* he is mild and timid, in *Take Me Out to the Ball Game* he is ruthlessly pursued by Betty Garrett, and in *On the Town* Miss Garrett has to get even tougher in order to drag him away from the innocent pleasures of New York as recommended by grandfather to 'My place'.

It was nice, but it was certainly limiting. Sinatra's early crooning style went out of fashion, his career dipped, and he was considered all washed up when he made a comeback with a strong dramatic role in *From Here to Eternity*. No more, either in public or in private life, the mild, droopy, helpless image of the early Sinatra, but instead a new wry, tough, wiry, even hell-raising personality hove into view. No one would have thought early Sinatra good casting for *Pal Joey*, but the new Sinatra fitted it like a glove, down to the splendid effrontery with which he threw off 'The lady is a tramp' and other cynical reflections on humanity, courtesy of Rodgers and Hart. As Nathan Detroit in *Guys and Dolls* he seemed less at home – the quaint humours of Damon Runyan hardly fit him – and his role as the angry newspaperman in *High Society* is the least rewarding of the three principals, though he gets the best Cole Porter song to sing, 'Mind if I make love to you'. In *Can Can*, his last real musical to date, he also makes the most of the best Cole Porter gave the score, rendering 'It's all right with me' as the embodiment of the cool and the flip, even in extremes of amorous adversity. The two screen Sinatras are equally compelling: equally definitive. It is just difficult to believe they are the same man.

Now we never have any difficulty believing that Fred Astaire is the same man, right from the age of thirty-three, when he had a small role partnering Joan Crawford in *Dancing Lady*, to approaching seventy, which is what he was when he made his latest, but perhaps not even now his last musical, *Finian's Rainbow* (1968). His range as an actor may be limited, but he has had the good sense never to step outside it too far, and his positive characteristics are absolutely reliable and constant, informing all he does with style, grace and polish, lightness, ease and, for want of a better term, gentlemanliness. And technically, as a dancer, his range is almost limitless. The only real competition the cinema has ever offered him is Gene Kelly, and there the competition is only on the most superficial level of standing: in every other respect their personalities, aims and styles are so different there is little useful point in trying to compare them. The only interesting point of contact is the question, urgent at the beginning of the 1940s, of how the old master would get on in the coming era, with its new stars – Kelly supreme among the men – and its new kinds of musical film. Especially since his partnership with Ginger Rogers had broken up in 1939 after six years and nine films.

The answer, as it turned out, was very well. He had already in the Ginger years had one outing with another leading lady, Joan Fontaine, to perfectly agreeable results in *Damsel in Distress*, though the pattern was much the same as for the Astaire-Rogers musicals. The first thing he did after parting from Ginger was to go into a lavish revue-type film with a very flimsy plot at MGM, *Broadway Melody of 1940*, in which he was coupled with the studio's reigning dance champion, Eleanor Powell, and a spanking new score by Cole Porter, including the classic 'I concentrate on you'. The highlight was a long dance routine for Astaire and Powell to 'Begin the Beguine' (not a number specially written for the film), in which Eleanor Powell's famous speed and precision inspired Astaire to new heights of sheer (if rather cold) virtuosity. Not all his films in the next thirty years would be equally effective, not all his new leading ladies equally compatible, but hardly ever was there a film to which he did not bring something new, or which did not bring out something new in him. He proved to have the power of constant re-creation, the ability always to revitalize his image, either by a new technical challenge or through meeting the personal challenge of a new co-star. He could adapt himself to the new sort of musical just as readily as he had adapted himself to the old kind of vehicle; it was all one more stimulus to fresh effort.

Most immediately, thinking of Astaire in these years, one thinks of him in terms of his leading ladies. Rita Hayworth, for instance, who starred with him in *You'll Never Get Rich* (to the music of Cole Porter) and *You Were Never Lovelier* (to the music of Jerome Kern). Hayworth was long, lanky, a beautiful athletic animal who brought out a new sprightliness, a warmer romantic tone in Astaire. In contrast, Lucille Bremer, who partnered him in *Ziegfeld Follies* and *Yolanda and the Thief*, was cold, remote, delicate, and his relations with her were much more stylized. The two big numbers in *Ziegfeld Follies* are 'This heart of mine', another bitter-sweet sophisticated anecdote about a gentleman thief at a ball and the society beauty he does not quite take in, and 'Limehouse blues'. The latter is mostly occupied with a dream sequence of formalized romance in a fantastic abstraction of willow-pattern China which runs through the mind of a dying Chinaman (Astaire) smitten by the mysterious beauty of a Chinese girl (Lucille Bremer) he has only glimpsed in passing: both are made up to look vaguely oriental, and their dance is complex, hieratic, with much play on the opening and closing of fans and the minimum of direct physical contact – a remarkable stylistic innovation for Astaire. *Yolanda and the Thief* was burdened with a heavily whimsical story which required Astaire to pose as the guardian angel of the extravagantly trusting and virginal Lucille, but sprang into life in an elaborate dream ballet (another) in which Astaire's mixed motivation in his charade – robbery or love – is dramatized in smartly surrealist terms, full of wild Freudian symbolism. Not the sort of thing one would normally have expected the suave Astaire to be involved in, but he coped wonderfully well and impressively extended his range.

With Garland in *Easter Parade*, as I have said, he did not get on so well, and a single return to Ginger in *The Barkleys of Broadway* (1948) added nothing new, though it was an agreeably nostalgic interlude. Vera-Ellen in *Belle of New York* was acceptable but colourless; the film came to life mainly in its moments of light-hearted fantasy as they danced on the rooftops. But a really great, made-to-measure partner for Astaire was about to appear, in the shape of Cyd Charisse, with whom he made two of his greatest films, *The Band Wagon* and *Silk Stockings*.

Cyd Charisse as a dancer has qualities quite new in any Astaire partner: tall, dark, statuesque, her particular speciality is a sort of smouldering, vibrant sexiness. First fully revealed in the unforgettable sleazy night-club sequence of *Singin' in the Rain*'s 'Broadway ballet', where she enters our field of vision shapely leg first, coils and curves in serpentine arabesques around the innocent Kelly and knowingly fans his ardour by cleaning his steamy spectacles on the inside of her thigh, this quality is given definitive expression in the climactic 'Girl hunt' ballet of *The Band Wagon*, a cool and witty burlesque in dance terms of a Mickey Spillane thriller with Astaire as a super-cool private eye and Charisse as the *femme fatale* who 'comes at him in sections' and with whom he ends up, philosophically observing: 'She was bad, she was dangerous, I wouldn't trust her any further than I could throw her. – But – she was my kind of woman.' Elsewhere in the film they share an idyllic romantic duet to 'Dancing in the dark' in Central Park at night, and Astaire has several classic numbers to himself: brushed aside as a hasbeen at the station he walks away, with the inimitable Astaire walk that is already at least halfway to dance, singing with cheerful resignation 'By myself', and before long he is galvanizing an amusement arcade with one of his brightest and breeziest tap routines, 'A shine on my shoes'. Once started on citing numbers from the film, though, it is hard to know where to stop: what about 'Entertainment', that effervescent concerted number that brings him together with his great British contemporary Jack Buchanan, the imperishable Oscar Levant and Nanette Fabray, or 'Triplets', in which Astaire, Buchanan and Fabray, done up as terrible infants, catalogue the horrors of being an inseparable threesome? For sheer variety and vitality, even Astaire has seldom surpassed this towering film.

Silk Stockings, the other film in which he starred with Cyd Charisse, is little inferior. A remake of *Ninotchka* with Cole Porter music, it casts Cyd Charisse as the lady commissar with hidden fires and Fred Astaire as the Westerner who convinces her they were 'fated to be mated'. This time Astaire is mellow, relaxed, romantic: with casual ease he sweeps aside the formidable lady's pat catalogue of pseudo-scientific theory about love and carries her into a dancing declaration of mutual attraction in 'All of you' – an attraction later wholeheartedly celebrated in the 'Fated to be mated' sequence danced by the two of them through a series of deserted film sets in a studio. Both stars also get important solo spots, Cyd Charisse succumbing completely to the lure of the 'Silk stockings' and Fred Astaire in (to date) his last and one of his most inventive 'white tie and tails' numbers, 'The Ritz roll and rock'. And the whole film is held together, comedy, romance and all, by the perfect control of Mamoulian's direction, with its flawless ease of transition – the last of the great musicals in the great tradition of MGM's heyday.

Through the years other partners came and went, other demands were made on Astaire and generally met. In *The Sky's the Limit*, for example, early in the 1940s, his partner was forgettable Joan Leslie, but the score contained two wonderful Arlen songs 'My shining hour' and 'One for my baby', the latter in particular giving a whole new dimension to Astaire's great powers, if not exactly as a singer, then certainly as a putter-over of songs. In *Daddy Long Legs*, early in the 1950s, the challenge was a partner, Leslie Caron, out of the ballet, and dance numbers that risked a head-on stylistic collision between Roland Petit and Astaire himself as choreographers – though not, mercifully, in collaboration. In the event, Astaire copes well with Petit in one of the dream ballets, where he appears with wings as Leslie Caron's guardian angel in a tender but not too serious *pas de deux*, and Leslie Caron manages to fit into 'Something's got to give' as choreographed by Astaire and David Robel almost as happily. But you get the feeling that the rapport between the stars was less than ideal.

Fred Astaire and Cyd Charisse / Silk Stockings

Not so in *Funny Face*, the other great peak of Astaire's later career. Here the leading lady is Audrey Hepburn, an actress who is also a trained dancer and enough of a singer to manage her Gershwin numbers touchingly. She has a special quality of waif-like vulnerability, allied with tough determination that brings out a new tenderness, and at times a new toughness, in Astaire. In the numbers it is nearly all tenderness: a curious, almost protective quality which irradiates the title number, sung and danced in the confined space of a photographic dark-room; a fragile, ethereal charm in the white-clad dance of love to 'He loves and she loves', in idyllic soft focus; a wry tenderness in 'Let's kiss and make up' where he engages in an elaborate bull-fighting mime with his coat as a cape to amuse her. The really energetic singing and dancing is reserved for Astaire in partnership with another artiste one could, ungallantly, call a veteran: Kay Thompson, with whom he does an all-out, stop-at-nothing version of 'Clap yo' hands' in order to convince a possibly hostile gathering of French intellectuals that they are not only genuine folk artists but a pair of the friendliest vibrations imaginable. And, of course, for Audrey Hepburn, in her athletic dance at the cellar club, where she lets rip to show just what her chosen philosophy, Empathicalism, can do to release inhibitions and repressions.

Again, like *The Band Wagon* and *Silk Stockings*, *Funny Face* is primarily a director's musical, an integrated whole in which the performers, even performers as magnetic as Fred Astaire, are only ingredients. He may be the ingredient which gives the dish its special savour, but that is all. And because he was able to adjust to this situation, to accept it, to do some of his best work within it, Fred Astaire remained triumphantly undated, as much at home in the forties and fifties as he had been in the twenties and thirties. And as he would be in the sixties for that matter: but *Finian's Rainbow* is an example of yet another approach to the musical, seemingly the only one possible in the era of the hard-ticket super-epic. And that is something else again, calling for separate consideration on its own terms.

Meanwhile, to tidy up our catalogue of stars, there are a few loose ends. Howard Keel and Kathryn Grayson I never did like much, and since their finest hour together, *Kiss Me Kate* (1953), was not all that fine, I shall ignore them. Howard Keel did marginally better with another lesser luminary among MGM singing ladies, Jane Powell, in *Seven Brides for Seven Brothers*, where his hefty physique and powerful voice were ideally employed; Jane Powell, too, fairness forces me to add, was in a number of enjoyable films, from *Rich, Young and Pretty* to *The Girl Most Likely*, but she was seldom if ever the principal reason for enjoying them. Debbie Reynolds had her moments of freshness in *Singin' in the Rain* and *Give a Girl a Break*, but did not mature attractively. Donald O'Connor, once a child actor, fought his way up from terrible teenage B-features to one or two films, like *Singin' in the Rain* (especially) and *Call Me Madam*, which showed he had gifts as a dancer and comedian that the screen seldom had the sense to use properly. There was Danny Kaye, whom I dislike too much even to write about, though nearly all his films were at least semi-musicals and one of them, *Merry Andrew*, was the only one to be directed by choreographer Michael Kidd. And now there is Elvis Presley, who turns out an endless succession of elementary low-budget vehicles which, except for the style of music played in them, might have been exactly the same thirty years earlier – and only one of which, a long way the best, was directed by a musical director of any standing: *Viva Las Vegas* (*Love in Las Vegas* for Britain), made by George Sidney.

Last but by no means least, there are the two great dancing ladies of MGM, Cyd Charisse and Ann Miller. Cyd Charisse's definitive moments have already been mentioned, but there have been few films to which she did not bring something special. *Deep in My Heart*, a soporific biography of Sigmund Romberg, was temporarily lifted by a richly erotic dance evocation of *The Desert Song* to 'One alone' from Cyd Charisse and James Mitchell. Her big number in a gym with a lot of pugilists was the best thing in *It's Always Fair Weather*. And isolated dance numbers in dramatic films like *Party Girl* and *The Silencers* (in which she had to shoot someone with a specially equipped brassière) can always be relied on to liven things up. But the enlivener *par excellence* was Ann Miller. Her speciality was – is, since she has recently made a big Broadway comeback in *Mame* – the fast tap routine. Tough, direct, all bounce, go and energy, she tapped her way through film after film, sometimes unforgettably, as when imperatively demanding a 'Prehistoric man' in *On the Town*, complaining (but convincing no one) that 'It's too darn hot' in *Kiss Me Kate*, or, in slightly less noisy form, explaining in *Hit the Deck* that as 'The lady from the Bayou' she prefers to dance with her shoes off.

Small pleasures, perhaps, but real and irreplaceable. There is always room – we hope and believe – for the really big talent, the star personality which imposes itself willy-nilly. Today there is plenty of room for Julie Andrews or Barbra Streisand, and with the great empty spaces of the giant screens to be filled up, the bigger the personality the better. But what of the lovely people who decorated the small corners of small films, did their number and went? What of the Ann Millers? What indeed. It is a sad thought for us all that they really don't make pictures like that any more.

'The Jets'/West Side Story (above); Shirley
Jones/The Music Man

Petula Clark and Fred Astaire/Finian's Rainbow

The Action to the Thought:
A Cabinet of Makers

The thing that has most bedevilled critical consideration of the film musical throughout its history is the constant comparison, direct or implied, with the theatre. Directly, there has been a regular tendency to judge film musicals in negative terms, depending on whether they are or are not freed from the proscenium arch, whether they do things which obviously could not be done in the theatre, and whether they are within their rights so to do. Sometimes, indeed, the films have come under fire for mutually contradictory reasons simultaneously: Busby Berkeley was blamed in *Forty-Second Street* and *Footlight Parade* both for keeping the film musical shackled to stage convention and for putting together numbers that no stage could contain, seen from angles no theatre audience could ever hope to see them.

Indirectly, and therefore insidiously creating much more confusion, there has been the comparison between the theatre's and the film's use of dance, often in the strange supposition that they are, ought to be, or even *can* be, the same. The film is often accused of illiteracy and barbarity by purists because it tends to use dance as raw material rather than an end in itself – and of course the complaint is justified if the ostensible aim of the film-maker is to produce a record of something already existing in its own integrity. There would be no point in recording a classic performance of *Swan Lake*, for example, unless you observed the coherence and purpose of its choreography in relation to the total stage picture, the intended angle of vision and so on; piecemeal selection of the immediately telling detail at the expense of the overall context just would not do. But then such a record, while a perfectly legitimate use of the camera, is hardly in any normal sense cinema at all, and certainly quite beside the point in any discussion of the musical film.

Of course, most musicals use dance to some extent, though it is surprising how many are not, as one tends to imagine, founded on dance. And a lot that seem to be are not. Very few of the classic Busby Berkeley numbers turn on dancing at all; their 'choreography' is almost entirely the manœuvring of the cameras in relation to static people and objects, or the manœuvring of people and objects into changing patterns for the camera. But always the camera is a vital participant, never a passive recorder; and the effect is produced by the succession of changing, self-transforming patterns rather than by the steps (if any) by which one becomes another. Dance, in fact, is kept to an absolute minimum, appearing at most as an interlude of relaxation for the camera, as when James Cagney and Ruby Keeler dance together on the bar top in the 'Shanghai Lil' number of *Footlight Parade* or Eleanor Powell goes into her tap routine in the midst of stunning transformation scenes in *Lady Be Good*. In some of Berkeley's more elaborate trick numbers, like 'Don't say goodnight' from *Wonder Bar*, a group of dancers may be the starting point, but even here what obviously interests him most is not the dance they do, and its qualities as dance to be captured on film, but the fancy work with mirrors that makes half a dozen couples look like hundreds. As he observed himself in interview with David Martin:

'If you will go through all of my pictures you will see very little actual dancing. It wasn't because I didn't know how to create it and do it, but I wanted to do something *new and different*. Something that had never been seen before. Had they ever seen seventy or a hundred pianos waltzing? Had they ever seen lighted violins before? The same goes for all my various formations. I wanted to do

74

something unusual and entertain an audience. As I have often said, it isn't particularly the steps but what you do with them. . . . Musical numbers are musical numbers. If they employ a certain amount of dance, fine, and if they employ other rudiments or props or something else that's effective, that's fine if it entertains an audience.'

All of which adds up to an important statement about the nature of the musical film. When you come down to it, dance becomes, and must be, only one of the raw materials at the director's disposal, to use as he thinks fit. Within this framework it may of course assume a greater or lesser importance, and the director will adapt his camera technique according to the importance he wishes dance as such (or more usually dancers as such) to have in his work. The possible variations and gradations are infinite, from no recognizable dancing in a whole musical number to a total, continuous, fully worked-out piece of choreography on film. But it can never be said on principle that one way of using dance, or not using it, is more right and proper, more essentially filmic, than another. The only test is the largely subjective, instinctive one of whether it works or not.

The point is cardinal, but it is a very difficult one to persuade *aficionados* of the dance or stage choreographers to accept. Agnes de Mille in her autobiography *Dance to the Piper* tells a sad story of her first contact with the cinema, when she was hired to arrange the dances for Cukor's version of *Romeo and Juliet*. She arranged a sequence of elaborate court dances for the ball scene, and was then outraged when Cukor photographed them piecemeal, from all the wrong angles, and used them merely as a background for his action. Her distress is understandable, but obviously Cukor was right: the function of the dances in this sequence was not to exist and be recorded as a self-sufficient, coherent entity, but to form part of a more complex filmic whole, to be, though a finished creation to their choreographer, merely so much raw material to the director.

Romeo and Juliet was not, of course, properly speaking a musical, but the lesson applies to musicals too, and forbids us to draw any too hard and fast line between what is a musical and what is not. The distinction, if it is worth making, is frequently one of feeling rather than form. A film like *Love Me or Leave Me* may have enough musical numbers in it to fill an LP, and yet not finally feel like a musical, but remain a dramatic film with incidental songs. On the other hand something like Douglas Sirk's *Has Anybody Seen My Gal?* can have the climate of a musical without a single fully-developed number, only a little teenage dancing in the local drug store, so that it comes as no surprise, no break in the mood, when heroine Piper Laurie breaks into a few lines of song quite casually with 'Give us a little kiss' as she and the hero drive away in his rickety car, even though the number is given no excuse on the purely realistic level.

More extremely, one can find films like George Sidney's *The Three Musketeers*, with Gene Kelly, which have no music of that kind in them at all, but which are yet made like musicals, invaded by the spirit of the dance even if no dancing is visible. This is because, ultimately, choreography and camerawork are not separable; choreography becomes a function of direction, even if the director has a specialist assistant to work out the details for him in this area. At its simplest choreography is merely the arrangement of movement – usually, though not necessarily, to music. In the cinema movement may take place within the frame, or the frame itself may move as the camera moves, or the

75

illusion of movement – which *is* movement if the illusion is effective – may be created by the dynamic juxtaposition of static compositions. Or all these kinds of movement may co-exist. All of them, therefore, are part of screen choreography, which cannot be artificially confined as a working concept to just the first sort, the arranged movement of dancers within the frame.

Screen choreography can be as simple as you like, or as complicated as you like. When Fred Astaire begins to walk along the station platform in *The Band Wagon* singing 'By myself' that is choreography, though he only walks and sings, because the walk and the camera movement are exactly timed to provide a sort of delicate visual counterpoint to the song, both being just that fraction faster than one ever quite remembers them, that fraction faster than the movement of the song itself, so that the apparent defeat of the song's words, with their suggestion of putting a good face on a bad situation, is subtly transformed into an affirmation of the character's unquenchable vitality. In *Funny Face* there is another very simple example. Everyone has been breathlessly awaiting Audrey Hepburn's promised transformation from ugly duckling to bird of paradise. When at last she appears on the fashion-parade walk, the last word in poise and beauty, the soundtrack is invaded by the strains of 'S'wonderful' and the camera, moving easily with her, backs away, rising ecstatically as it pulls out to reveal the delighted, thunderstruck spectators, and then sinks to rest again behind them with a sort of visual sigh of satisfaction. It is a dramatic scene, but the handling is entirely musical, in the broadest cinematic sense choreographic.

It is not for nothing that the genre is called 'musical', rather than dance film or anything more limiting. What the director needs is in fact much more importantly a musical sense than a detailed technical knowledge of the dance, for every musical film, whatever the relative importance of its various constituents, must be essentially 'a city built to music'. A number of the best musical directors have been dancers and dance directors in their time as well, but it is not necessary. Berkeley has, though he disclaims any formal knowledge of the dance; Kelly and Donen were, and so was Charles Walters. Mamoulian and Minnelli have never been, nor has Francis Ford Coppola, though he arranged the movements of all the numbers in *Finian's Rainbow* except for Fred Astaire's solo dance amid the packing cases. George Sidney has been an amateur and professional musician, but never a dancing man. Obviously, if the dancing as such is to bulk large in a film's total effect, a very close rapport is necessary between choreographer and director, so it may help if the choreographer is also the director. On the other hand, it may hinder if his point of view in making the film remains first and foremost that of the choreographer, leaving the director part only as a subservient functionary (in just the same way that an actor may successfully direct a film in which he himself appears, but only if the director takes precedence over the actor in the actual process of film-making). The director may, of course, choose to give the choreographer (himself or another) his head, and so place the main emphasis of a musical sequence on the dancing, but we should always feel that it is a *choice*, not something imposed or a foregone conclusion. And the camera must always, somehow, seem to be an active participant, never merely a passive or at best sluggish recording instrument.

That was rather the trouble, for instance, with Bob Fosse's début as a fully-fledged director in *Sweet Charity*. From his earlier work in the cinema, as an inventive collaborator with directors such as Richard Quine (*My Sister Eileen*) and Stanley Donen (*Damn Yankees*), one can imagine that he should be able to make an effective musical film of his own. But here the problem of having a pre-established stage success of his own to rethink in film terms has been too much, and most of the numbers are awkwardly, if at all, integrated into the flow and movement of the whole, remaining all too evidently set-pieces lovingly but too reverentially recorded by the camera. Probably it would have been more effective all round to confide the filming of *Sweet Charity* to another director altogether, who could take a more cavalier view of its staging, even with Fosse as choreographer, and give Fosse a subject entirely new to him to direct as a film musical. Wiser, in general, was the début in film direction of another dance director, Herbert Ross, who chose the rather extreme course of making, in *Goodbye Mr Chips*, a musical with no dancing and no numbers (apart from a couple of incidental stage interludes) which are visibly sung at all, but which yet is pervaded by the spirit of music. Too extreme, perhaps: the film is quite enjoyable, but one is left with the feeling that now Ross has proved his point, to himself and everyone else, by establishing that as a director he is not just a singing, dancing fool, it would be nice if he were to make a rather more overtly musical musical.

Of course, we should remember that if we look back into the history of the musical, we find that it did not begin as primarily, or even importantly, a dancing form. The first Hollywood musicals to establish a viable film form, as soon as the sound camera was liberated from its soundproof box and enabled to cut free from the static recording of stagy scenes and routines, were the early sound films of Ernst Lubitsch. *The Love Parade* in 1929, *Monte Carlo* in 1930 and *The Smiling Lieutenant* in 1931 were the first Hollywood musicals to break out of a circumscribed stage-revue type of structure and tell their stories freely and gaily in terms of music. But not at all in terms of dance.

The Love Parade may open with a row of kicking legs, but thereupon Maurice Chevalier's comic valet goes straight into his song, and from there on it is all song and incidental music, with nothing you could properly call dance. On the other hand, no one would notice this as a lack: the whole film, its form and tempo, is dictated by the music, and even in those rare moments when there is no music of any sort on the soundtrack the performers seem to be keeping time with an inaudible music in their inner ear. The story is minimal, a sort of Ruritanian *Taming of the Shrew* in which the infectiously charming Maurice Chevalier manages to marry and subdue an over-spirited royal bride, Jeanette MacDonald, and the songs follow one another thick and fast, dialogue flowing into song into wordless action-to-music into song into dialogue with the utmost unselfconscious ease. Lightness and speed were of the essence, and complete freedom for the camera to roam as it would, the spectator's eye skipping from object to object in a constant sparkle of effect. This was the famous Lubitsch touch, and by finding at once a way of using sound and music to help in its creation, as in his silent films he had done with visuals alone, Lubitsch created the first true screen musical in the world.

Monte Carlo showed some advance, and its big set-piece, the montage of train wheels, peasant bystanders in the fields, and Jeanette MacDonald singing 'Beyond the blue horizon' out of a train window, accompanied by a sound montage of train whistles, hisses of steam and so on, was immediately famous, though not, perhaps because of its intricacy, much imitated by anyone else. The trilogy was completed with *The Smiling Lieutenant* (Chevalier again), a very free version of Oscar Straus's *Waltz Dream* about yet another imperious princess in pursuit of a mate; and by this time the technique was perfected. From beginning to end everything, sights and sounds, words and music, are in a constant state of flow, with the smoothest possible, most nearly invisible transitions from one state to another. Dialogue itself is in fact kept to an epigrammatic minimum, and hardly a word is spoken in the whole second half of the film, everything being done in action to music, rising occasionally, as though by some irresistible inner necessity, to song.

Even the great Rouben Mamoulian, making his first fully-fledged musical the following year, *Love Me Tonight*, could do little to improve on the technique, and obviously had it in mind, since he was working for the same studio and with Lubitsch's two favourite stars, Maurice Chevalier and Jeanette MacDonald, fresh from making Lubitsch's fourth film to star one or other, second to star them both, *One Hour With You*. *One Hour With You*, a *risqué* modern marital comedy for a change, even anticipated *Love Me Tonight*'s use of very formalized speech and rhymed spoken dialogue between songs. On the other hand *Love Me Tonight* has passages of sheer poetry and romantic invention beyond Lubitsch, and the two directors are temperamentally very different, so that despite a few superficial similarities of method one would never mistake a scene like the opening, with its orchestration of everyday morning sounds, or the elaborate transition in the course of 'Isn't it romantic?' from Maurice Chevalier in Paris across time and space to Jeanette MacDonald in her castle in the country, for the work of Lubitsch.

Strangely enough, this first extraordinary flowering of the completely integrated, musical musical had almost no follow-up. Lubitsch's only later musical, *The Merry Widow* of 1934, was a deliberate harking-back to the style, and, reuniting Chevalier and MacDonald for the last time under Lubitsch's direction, almost a conscious farewell to it. (In 1947 Lubitsch began shooting *That Lady in Ermine*, an operetta in virtually the same style with Betty Grable and Douglas Fairbanks Jr., but died after eight days of shooting and the film was completed by Preminger. It is intriguing to consider what Lubitsch would have made of it at that distance of time.) In 1936 Sternberg made his only attempt at the musical genre in *The King Steps Out*, a vehicle for Grace Moore which seemed to be aiming at the same sort of Ruritanian glamour and artificiality that Lubitsch had created; but he lacked the lightness of touch (and so, even more, did Grace Moore), so despite a few enchanting visual effects the old magic was not recaptured.

Meanwhile, other things were happening in the musical; after a lagging of interest in the genre other approaches were being tried. To begin with, people began to worry about liberating the screen musical from the proscenium arch, whether it should be so liberated, and if it could be, how. The proscenium arch had never seemed an important consideration in Lubitsch's films, or in *Love Me Tonight*, which were unmistakably of the cinema, cinematic. But when musicals made a big popular comeback with *Forty-Second Street* and its successors the issue was raised again, largely because the background to such stories as they had was nearly always theatrical, involving the putting-on of a show. The numbers, consequently, were nearly always 'realistically' motivated: an audition or a rehearsal would be in progress (even if an unseen orchestra joined in after the first few chords from the accompanist), or it was opening night, in an atmosphere of hushed expectancy the curtains parted and. . . .

Now needless to say – or it should be needless to say – this was a convention just like any other. If Busby Berkeley was directing the musical sequences, the curtains parted and what happened next could happen on no stage, nowhere. But people worried; worried because of the theatrical connection to begin with, then worried because it was not realistically followed through. Consequently, it was hardly noticed that what Busby Berkeley was doing was to create an entirely new, entirely cinematic form of musical extravaganza, in which the camera and its techniques played a crucial role by creating a new sort of space-time continuum which could not conceivably exist anywhere else but on a cinema screen. With his patterns of chorines shot from above or below, his dreamlike unprepared transitions from air to land to water, he abolishes gravity and annihilates time. As we have observed this would be fatally destructive of a dancer's effectiveness because it is a kinesthetic necessity that we remain aware of the dancer's struggle against the physical limitations of his art; but Berkeley is rarely much concerned with dancers, he is an abstractionist constructing an abstract universe of objects patterned to music.

In this Berkeley had any number of imitators at the time, though none of them seemed to grasp the principles by which his effects were achieved – largely, no doubt, because as he has many times told us, there were no principles, only instinct. There is hardly a single example in the cinema of his direct influence, fruitful or otherwise; he was and remains a creator right out on his own. In the musical, at least: funnily enough, his ideas can be seen bearing fruit on some strange branches of film art – in the abstract experimental films of the 1930s and 1940s, for example, and in the montage sequences once a regular feature of Hollywood dramas, showing the passage of time by leaves torn from a calendar, pages of newspaper flicking over, or evoking a period by flashes of charleston-ing legs, cops axing open barrels of bootleggers' liquor, and stock-brokers jumping out of Wall Street windows. But in general his way of abstraction remained solitary, in the musical anyway, for the day of the dancer was coming and Berkeley had little to offer dancers except competition. 'Why did I have Ann Miller dance through the musicians with their arms sticking up through holes cut in the floor (in *Small Town Girl*)?' he asks rhetorically in the interview already quoted. 'Because it was interesting to watch, interesting to photograph. I could have had her come out on the stage and do a buck-and-wing routine, call it a dance, and that would have been it. But I always try for the new and different. They don't think about these things today.'

Perhaps not entirely true, but true enough in the way he means it. When the dance invaded the film musical, there was little room left for the musical routine as Berkeley conceived it, and it must have seemed to him that all too often the camera was becoming the dancer's slave instead of his master. Ideally, of course, it should be neither, but his valued collaborator in the production of a whole which is bigger than, and different from, the sum of its parts. Which is where Fred Astaire comes in. It is not easy to assign precise responsibility for what happens in the Astaire-Rogers films. That Fred Astaire dances we can see for ourselves. But the credits list Hermes Pan alone as choreographer, and the directors are mostly familiar names of the time, prolific technicians who were well able to turn a hand to anything; and, in the case of the director most frequently involved, Mark Sandrich (who directed five of the nine) actually specialized in musicals. And yet there is really no doubt that in every important respect Astaire was the creator of his own screen style and image: he devised his routines, with the assistance of Hermes Pan, and he laid down in outline how they should be filmed. Basically, he insisted that since the essence of his dance was that it used the whole of his body, and that body in relation to its environment, it must be given space to work. There are sometimes close-ups in his routines, but on the whole they are filmed in medium shot, showing him at full length (and Ginger Rogers when present) and following the dancers round with a very mobile camera.

Was this 'stagy'? The question is artificial, since it presupposes that the techniques of dancing on stage and in films are mutually exclusive (when in fact it is obvious that there is a very considerable area of overlap) and that the film justifies itself and defines itself by exclusion, so that it can do 'legitimately' only what no other medium can in any circumstances do. On the contrary: movement within the frame is not necessarily any less filmic than movement of the frame, and interplay of the two movements may produce something completely unprecedented in the dance. So it does in the Astaire-Rogers films. In Mc-Luhan's terms, the Astaire conception of musical cinema is cool, as against the hot conception of Busby Berkeley, but neither is any less proper than the other. And the use of a very mobile camera brings its own bonuses: by bringing back the sense of gravitational limitation to human movement in a musical number, Astaire sets screen dance free in another way: its powers of movement vertically may be limited to the height of a dancer's highest leap, but horizontally it is completely untrammelled, free to move without limitation in any natural or constructed world. Only very rarely has Astaire broken the rules of his own art-definition – the slow-motion passage in 'I used to be colour blind' (Carefree) and the dancing on walls and ceiling sequence in Royal Wedding (Wedding Bells in Britain) are the only examples I can think of – and in general his physical reality and presence is the most powerfully felt element in his films. Happily so, for he if anybody is one of those whom the camera loves.

The popularity of the Astaire-Rogers films led inevitably to the appearance of other dancing pairs, like Carole Lombard and George Raft in Bolero (which was foolish but fun) and Rumba (which was terrible), or Eleanor Powell and James Stewart (of all people) in Born to Dance, which did not do much to suggest that Eleanor Powell was specially suited to being half of a double act. But just as Busby Berkeley films, often imitated, remained defiantly inimitable, so did the Astaire-Rogers films, not so much because there were no such good dancers around (though there weren't) as because no one else but Fred Astaire had really mastered the art of making dance look natural on screen. No one would until Gene Kelly came along.

By the end of the 1930s the musical had reached a decisive point. The older order was changing, Astaire and Rogers had split, Berkeley had left Warner Brothers, and nothing was quite the same. One solitary masterpiece, Mamoulian's dramatic musical *High, Wide and Handsome* (1937) with its classic Kern score and integration of music and story, had come and gone, arriving apparently out of nowhere and disappearing again without progeny. At which point along came *The Wizard of Oz*. It was not an out-and-out masterpiece: the colour was inappropriately garish, the set designs lacking in style, and the direction merely competent, with no sign of any particular feeling for music and movement. But it did signify a new way of looking at the musical genre, one for which its associate producer, Arthur Freed, was above any other single person responsible. Freed had been for some time a successful lyric-writer, particularly for Nacio Herb Brown (the score of *Singin' in the Rain* is made up of their vintage songs); as a film producer he was rapidly to become the Diaghilev of the musical.

It was his idea that musicals should not be random assemblages of numbers, slim stories interrupted every so often by a cue for song or dance. Not all musicals before him had been that, but he seemed to be the first person to work out a theory from which consistent practice might be achieved, not so much a rigid formula as an informing idea of what a musical should be and what it should not. What he wanted was to produce films which would make sense as a whole. On *The Wizard of Oz* his responsibilities were mainly musical, and so he started there, by encouraging composer and lyricist, Harold Arlen and E. Y. Harburg, to write a fully developed, integrated score in which all the songs and musical episodes advanced the story and created dramatic atmosphere. So strongly did they all feel this that one of the most popular songs from the score, 'The jitterbug', was eliminated from the final print because it was felt to represent a dispensible excrescence – though they all fought like tigers for the retention of the other big hit in the film, 'Over the rainbow', feeling quite correctly that its presence at the beginning was a dramatic necessity.

It was a start, anyway. But to make a whole film work the way he wanted it, Freed appreciated that more than just an idea was needed: he needed to function as a true impresario, bringing together teams of collaborators with similar ideas and, more importantly, talents and personalities which would combine in the right stimulating way. While he produced relatively conventional films, among them several of the Judy Garland films directed by Busby Berkeley – *Babes in Arms, Strike Up the Band, Babes on Broadway, For Me and My Gal* – he laid his plans carefully. He found new stars and let them develop their talents in various directions – Gene Kelly and Stanley Donen came to MGM at this time, to work for him – and he invested in the future by bringing to Hollywood a successful Broadway designer-director of spectacular revue, Vincente Minnelli, so that he would just haunt the studios for a year or so, picking up all he could about the art and craft of film-making. In 1943 he was ready to let Minnelli, after trying his hand on sequences of *Babes on Broadway* and *Panama Hattie*, direct his first complete film, *Cabin in the Sky*.

Minnelli's next musical, *Meet Me in St Louis* (1944), is the really decisive beginning of the total musical in Hollywood, but *Cabin in the Sky* is a significant and highly honourable precursor. As it happened, the show on which it was based was one Minnelli had directed on Broadway, but for the film it was completely refashioned, with more new songs imported than originals were left from Vernon Duke's stage score. But the overall shape of the original, with its dream structure (good and evil fight for the soul of a dying man in a fantasy world), was preserved in order that the film might be a piece of sophisticated *faux-naif*, a highly conventionalized evocation of a primitive Negro heaven and hell, and mostly what is in between, a world whose angels and devils may every so often have converse with men. The film is a triumph of visual stylization, a parable set to music with an unfailing sense of visual elegance and a surprisingly true instinct for rhythm in music and in film staging and editing. There are moments, though, where the material seems not wholly assimilated to its new medium; touches of stiffness and uncertainty which will not recur in Minnelli's work.

In *Meet Me in St Louis* we can for the first time see Minnelli's talents whole, or nearly whole (there is still almost no real dance for him to cope with). And it is here that we begin to appreciate the secret of his special way with musicals. It is that while phenomenally sensitive to music (he is a pianist himself) and to the subtlest variations of pacing and rhythm within a scene, a movement, and possessed of a great visual flair, the consequence no doubt of his having begun his professional life as a painter and designer, he is fundamentally a *dramatic* director. *Meet Me in St Louis* is conceived and directed as a coherent drama, a story about believable people in a believable situation. The musical episodes are all judged from the outset according to their power to advance the story, epitomize a situation, intensify a mood; they all have to have their dramatic *raison d'être*.

But music and choreographed movement are not confined to the obviously musical episodes, the outbursts of sheer *joie de vivre* like the 'Trolley song' or the moments of intense introspection like 'Have yourself a merry little Christmas'; music and drama interpenetrate each other. The sequence in which the youngest daughter, Tootie (Margaret O'Brien), is dared into playing a Hallow-e'en joke on a frightening neighbour, scared out of her wits but too inflexibly proud to admit it, is handled with choreographic precision as the camera tracks away in front of her, holding her face in centre frame as she leaves the light and warmth of the bonfire for the menacing, dangerous dark. And the principal dance episode, the cakewalk Esther (Judy Garland) and Tootie dance for their family to 'Under the bamboo tree', is given just the right fragile charm by Minnelli and his choreographer Charles Walters. The collaboration, under Arthur Freed's watchful eye, is incomparable in other respects, the casting down to the smallest roles is just right, and how beautifully the scene of father and mother (Leon Ames and Mary Astor) singing 'You and I' at the piano catches the required Victorian familial warmth and intimacy – not only because of the performances, but because of the impeccable re-creation by designers Lemuel Ayers and Jack Martin Smith of a turn-of-the-century middle-class domestic environment. *Meet Me in St Louis* is one of those films where just everything goes right, seemingly effortlessly, by chance. But chance, one may guess, like goodness, had nothing to do with it.

From this delicate, nostalgic piece of Americana one might sense, though it is not obtrusively on display, the quality which is the most clearly distinguishing feature of Minnelli's artistic personality: his sophistication. One could sense it because it is the crowning achievement of sophistication to be truly simple. But it was not until his next musical, *Ziegfeld Follies,* that it was strikingly, shamelessly put on display. Naturally, sophistication is not in itself a good thing or a bad thing. Many of the greatest achievements of the American cinema, from Griffith and Ford down, have sprung from a certain species of grand simplicity and innocence; so too, from a less grand variety, have come many of its most notorious disasters. Nor will sophistication alone avail its possessor much in the absence of other vital talents and qualities. But, for better or for worse, Minnelli has it, and it is usually in his case for better.

Ziegfeld Follies is the perfect field for its deployment: evoking the pre-war splendours of the mammoth Ziegfeld shows – the last edition of which Minnelli had directed on stage before coming to Hollywood – the film is constructed in a series of twelve separate episodes, linked by no more than the idea, stated in the prologue, of paying tribute to the showmanship of Ziegfeld (to whom MGM had already paid lavish tribute in *The Great Ziegfeld* and *Ziegfeld Girl*). Not all the episodes are directed by Minnelli: three numbers in the middle are directed by other hands, and so is the spectacular opening number 'Bring on the beautiful girls', which, though it has been hailed by some unwary critics as the quintessence of Minnelli, was actually directed by George Sidney. Three Minnelli numbers in particular do stand out, though: an aria from *La Traviata*, staged as a highly stylized ball scene with wild surrealistic costumes by Irene Sharaff, all in black and white except for the scarlet of the prima donna, and geometrically formal choreography by Eugene Loring, a refined stylist and one of Minnelli's happiest dance collaborators; Fred Astaire and Lucille Bremer in 'This heart of mine'; and the same pair in 'Limehouse blues', both of which I have already described.

All these three show a fusion of design, choreography and camerawork at a level of intricacy and cool virtuosity which takes the breath away. In the first, for instance, there is the lead-in to Violetta's entrance by way of a sudden dramatic crane shot starting above the heads of the dancing couples, looking down on them, and then diving giddily to capture the first glimpse of this stunning vision in red, the only touch of strong colour in a nearly monochromatic scene, from right down at ground level, framed in a retreating vista of dancers. 'This heart of mine' also makes play with strong colour contacts: the rich red of the ballroom against the pale blues of the background and the white of the moving shell which shuts the interior from our view; the matching red of the outdoor dancers' costumes against Astaire's black and white, and Lucille Bremer's glittering pastel shades. 'Limehouse blues', also choreographed by Eugene Loring, features a remarkable contrast of moods and atmospheres between the prologue and epilogue, shot in a misty, romantic evocation of Limehouse in the nineties, and the abstract formalizations of the main danced episodes (apparently the frame, which makes the dance itself a dream, a favourite Minnelli device, was thought up by Minnelli himself because he regarded Loring's treatment of the music as too divorced from the spirit of the song itself – a very tangible example of his taste and intelligence).

During the years which followed, Minnelli devoted at least half his time to non-musical subjects, and finally, when the great era of musicals at MGM passed, seemed to abandon them altogether (though in 1969 he returned to the genre with *On a Clear Day You Can See Forever*). Ironically, but perhaps only properly, while in his musicals (the non-narrative *Ziegfeld Follies* is an obvious special case) his greatest strength seemed to be in his abilities as a dramatic

director, holding all the other elements very firmly in place, in his straight
dramas one is always conscious of his musical gifts: in the grand symphonic
structures of his heavier dramas like *Some Came Running* and *Two Weeks in
Another Town*, with their stunning all-out climaxes, in the visual sparkle and
rhythmic variety of comedies like *Father of the Bride*, *The Long, Long Trailer* and
The Courtship of Eddie's Father. Which is only to say, I suppose, that he is consistent in style and approach, whatever he is doing. The feeling remains, though,
that it is the musicals which really bring out the best in him, because they bring
out the most, they use the widest range of his talents. Not so much *Brigadoon* and
Kismet, where the material is unsuitable to him (phony-folksy in the first, ersatz
oriental in the second) and inferior in itself. Not even in *Gigi*, which though it
is made with taste is rather airless and lacks sparkle. *Bells are Ringing*, on the
other hand, I rather like: a strange and isolated throwback (from 1960) to the
great days in the early 1950s, and very amusingly written by a couple, Betty
Comden and Adolph Green, whose tough urban wit nicely complements Minnelli's more delicate but equally urban variety.

Comden and Green had already collaborated with him on one of his finest
works, *The Band Wagon*, which I am inclined to put among the outright masterpieces of the whole musical genre. It is the putting-on-a-show musical to end all
putting-on-a-show musicals and one would probably not be far wrong in seeing
the writers played by Oscar Levant and Nanette Fabray as self-portraits. The
film is again, like – but how different from – *Meet Me in St Louis*, a perfect meeting of minds and talents at the behest of Arthur Freed: the polish and sophistication of Minnelli meeting the polish and sophistication of Fred Astaire, the

Gene Kelly/The Pirate

energy and humour of Comden and Green coupled with the energy, humour and sometimes deeper feeling of song-writers Dietz and Schwartz, the lithe grace of Fred Astaire set off by the fine athletic line of Cyd Charisse, the face and figure of Cyd Charisse matched with the vibrant voice of India Addams, the long-delayed meeting of Fred Astaire with his British equivalent, Jack Buchanan, the sets of Oliver Smith (for the numbers) and Preston Ames, the vital choreography of Michael Kidd, all fused into a satisfying whole by the roving camera of Minnelli – whichever way one maps out the correspondences, it is impossible to find a false agreement, and the film's unflagging zest testifies to the perfect concurrence of all those involved, and their ability, individually and collectively, to inspire Minnelli perhaps beyond compare.

His other three musicals, anyway, do not quite hold their own in comparison. *Yolanda and the Thief* is a rather heavy piece of nonsense which evidently, on a human and plot level, Minnelli could not begin to take seriously, and it shows – only the surrealistic dream sequence (dreams again!) and the joyously anachronistic finale come alive. *The Pirate* and *An American in Paris*, on the other hand, are masterpieces all but a hair's-breadth – that hair's-breath being, I think, the failure of Minnelli and Gene Kelly to achieve perfect artistic harmony. There is, it seems to me, something slightly coarse and obvious about Kelly's personality, his approach, his choreography (they are not necessarily any the worse for it) which is at odds with Minnelli's extreme refinement and fastidiousness. Minnelli and the basic material of *The Pirate* – a play by the sophisticated S. N. Behrman, songs by that arch-sophisticate Cole Porter – are in perfect harmony, but his stars, Kelly and Judy Garland, are not: true sophistication is beyond the range of either. In spite of this the film works quite well, and sometimes superbly, as in the dazzling crane shots of Kelly's arrival at the quay and the fantastic virtuoso shooting of his fantastic virtuoso number of 'Niña', with the camera prowling after him up and down, in and out, over every inch of Jack Martin Smith's Caribbean street-scene set. And in overall shaping *The Pirate* is just about perfect.

The same cannot be said for *An American in Paris*, where the title ballet, exhilarating as it is, dangerously overbalances the second half. The film as a whole breaks down into any number of glittering fragments, many of them to be treasured: like Oscar Levant's dream(!) rendering of parts of Gershwin's Piano Concerto with himself as soloist, orchestra, conductor, and finally the solitary, but enthusiastic, member of the audience; the happy-go-lucky improvisation of Kelly's and Levant's rendering of 'Tra-la-la-la'; and of course, the big ballet itself. Obviously, the condition of the film's creation was the participation of Kelly and his great chance as a choreographer in the ballet, so there is little point in wondering whether it would have been stylistically more coherent if he had directed the film himself, or if Minnelli had made it with another star, say, Fred Astaire. For that matter, even if it had been more coherent, either way, would its heights have been so exhilarating? Would we finally, on balance, have been the gainers? In any case, parts of *An American in Paris* certainly count among Gene Kelly's finest hours, and if I happen to look for Minnelli's finest hours rather in *Meet Me in St Louis* or *The Band Wagon* (or *The Bad and the Beautiful*, but that's another story . . .), well, it is only one man's opinion, and of finest hours Minnelli has enough and to spare.

After Minnelli the greatest discoveries in Arthur Freed's great gallery are surely Gene Kelly and Stanley Donen, considered first as a two-headed monster and then individually. Kelly's gifts as a musical director are difficult to assess because we do not know, it is impossible to know, and no doubt they themselves did not and do not know, exactly how much Kelly and Donen each contributed to their joint works *On the Town*, *Singin' in the Rain* and *It's Always Fair Weather*, plus however much they were responsible for in *Take Me Out to the Ball Game*. The question is important because two of these films, *On the Town* and *Singin' in the Rain*, are masterpieces of the genre, and left on his own as a director Kelly has tended to steer clear of it, returning only in *Hello Dolly!*, a glittering anthology of magic moments from old MGM movies which is fun but hardly magisterial. Donen, on the other hand, has made several very good musicals without Kelly – *Give a Girl a Break*, *Seven Brides for Seven Brothers*, *The Pajama Game*, *Damn Yankees* – and one masterpiece, *Funny Face*. It is tempting, therefore, to suppose that Donen was all along the directorial talent: tempting, but probably quite unfair, since if only by the power of his physical presence, Kelly was obviously in many respects the dominant partner.

Best then to take the films for what they are, without looking too closely into their origins. All three of them were written by Comden and Green, *On the Town* from their Broadway success but completely transformed along the way, the other two as originals. *It's Always Fair Weather* is far and away the least of these: the script has a ring of truth in the picture of an old-pals' rather sour and embarrassed reunion, but it is not matched by the film, which tries desperately to put a good face on things, and comes near to desperation also in its frantic search for new gimmicks – Gene Kelly's roller-skate solo, Cyd Charisse's dance with the boxers, the trio of the three old pals, Kelly, Dan Dailey and Michael Kidd, with dustbin lids. It is true that the film makes rather good use of Dolores Gray's forceful talents, but the only other real musical chance she got was enlivening stretches of the otherwise deadly *Kismet*, so that was more or less that.

The various delights of *On the Town* and *Singin' in the Rain* have probably been sufficiently hinted at already, but it is worth taking a closer look at the way they are made. The most noticeable thing about them is their extreme technical variety and dexterity: compared with the meticulous Minnelli, Kelly and Donen show an extraordinary and really very fetching willingness to try anything at least once, hit or miss. Since they were both dancers and are both choreographers, one might expect their films to be dance-dominated, but not at all. *On the Town* especially makes very frequent use of montage effects, and of camera movement rather than movement within the frame. The three sailors' first hour or so in New York, for example, is described in a fast montage of shots here, there and everywhere in the city, cut to the urgent, bustling music of 'New York, New York'. Ann Miller is given a chance to tap away in the 'Prehistoric man' sequence, but it is not a straight dance-to-camera number, since it makes use also of cut-in non-consecutive close-ups of the gang running delightfully wild with the museum's ethnological exhibits.

There are two ballets in the film, but only one of them, the 'A Day in New York' ballet, is handled straightforwardly from a directorial point of view; the shorter ballet in which the qualities attributed to 'Miss Turnstiles' in a subway poster are wittily burlesqued is chopped up, mixing various locales, sets and costumes with throwaway ease and includes an elaborate split screen effect which shows her in her various characters – home body, sports girl, student of the ballet – all at once. The title number uses the geographic flexibility of the cinema to the full by starting on the top of the Empire State Building, bringing the six principals down in the elevator and continuing as they stride off, arms interlinked, still singing. The musical battle of wills between Frank Sinatra and Betty Garrett as she tries to drag him up to 'My place' goes to the opposite extreme; it is staged entirely in the front seats of the cab she drives.

On the Town is the perfect example of *flow* in the modern musical, with everything advancing the plot and no slack passages, no cues for song, no overelaborate balletic interludes (even 'A Day in New York' is kept within reasonable limits, and serves its function in character-development). *Singin' in the Rain* is built more like a rich plum pudding, stuffed full of goodies. The means it uses range from the simplest to the most elaborate. Very simple, very 'dancery', are the title number, sung and danced by Kelly almost alone in a rainy street with a very mobile camera following him, darting ahead or craning up high, and with all the cuts made on movements or at moments of transition in the music so that they are unnoticeable. The same with the big romantic *pas de deux* on a deserted sound stage to 'You were meant for me': the song is done in one shot, then at the transition from song to dance comes an unnoticeable cut to a looser medium shot. As the movements of the dance become more generous, covering a wider sweep of the floor, we cut on a turn to a long shot starting high up on a crane and hovering over the dancers until as the music and movement die down again we cut on another turn back to the medium shot, which moves into close-up as Kelly sings the last words of the song, and then cranes away to long shot from above as the music fades away.

At the other end of the scale are the elaborate 'Broadway Ballet', telling the story of a boy from the sticks who achieves fame and fortune as a dancer but never manages to possess his elusive *femme fatale* (Cyd Charisse – who else?) in a series of spectacular transformation scenes, and, even more directly a tribute to Busby Berkeley's cinema, the big montage to show the coming of sound to the cinema with an ever-more frantic alternation, visual and aural, of 'The wedding of the painted doll' and 'I've got a feelin' you're foolin''. This leads at last into a full dress version of 'Beautiful girl', complete with twenties fashion parade, beginning and ending with a typical Berkeley vertical crane shot, suggested perhaps by similarly composed shots in the 'Shadow Waltz' number of *The Golddiggers of 1933*, variations on the 'human flower' effect which shows symmetrically arranged figures from directly above with one face at the centre.

It would be difficult to choose between *On the Town* and *Singin' in the Rain*. *On the Town* is probably more formally satisfying, better integrated as a whole, but *Singin' in the Rain* is more lovable, and for a film buff irresistibly loaded with nostalgia as well. Only one of Donen's solo musicals come up to this standard, though all of them have splendidly inventive passages: there are split screens again in *Give a Girl a Break*; and later on, in *The Pajama Game* and *Damn Yankees*, he manages marvellous visual effects with location shooting, dazzle from shooting into lights and so on. (On these last two films he shares director credit with the stage veteran George Abbott, but presumably the camera direction and visual detailing were entirely his.) In so far as these films, and his other, non-musical films enable one to distinguish his personality from Kelly's, he emerges as a gentler, less assertive, more style-conscious personality. In *Seven Brides for Seven Brothers* he handles the famous square-dance with the requisite gusto and split-second timing, but elsewhere seems a trifle embarrassed by the hearty humours of the story, a Western version of the Rape of the Sabines, and not truly at home in the wide open spaces (especially as represented in rather dowdy studio sets).

The one film which does match up completely to the best of Kelly-Donen films in Donen's later work is of course *Funny Face* (1956). In that, you might say, he almost out-Minnellis Minnelli, achieving all the visual glamour, the outrageous sophistication that Minnelli could supply, and colouring it with a special, melting romanticism of his own which has never been quite in Minnelli's range. If his camera was ever in danger of seeming like a smart little girl with no heart, the danger has now passed, as it is in love with a wonderful guy – Fred Astaire. And a wonderful co-star, Audrey Hepburn, whose chemistry with Astaire is unbelievably right. It is hard to know what to praise most in the film. The sheer verve with which it launches into Kay Thompson's great opening number, 'Think pink', in which, at the behest of the most powerful fashion editress in the world, everything is turned pink before our eyes, at times in Donen's favourite split-screen technique as we are encouraged to thrive in pink, dive in pink, come alive in pink? Or the exquisitely reticent and intimate feeling of Audrey Hepburn's 'How long has this been going on?' shot all in dark, sombre colours until at the climax of the music a sudden twirl of glowing colour

is matched by a musical swirl on the soundtrack before it all fades away, sight and sound, with a forlorn might-have-been image of Audrey Hepburn in a pool of clear cold light? Or the wonderfully simple idea of staging 'Funny face' in Astaire the photographer's dark-room, deliberately confining the space in which they can dance and reducing the colour range to various shades of red until the very end, when again Audrey Hepburn is caught in a pool of light but this time bright and searching, promising an optimistic future?

It would be possible to go on in this way throughout the film: it does not have one poor number, one passage of fluffing and uncertainty. From every point of view it is about as near flawless as one can hope for in an imperfect world. It should have been produced by Arthur Freed, but it wasn't: instead it was master-minded by one of the great unsung backroom boys of the musical cinema, Roger Edens, long an assistant to Freed, song-writer, arranger and all-purpose Man Friday around Culver City. Kay Thompson was also a refugee from MGM, where she had written, arranged and generally smoothed things behind the scenes (she wrote, for example, 'A great lady has an interview' in *Ziegfeld Follies*). The choreography is by Eugene Loring, whom we have also met at MGM, and Astaire himself. And the great photographer Richard Avedon is special visual consultant. Artistically it is a famous victory; it is also virtually the death-knell of the intimate, integrated musical as we have come to know and love it in the previous dozen or so years.

Not quite. There is one more masterpiece to come from MGM, and who is going to give it to us? Not Minnelli, not Kelly, not the quiet Charles Walters, whose gentle *Lili* had charmed us and whose *Jumbo* in 1962 will be just about

the best of the new super-musicals, nor George Sidney, who did well by *The Harvey Girls* and *Kiss Me Kate* and is now off at another studio turning *Pal Joey* into a vehicle for Frank Sinatra and Rita Hayworth. But, surprise, surprise, the old master himself, Rouben Mamoulian, in his latest film, *Silk Stockings*. In his whole long career Mamoulian has made only four real musicals: *Love Me Tonight*, *High, Wide and Handsome*, *Summer Holiday* and *Silk Stockings*. And each one has been extraordinary. *Summer Holiday* was made in 1946, for Arthur Freed at MGM after Mamoulian had been inactive in the cinema for five years, and one could write a book about its brilliant inventions: the informal staging of numbers in the open air, with the kids dancing around on real lawns under real trees; the half parody, half affectionate re-creation of famous American paintings during the graduation day ceremony; the erotic nightmare in which Marilyn Maxwell, as a statuesque saloon girl Mickey Rooney has unwisely tangled with, gets bigger and bigger, redder and redder as he gets drunker and drunker, less and less sure of himself. Or for that matter the passages in which nothing apparently remarkable is done, but the atmosphere of small-town American life in the 1900s, the spirit of a disparate but united, loving family, is captured in vivid, affectionate detail.

The film was a failure, critical and commercial, and it was ten years before Mamoulian returned, again for Freed, again at MGM, with *Silk Stockings*.

Fred Astaire/Silk Stockings

Curiously, this was his first dancing musical: *Love Me Tonight* and *High, Wide and Handsome* have no real formal dancing, and neither does *Summer Holiday* when you look at it closely, though one carries away the impression that it has. In *Silk Stockings*, though, Mamoulian has made the ultimate dancer's musical. Seldom has dance on the screen been so perfectly used as an extension of dialogue, to take over when words fail and tell us things no words can. Fred Astaire and Cyd Charisse are at the top of their form, Mamoulian's invention is as prodigious as ever, and the film, also at the time a critical and commercial disappointment, has now established itself unmistakably as a classic. For a lesson in how things should be done, it seems, you can't beat an old master; and if the musical as the 1940s made it had to go out at all, it could hardly have gone out in grander style.

*Margaret O'Brien and Judy Garland/Meet Me in
St Louis*

The end of the 1950s looked like the complete eclipse of the Hollywood musical. Costs, they said, were rising by leaps and bounds, the foreign market was becoming increasingly important for balancing books, musicals could not be dubbed and foreigners didn't like them anyway: the great musical stars were getting older and no new talents seemed to be coming along to take their places. So why bother? So for several years hardly anybody did. After the last great flare-up of 1956–7 with *Funny Face* and *Silk Stockings* something seemed to have died. Minnelli's *Gigi* of 1958 was tame, his *Bells are Ringing* of 1960 game, but thereafter he gave up musicals altogether for a decade. The following year there was *Damn Yankees*, which was silly and stylish, and *South Pacific* which was afflicted with elephantiasis and green sickness, and was dire. 1959 and 1960 had a *Porgy and Bess* which should have been shot by Mamoulian but was, alas, finally handed over to Preminger, a flat *Can Can*, and the intermittently lively but style-less *Li'l Abner*. Not much there to encourage. But salvation, of a sort, was at hand: in 1961 along came *West Side Story*. It was big, it was pretentious, it was cultural, the Continentals liked it (even with subtitles) and it made a lot of money. The answer was obvious, or seemed so. To make money with musicals, you gotta think big. And the corollary of think big was, apparently, think Broadway.

Thinking Broadway was not a new phenomenon in the musical. Hollywood producers with their pathetic longing for the 'safe' property, the 'pre-sold' audience, had always been just as likely to buy up the rights of the latest Broadway musical hit as the latest sell-out straight play, the latest best-selling novel and so on. But rarely did they follow the lead of the original very closely once they had got it: as we have seen, they were most likely to junk most of the score, rewrite the story and start again, probably retaining only the title. Most of the great screen musicals have been originals: *Meet Me in St Louis*, *Singin' in the Rain*, *Love Me Tonight*, *Summer Holiday*, *High, Wide and Handsome*, *The Pirate*, *An American in Paris*, *The Love Parade*, *Monte Carlo*, *One Hour With You*, *A Star is Born*, and even, despite their titular reference to Broadway shows, *The Band Wagon* and *Funny Face*. The only important exceptions I can think of, *The Smiling Lieutenant* and *On the Town*, transform their material so completely as to become in effect original screen musicals. 'Fidelity' to the original was something almost unknown – until, that is, they started making films of Rodgers and Hammerstein's great Broadway successes: *Oklahoma!* in 1955, *Carousel* and *The King and I* in 1956, *South Pacific* in 1958. The stage success of these had been so legendary that (a) it allowed Rodgers and Hammerstein to make their own terms for the filming, (b) even if they had permitted it, which they probably wouldn't have, no one was likely to tamper too much with what looked like a guaranteed gold mine. So the films were faithful to a fault: little more than records of the stage productions, pruned here and there and using the odd location in order to 'make the most of the medium', but taking over the big numbers, the dream ballets and so on pretty well intact from the stage.

It was not a good idea, artistically; not surprisingly, the most effective of the films as a film was *The King and I*, largely because the show was the most intimate, the most firmly centred on the two principal performers. But all the films made money, and *South Pacific* in particular made a very great deal of money indeed. This gave confidence just when confidence was what the screen musical most sorely needed. But confidence in the wrong things; confidence in the inherent power of the Broadway hit to pull in a large cinema audience more or less however it was filmed, but preferably keeping as closely as possible to the original. *West Side Story* only reconfirmed this. Directed by its original Broadway director, Jerome Robbins, in collaboration with Robert Wise, a Hollywood professional who had never made a thorough-going musical, it stuck very closely to Broadway. Essentially, that is: the famous dance routines were carried over to film almost intact, though a lot of rather strenuous ingenuity was brought to bear on recording them in obtrusively 'cinematic' terms, and all the songs were still there embedded in the same dramatic structure. The only radical difference was that while the stage musical all took place in boldly stylized sets, and was consistent in its degree of removal from reality, the film chose to shoot as much as possible on the streets where it had never really happened. The result was a terrible grinding of gears as a collection of likely enough looking slum louts strolling along in a real New York street began to skip and jump in an obviously balletic fashion and then went into their dance. Far from easing the transitions, which in the stage show had been impeccably smooth and well-ordered, from speech to song or dance and back again, it made them stick out like a sore thumb.

All the same, the film had an international success and determined the screen musical of the 1960s in more ways than one: not only in its subservience to a Broadway original, and in its insistence on sheer size – the enormously wide screen, packed as often as not just like a proscenium arch at grand finale time – but in another, even less happy characteristic, its use of non-singing, non-dancing film stars in the leads. For star power, in film terms, was usually the one thing a Broadway show lacked, the one thing that the cinema could effectively add by dropping in a name or two. And since there were not that many musical names around any more, well, any old name would do. A singing voice could be dubbed on, often with little detailed care that it should match the performer it was attributed to, or it might be fun to let the performer have a go at singing for himself, and let the audience share the joke. Dancing was more difficult – it is always easier to make a straight actor look like a singer than a dancer – but that too could probably be arranged by modifying the choreography so that the real dancers could just dance around him or her, and with the magic of the cinema you would never notice. Or would you? In most cases, yes. There may be little harm in giving a dancer a singing voice – Rita Hayworth and Cyd Charisse are outstanding examples of dancers who have always had someone else sing for them on screen – because such performers have at least one outstanding musical talent. But when all the musical talents are borrowed or faked it really shows.

With this pattern in mind, most of the big film musicals of the 1960s can be accounted for: *Bye Bye Birdie, Gypsy, The Music Man, The Unsinkable Molly Brown, My Fair Lady, The Sound of Music, How to Succeed in Business Without Really Trying, Camelot, Half a Sixpence, Oliver!, Sweet Charity, Funny Girl, Hello Dolly!* All straight from recent stage hits, often much too straight. All big screen extravaganzas, giving as it is hoped size and stature to the film by sheer magnification. Many of them putting film stars who cannot do justice to the songs in the lead – Rosalind Russell in *Gypsy*, Audrey Hepburn in *My Fair Lady*, Vanessa

Redgrave and Franco Nero in *Camelot* – so that they need to be dubbed or lean heavily on the assistance of modern recording techniques. And even those which did retain the original stage stars, like Robert Preston in *The Music Man* and Rex Harrison and Stanley Holloway in *My Fair Lady*, did so only it seems after terrible soul-searching and desperate attempts to cast someone else. There have also been some musicals based on past hits, which were not so obvious as surefire material and therefore not so sacrosanct – *Jumbo*, *Finian's Rainbow*, *Paint Your Wagon* – and these have in general turned out better, their makers revelling in their freedom from the shackles of audience expectations and established patterns. And there have been a handful of originals, ranging from the good – *Thoroughly Modern Millie* – through the moderate – *Goodbye Mr Chips*, *Mary Poppins* – to the awful – *Dr Dolittle*, *Chitty Chitty Bang Bang*.

Totting up the record, it is not encouraging. Of all these films I can whole-heartedly approve only of *Finian's Rainbow*, *Jumbo*, *The Sound of Music* and *Funny Girl*, with a milder recommendation for *How to Succeed*, *Half a Sixpence*, *Hello Dolly!* and *Paint Your Wagon*. If there is a pattern in that, it is difficult to see, but some possible lines of consistency suggest themselves. First, that musical films are best if they are created directly for the screen, or, if they must come from stage shows, they should come from stage shows which were flops, or a long time ago, or are in some other way not held in reverence by those making the film. Second, the presence of a musical specialist somewhere around does help. A real musical star, like Julie Andrews or Barbra Streisand, or the unsinkable Fred Astaire, can make all the difference; a director who knows his music and/or his dance, like Charles Walters on *Jumbo*, George Sidney on *Half a*

Fred Astaire/Finian's Rainbow

Sixpence, Gene Kelly on *Hello Dolly!*, is an inestimable asset. Not that musicals absolutely must be made by musical specialists. One or two really great musicals in the past have been made by people with no special experience in the genre – Cukor's *A Star is Born* above all, though he was not able to pull it off again in the stiff, stagy *My Fair Lady*. And other non-specialists have turned a hand to the musical with pleasing results – George Marshall in that sophisticated, stylized Western send-up *Red Garters*, Howard Hawks in the splendidly brash and vulgar *Gentlemen Prefer Blondes*. In the 1960s more and more musicals have been confided to the direction of non-musical directors, sometimes, admittedly, with happy results, more often disastrous.

I remember, for instance, Richard Fleischer saying in the second week of shooting on *Dr Dolittle* that though he had never made a musical before, no one had seen a foot of the film he had shot, and no one, least of all himself, could say whether he would have any talent for it or not, he had already been offered no fewer than five other big musical properties to direct. As it turned out, it was just as well he had enough sense not to accept any of them, since *Dr Dolittle* proved to be one of his unhappiest ventures, with no noticeable feeling for the musical form at all. But the same kind of thinking, if one may dignify it with the name, has given *Oliver!* to Carol Reed, *Chitty Chitty Bang Bang* to Ken Hughes, *Funny Girl* to William Wyler, *The Sound of Music* to Robert Wise, *Camelot* and *Paint Your Wagon* to Joshua Logan (on the strength, presumably, of *South Pacific*, which one would have thought no recommendation, except financially). Sometimes the gamble has paid off: *Paint Your Wagon* is a moderate success (though *Camelot* isn't); *Funny Girl* and *The Sound of Music* triumph by virtue of their stars, Barbra Streisand and Julie Andrews respectively, and because Wyler and Wise show a flair for music, and a gift for sheer film-making, which leaves them unintimidated by the reputation of the properties they are handling and untrammelled by timid subservience to the conventions of the genre they are tackling for the first real time. (One might hope this would be true of most established directors new to the genre, but it seems to work the other way: they usually fall into ineffectual attempts to cover their insecurity by following set patterns.)

And meanwhile, as people in Hollywood, faced with disappointing returns from most of the latest crop of super-musicals, begin again, for perhaps the dozenth time in the musical's brief history, to pronounce the genre dead beyond hope of resuscitation, a new variation has been quietly making its appearance. This might be best defined as the scopitone musical. Scopitone is a sort of juke-box originating in Italy which accompanies records with filmlets of the singer singing or atmospheric three-minutes of landscape and happening, or psyche-delic play of coloured lights. And unobtrusively the same principle seems to be cropping up in large-screen entertainments. In films such as *Easy Rider*, or *Alice's Restaurant*, or *Zabriskie Point*, or, given a slightly more classical bias in the choice of music, *2001*, the soundtrack is the essential; the musical selections dictate the shape of the film, while the visuals provide merely a lively or restful accompaniment, something to keep our eyes agreeably occupied while we listen. Even more obviously dramatic, plotty films like *Midnight Cowboy* and *Butch Cassidy and the Sundance Kid* are liable to take time out every now and then for a sequence which works in this way: the pretty high-jinks on a bicycle which

accompany the wholly irrelevant and anachronistic song 'Raindrops are falling on my head' in *Butch Cassidy* are a striking case in point. Perhaps the responsibility for this development can be placed at the door of Francis Ford Coppola, who made the first and in most ways the most satisfactory scopitone film, *You're a Big Boy Now* (1967). But at least that was made with a real musical feeling (which of course Coppola has since confirmed by making a fully-fledged musical in *Finian's Rainbow*). It is almost a musical with no musical numbers delivered directly from the screen; it has the distinct feel of a musical to it. As much can hardly be said of the rest: at best they seem to have found a neat, unobtrusive way of supporting the flagging cinema with infusions from the booming pop record business. But not, unfortunately, a way that qualifies them as true musicals.

Is there a hope, then, for the musical, as we enter the 1970s? Well, there is always hope. Hope wells up in one as the camera at the opening of *The Sound of Music* soars high over the Austrian Alps and then plunges magically downwards the tiny figure of Julie Andrews as she launches into the title song. Hope is triumphantly confirmed as the first half of *Funny Girl* ends with the most shattering helicopter shot in the history of the cinema, conveying limitless elation as we descend from way above the Statue of Liberty right into close-up of Barbra Streisand in full song and perfect synchronization, only to whirl away again until she is only a tiny orange speck in the midst of the great grey sea. While such things can be done with such stars it is impossible not to go on hoping. And if a new director, a young director, can make one of the most sheerly inventive, most free, invigorating film musicals we have had in years, as Francis Ford Coppola has with *Finian's Rainbow*, then the hope seems more

98

Sarah Seegar, Mary Wickes, Hermione Gingold,
Adnia Rice and Peggy Mondo/The Music Man

than ever justified. It is strangely appropriate, though, that if we usher in the new era with a film by a new director still in his thirties, primed with all the latest techniques of film-making from Godard to Lester, we should at the same time be ending more or less where we began: with the great, the imperishable, the one-and-only Fred Astaire.

85590

Gene Kelly and Judy Garland/The Pirate

Selected Filmographies

Researching films made six thousand miles away and anything up to forty years ago is no sinecure, even for a confirmed filmophile. Many of the original production companies are no longer in existence; those which are have usually sold the greater part of their output to TV along with the appropriate files, cue sheets, and publicity material.

Fortunately, I have had some assistance from enlightened individuals in the industry, most notably Max Bercutt of Warner Brothers at Burbank, who showered cast lists, musical programmes and stills on me with an enthusiasm above and beyond the call of duty. Thanks are also due to people in London offices including John Fairbairn and Jo Wimshurst of 20th Century-Fox; Tony Owttrim and Geoffrey Romney of Walt Disney Productions; Clive Sutton of United Artists; Edward A. Patman of MGM; Michael Buist of Universal and Gerry Lewis of Paramount, all of whom responded to various degrees.

Acknowledgements also to my good friend Robert Farnon; to Albert Hand and his Elvis Presley Fan Club; Decca Records for their store of information on Bing Crosby films; Roy Bower who added to my knowledge of jazz films, as did Reg Cooper of *Jazz Journal* and various of his readers who volunteered information.

Above all, many thanks to the British Film Institute. Without the resources of their Research and Information Library I could not have started. Since every new reference book must necessarily be a collation of what has gone before, I must credit some of the existing literature which has confirmed the accuracy of many of the facts herein: the *Film Daily* Year Books, *The ASCAP Biographical Dictionary*, *Popular Music* (Nat Shapiro), *The Filmgoer's Companion* (Leslie Halliwell), *The Blue Book Of Hollywood Musicals* (Jack Burton), *Immortals Of The Screen* (Ray Stuart), *Leo's Encyclopedia* (MGM Publicity Dept.), together with more biographies than I can possibly name.

Inevitably, taking into consideration the sparseness of credits for films of the twenties and thirties, there are gaps. But not, I hope, inaccuracies.

In one sense the book is incomplete. The reader will search in vain for the Western 'musicals' of Roy Rogers, Gene Autry and others; nor have I found it expedient to include dramatic films with occasional interpolated songs or night-club sequences. In one way or another, every one of the 1,443 films listed in this book can, to the best of my knowledge, be classed as a 'Hollywood Musical'.

In the final analysis, the greatest guarantee of veracity must come from personal viewing of the films concerned. Having watched and annotated Hollywood musicals ever since I was first hooked on the Astaire-Rogers films, my own experience has been strengthened in recent years by the willingness of the television companies to remind us of the many great musicals of the past.

But even this doesn't tell the whole story. Many historic musicals have gone beyond recall, or emerge occasionally to limited viewing at the National Film Theatre. It was my original intention, and is still my hope, that the names in this part of the book will not only trace the history of the Hollywood Musical and all those talented people involved with the *genre*, but will ensure that these films which formed such an important part of the screen industry will not be completely forgotten.

Carlyon Bay   A. J.
1971

Abbreviations:

Prod Producer (also indicating associate producer where known).

Dir Director.

Scr Author(s) of screenplay, also listing original source, i.e. novel, Broadway show, or original screen story.

Ph Director of photography, or, in some instances, cameraman.

Art Art director. In some cases may include set designer. Some recent films have referred to production designer.

Sets Set designer or, in some cases, decorator.

Ed Film editor.

Cost Costume designer.

Sound Sound engineer or director. Credit sometimes given only to Head of Sound Department.

Dances May refer equally to dance director, director of musical sequences, or – as in some recent ventures – choreographer.

Mus Musical director. Credits of older films were often vague, giving one generic 'Music' credit to either the composer of the background score, the conductor of the studio orchestra (not necessarily the same person) or, more often than not, the Head of the Music Department who received the credits (and the Oscars) for the work of his staff.

Songs As a general rule the composer is listed before the lyricist, but occasionally the writers may collaborate on both music and lyrics.

The lettering (*a*), (*b*), (*c*), etc. after artists' names is reflected in the song titles, indicating the performers of each musical number (space has not permitted any distinction between vocal and dancing performances). I have tried to make this section as complete as possible, to settle any arguments about who sang what in which film, but positive identification of performers, especially in older films, has not always been possible.

Alexander's Ragtime Band Fox 1938 (*105 mins*) *with* Alice Faye (*a*), Tyrone Power (*b*), Ethel Merman (*c*), Don Ameche (*d*), Jack Haley (*e*), Dixie Dunbar (*f*), Chick Chandler (*g*), Wally Vernon (*h*). *Prod* Darryl F Zanuck, *assoc* Harry Joe Brown. *Dir* Henry King. *Scr* Kathryn Scoular, Lamar Trotti. *Ph* J Peverell Marley. *Art* James Basevi, Bernard Herzbrun. *Sets* Boris Leven. *Ed* Barbara McLean. *Dances* Seymour Felix. *Mus* Alfred Newman. *Songs* Irving Berlin: Alexander's ragtime band (*a*); Easter parade (*d*); Blue skies (*a-c*); My walking stick (*c*); Remember (*a*); Everybody's doing it (*a-f-h*); When the midnight choo-choo leaves for Alabam (*a*); Everybody step (*c*); A pretty girl is like a melody (*c*); Heat wave (*c*); We're on our way to France (*chorus*); All alone (*a*); What'll I do (*chorus*); Say it with music (*c*); Ragtime violin (*trio*); International rag (*a-g-b*); This is the life (*b*); Go to the devil (*c*); Oh how I hate to get up in the morning (*e and chorus*); Now it can be told (*a-d*); I can always find a little sunshine in the YMCA (*e*). *Background music* Marie; Cheek to cheek; When I lost you; Lazy.

An American In Paris MGM 1951 (*113 mins*) *with* Gene Kelly (*a*), Leslie Caron (*b*), Georges Guetary (*c*), Oscar Levant (*d*), Nina Foch. *Prod* Arthur Freed. *Dir* Vincente Minnelli. *Scr* Alan Jay Lerner. *Ph* John Alton, Alfred Gilks. *Art* Cedric Gibbons, Preston Ames. *Sets* Edwin B Willis, Keogh Gleason. *Ed* Adrienne Fazan. *Cost* Orry-Kelly, Irene Sharaff, Walter Plunkett. *Sound* Douglas Shearer. *Dances* Gene Kelly. *Mus* Johnny Green, Saul Chaplin. *Songs* George and Ira Gershwin: I got rhythm (*a and children*); 'S wonderful (*a-c*); Love is here to stay (*a-b*); I'll build a stairway to paradise (*c*); Tra-la-la-la (*a-d*); By Strauss (*a-c-d*); Embraceable you (*b*); Concerto in F (*d*); An American in Paris – ballet (*a-b*).

Anchors Aweigh MGM 1944 (*140 mins*) *with* Frank Sinatra (*a*), Gene Kelly (*b*), Kathryn Grayson (*c*), Jose Iturbi (*d*), Sharon McManus

(*e*), Carlos Ramirez (*f*), Dean Stockwell, Pamela Britton. *Prod* Joe Pasternak. *Dir* George Sidney. *Scr* Isobel Lennart from a story by Natalie Marcin. *Ph* Robert Planck, Charles Boyle. *Art* Cedric Gibbons. *Sets* Edwin B Willis, Richard Pefferle. *Ed* Adrienne Fazan. *Cost* Irene, Kay Dean. *Sound* Douglas Shearer. *Dances* Stanley Donen, Gene Kelly. *Mus* Georgie Stoll. *Songs* Jule Styne and Sammy Cahn: What makes the sunset (*a*); I begged her (*a-b*); I fall in love too easily (*a*); The charm of you (*a*); We hate to leave (*a-b*); *Interpolated songs* Jealousy (*c*); The donkey serenade (*d*); My heart sings (*c*); Another kiss (*a-c*); The worry song (cartoon dance) (*b*); Mexican hat dance (*b-e*); Brahms' lullaby (*a*); Hungarian dance no. 2 (*d*); If you knew Susie (*a-b*); Waltz serenade (*c*).

And The Angels Sing Paramount 1944 (*96 mins*) *with* Betty Hutton (*a*), Dorothy Lamour (*b*), Fred MacMurray (*c*), Mimi Chandler (*d*), Diana Lynn (*e*), Eddie Foy Jnr, Harry Barris. *Prod* E D Lestin. *Dir* George Marshall, Claude Binyon. *Scr* Norman Panama, Melvin Frank. *Ph* Karl Struss. *Art* Hans Dreier, Hal Pereira. *Sets* Ray Moyer. *Ed* Eda Warren. *Cost* Edith Head. *Sound* Gene Merritt, Joel Moss. *Dances* Danny Dare. *Mus* Victor Young. *Songs* Jimmy Van

An American in Paris/Gene Kelly; Anchors Aweigh/Frank Sinatra and Gene Kelly

Heusen and Johnny Burke: Bluebirds in my belfry (*a*); For the first hundred years (*a-b-d-e*); His rocking horse ran away (*a*); It could happen to you (*b-c*); Knocking on your own front door (*a-b-d-e*); My heart's wrapped up in gingham (*c*); How does your garden grow (*a-b-d-e*); Elman and Mercer: And the angels sing (*a-b-c-d-e*).

Annie Get Your Gun MGM 1949 (*107 mins*) *with* Betty Hutton (*a*), Howard Keel (*b*), Keenan Wynn (*c*), Louis Calhern (*d*), Edward Arnold, J Carroll Naish, Benay Venuta. *Prod* Arthur Freed. *Dir* George Sidney. *Scr* Sidney Sheldon from the book by Herbert and Dorothy Fields for the Broadway show. *Ph* Charles Rosher. *Art* Cedric Gibbons, Paul Groesse. *Sets* Edwin B Willis. *Ed* James A Newcom. *Sound* Douglas Shearer. *Dances* Robert Alton. *Mus* Adolph Deutsch, *assoc* Roger Edens. *Songs* Irving Berlin: Colonel Buffalo Bill (*chorus*); I got the sun in the morning (*a*); You can't get a man with a gun (*a*); They say it's wonderful (*a-b*); My defences are down (*b*); Doin' what comes naturally (*a*); There's no business like show business (*a-b-c-d*); The girl that I marry (*b*); Anything you can do (*a-b*); I'm an Indian too (*a*); Let's go West again (deleted before release).
Note Film was started under Busby Berkeley's direction, starring Judy Garland, who recorded the soundtrack before illness called a halt to the production.

Anything Goes Paramount 1936 (*92 mins*) *with* Ethel Merman (*a*), Bing Crosby (*b*), Grace Bradley (*c*), Charles Ruggles, Ida Lupino, Chill Wills (*d*), The Avalon Boys (*e*), Arthur Treacher. *Prod* Benjamin Glazer. *Dir* Lewis Milestone. *Scr* Guy Bolton, Howard Lindsay and Russell Crouse from their book for the Broadway show by Cole Porter. *Ph* Karl Struss. *Art* Hans Dreier. *Ed* Eda Warren. *Mus* Victor Young. *Songs* Cole Porter: Anything goes (*a*); You're the top (*a-b*); There'll always be a lady fair (*a-d-e*); I get a kick out of you (*a*). Hoagy Carmichael: Moonburn (*b*). Frederick Hollander and Leo Robin: Am I awake (*b*); My heart and I (*b*); Hopelessly in love; Shangai-de-ho. Richard Whiting and Leo Robin: Sailor beware (*b*).
Note Shown on TV as *Tops Is The Limit*.

Anything Goes Paramount 1956 (*106 mins*) *with* Bing Crosby (*a*), Donald O'Connor (*b*), Mitzi Gaynor (*c*), Zizi Jeanmaire (*d*), Phil Harris, Kurt Kasznar. *Prod* Robert Emmett Dolan. *Dir* Robert Lewis. *Scr* Sidney Sheldon from the book by Guy Bolton, P G Wodehouse, Howard Lindsay and Russell Crouse for the Broadway show by Cole Porter. *Ph* John F Warren. *Art* Hal Pereira, Joseph MacMillan Johnson. *Ed* Frank Bracht. *Cost* Edith Head. *Dances* Nick Castle, Roland Petit. *Mus* Joseph J Jilley. *Songs*

Cole Porter: Anything goes (*c*); I get a kick out of you (*d*); You're the top (*a-c*); Dream ballet – Let's do it / All through the night (*d*); It's de-lovely (*b-c*); All through the night (*a*); Blow Gabriel blow (*a-b-c-d*). Jimmy Van Heusen and Sammy Cahn: Ya gotta give the people hoke (*a-b*); A second-hand turban and a crystal ball (*a-b*); You can bounce right back (*b*).

April In Paris Warner 1952 (*101 mins*) *with* Doris Day (*a*), Ray Bolger (*b*), Claude Dauphin (*c*), Eve Miller, George Givot, Herbert Farjeon. *Prod* William Jacobs. *Dir* David Butler. *Scr* Jack Rose, Melville Shavelson. *Ph* Wilfrid M Cline. *Art* Leo K Kuter. *Sets* Lyle B Reifsnider. *Ed* Irene Morra. *Cost* Leah Rhodes. *Sound* C A Riggs, David Forrest. *Dances* Le Roy Prinz. *Mus* Ray Heindorf. *Songs* Vernon Duke and Sammy Cahn: April in Paris (lyr: E Y Harburg) (*a-c*); Isn't love wonderful (*a-b-c*); I'm gonna ring the bell tonight (*a-b and ensemble*); I know a place (*a*); That's what makes Paris Paree (*a-c*); Auprès de ma blonde (trad.) (*a-c*); Give me your lips (*c*).

At The Circus MGM 1939 (*87 mins*) *with* Groucho Marx (*a*), Harpo Marx (*b*), Chico Marx (*c*), Kenny Baker (*d*), Florence Rice (*e*), Eve Arden, Margaret Dumont. *Prod* Mervyn Le Roy. *Dir* Edward Buzzell. *Scr* Irving Brecher. *Ph* Leonard M Smith. *Art* Cedric Gibbons. *Sets* Edwin B Willis. *Ed* William Terhune. *Cost* Dolly Tree, Valles. *Sound* Douglas Shearer. *Dances* Bobby Connolly. *Mus* Franz Waxman, *assoc* Roger Edens. *Songs* Harold Arlen and E Y Harburg: Lydia the tattooed lady (*a*); Step up and take a bow (*d-e*); Two blind loves (*d-e*); Swingali (*b and Negro chorus*). Rodgers and Hart: Blue moon (*b*). Vejvoda and Brown: Beer barrel polka (*c*).

Babes In Arms MGM 1939 (*96 mins*) *with* Judy Garland (*a*), Mickey Rooney (*b*), Douglas McPhail (*c*), Leni Lynn (*d*), June Preisser (*e*), Betty Jaynes (*f*), Charles Winninger. *Prod* Arthur Freed. *Dir/Dances* Busby Berkeley. *Scr* Jack McGowan and Kay Van Riper from the Broadway show by Rodgers and Hart. *Ph* Ray June. *Art* Cedric Gibbons. *Sets* Edwin B Willis. *Ed* Frank Sullivan. *Cost* Dolly Tree. *Sound* Douglas Shearer. *Mus* Georgie Stoll, *assoc* Roger Edens. *Songs* Richard Rodgers and Lorenz Hart: Babes in arms (*a-b-c-f and ensemble*); Where or when (*a-c-f*); The lady is a tramp (*e*). Nacio Herb Brown and Arthur Freed: Broadway melody (*orch.*); Broadway rhythm (*a*); Good morning (*a-b-c and chorus*); You are my lucky star (*f*). Freed, Lyman and Arnheim: I cried for you (*a*); Harold Arlen and E Y Harburg: God's country (*a-b-c-f and chorus*). Roger Edens: Figaro (*a*). *Minstrel show interpolating* Oh Susanna (*a-b*); Ida

(*b*); By the light of the silvery moon (*a-b*); I'm just wild about Harry (*a-b and chorus*); Ja-da.
Babes On Broadway MGM 1941 (*118 mins*) *with* Judy Garland (*a*), Mickey Rooney (*b*), ·Virginia Weidler (*c*), Ray MacDonald (*d*), Richard Quine (*e*), Fay Bainter. *Prod* Arthur Freed. *Dir/Dances* Busby Berkeley. *Scr* Fred Finklehoffe, Elaine Ryan. *Ph* Lester White. *Art* Cedric Gibbons. *Sets* Edwin B Willis. *Ed* Frederick Y Smith. *Sound* Douglas Shearer. *Mus* Georgie Stoll, *assoc* Roger Edens. *Songs* Burton Lane and Ralph Freed: Babes on Broadway (*chorus*); How about you (*a-b*). Lane and E Y Harburg: Chin up, cheerio, carry on (*a*); Anything can happen in New York (*b-d-e*). Roger Edens: A bombshell from Brazil (*b*). Edens and Freed: Hoe down (*a-b-d*). Harold Rome: F D R Jones (*a*). *Interpolated songs* I belong to Glasgow (Lauder) (*b*); Mama yo quiero (Paiva and Stillman) (*b*); Waiting for the Robert E. Lee (*a*).
The Band Wagon MGM 1953 (*112 mins*) *with* Fred Astaire (*a*), Cyd Charisse (*b*), Nanette Fabray (*c*), Jack Buchanan (*d*), Oscar Levant (*e*), James Mitchell. *Prod* Arthur Freed, *assoc* Roger Edens. *Dir* Vincente Minnelli. *Scr* Betty Comden and Adolph Green, based on the 1931 Broadway show. *Ph* Harry Jackson. *Art* Cedric Gibbons, Preston Ames. *Sets* Edwin B Willis, Keogh Gleason. *Ed* Albert Akst. *Cost* Mary Ann Nyberg. *Sound* Douglas Shearer. *Dances* Michael Kidd. *Mus* Adolph Deutsch. *Songs* Arthur Schwartz and Howard Dietz: A shine on your shoes (*a*); By myself (*a*); Dancing in the dark (*a-b*); Triplets (*a-c-d*); A new sun in the sky (*b*); I guess I'll have to change my plan (*a-d*); Louisiana hayride (*c*); I love Louisa (*a-e*); That's entertainment (*a-b-c-d-e*); Beggar's waltz (*a*); Girl hunt ballet adpt. Roger Edens, narr. written by Vincente Minnelli (*a-b*); *Background music* You

and the night and the music; High and low; Something to remember you by.
Note Vocals for Cyd Charisse dubbed by India Adams.
The Barkleys Of Broadway MGM 1948 (*109 mins*) *with* Fred Astaire (*a*), Ginger Rogers (*b*), Oscar Levant (*c*), Gale Robbins, Billie Burke, Alan Marshall. *Prod* Arthur Freed. *Dir* Charles Walters. *Scr* Betty Comden, Adolph Green. *Ph* Harry Stradling. *Art* Cedric Gibbons, Edward Carfagno. *Sets* Edwin B Willis. *Ed* Albert Akst. *Cost* Irene. *Sound* Douglas Shearer. *Dances* Robert Alton, Hermes Pan. *Mus* Lennie Hayton. *Songs* Harry Warren and Ira Gershwin: They can't take that away from me (mus: George Gershwin) (*a-b*); Shoes with wings on (*a*); Swing trot (*a-b*); Manhattan downbeat (*a-b*); My one and only Highland fling (*a-b*); You'd be so hard to replace (*a*); A weekend in the country (*a-b-c*); Tchaikovsky Piano Concerto (*c*); Sabre dance (Khachaturian) (*c*).
The Belle Of New York MGM 1952 (*82 mins*) *with* Fred Astaire (*a*); Vera-Ellen (*b*), Alice Pearce (*c*), Gale Robbins, Keenan Wynn, Marjorie Main. *Prod* Arthur Freed. *Dir* Charles Walters. *Scr* Chester Erskine, Robert O'Brien and Irving Elinson from the operetta story by Hugh Morton. *Ph* Robert Planck. *Art* Cedric Gibbons, Jack Martin Smith. *Sets* Edwin B Willis. *Ed* Albert Akst. *Sound* Douglas Shearer. *Dances* Robert Alton. *Mus* Adolph Deutsch, *assoc* Roger Edens. *Songs* Harry Warren and Johnny Mercer: The belle of New York (*chorus*); Baby doll (*a*); Bachelor dinner song (*a*); Naughty but nice (*b-c*); Oops (*a-b*); Thank you Mr Currier, thank you Mr Ives (*b*); I love to beat a big bass drum (*a*); Bride's wedding day; Seeing is believing. Burton Lane and Alan Jay Lerner: I wanna be a dancin' man (*a*). Roger Edens: Let a little love come in (*b-c*)

Note Vocals for Vera-Ellen dubbed by
Anita Ellis.

Bells Are Ringing Freed/MGM 1960 (*127 mins*)
with Judy Holliday (*a*), Dean Martin (*b*), Eddie
Foy Jnr (*c*), Hal Linden (*d*), Fred Clark, Jean
Stapleton, Frank Gorshin. *Prod* Arthur Freed.
Dir Vincente Minnelli. *Scr* Betty Comden and
Adolph Green from their book for the Broadway
show. *Ph* Milton Krasner. *Art* George W Davis,
Preston Ames. *Ed* Adrienne Fazan. *Sound*
Wesley C Miller. *Dances* Charles O'Curran.
Mus André Prévin. *Songs* Jule Styne, Betty
Comden and Adolph Green: Bells are ringing
(*chorus*); Drop that name (*a*); Do it yourself (*b*);
I'm going back (*a*); I met a girl (*b*); It's a perfect
relationship (*a*); A simple little system (*c*); Better
than a dream (*a-b*); The party's over (*a*); The
Midas touch (*d*); Just in time (*a-b*).

The Best Things In Life Are Free Fox
1956 (*104 mins*) *with* Gordon Macrae (*a*),
Dan Dailey (*b*), Ernest Borgnine (*c*), Sheree
North (*d*), Jacques D'Amboise (*e*), Norman
Brooks (*f*), Roxanne Arlen (*g*), Murvyn Vye.
Prod Henry Ephron. *Dir* Michael Curtiz.
Scr William Bowers, Phoebe Ephron from a
story by John O'Hara. *Ph* Leon Shamroy.
Art Lyle R Wheeler, Maurice Rainsford.
Sets Walter M Scott, Paul S Fox, John De Cuir.
Ed Dorothy Spencer. *Dances* Rod Alexander.

Mus Lionel Newman. *Songs* B G De Sylva,
Lew Brown and Ray Henderson: The best
things in life are free (*a-b-c-d*); Sonny boy
(*a-b-c-f*); You try somebody else (*d-g*); Without
love (*d*); Button up your overcoat (*a-b-c-d*); This
is the Mrs (*c*); Black bottom (*d-e and chorus*);
This is my lucky day (*b*); One more time (*a*);
Sunny side up (*a-b-c-d*); Together (*b*); Good
news (*a-b-c*); The birth of the blues (*a-d-e*); It
all depends on you (*a-b-c-d*); Just a memory
(*d and chorus*); Broken hearted (*a-b-c*); If I had
a talking picture of you (*unidentified singer*);
Don't hold everything (*d*).
Biography of song-writers De Sylva, Brown
and Henderson.

The Big Broadcast Paramount 1932 *with*
Bing Crosby (*a*), The Mills Brothers (*b*), The
Boswell Sisters (*c*), Cab Calloway and his
Orchestra (*d*), Kate Smith (*e*), Arthur Tracy
(The Street Singer) (*f*), Eddie Lang (*g*), Vincent
Lopez and his Orchestra (*h*), Stuart Erwin.
Dir Frank Tuttle. *Scr* George Marion Jnr
based on 'Wild Waves' by William Ford
Manley. *Ph* George Folsey. *Songs* Ralph
Rainger and Leo Robin: Here lies love (*a-f-h*);
Please (*a-g*). Harold Arlen and Ted Koehler:
Kicking the gong around (*d*). *Interpolated songs*
Dinah (*a-b*); Where the blue of the night meets
the gold of the day (*a*); Crazy people (*c*); When
the moon comes over the mountain (*e*); Tiger
rag (*b*); Minnie the moocher (*d*); It was so
beautiful (*e*); Hot toddy.

The Big Broadcast Of 1936 Paramount 1935
(*97 mins*) *with* Bing Crosby (*a*), Ethel Merman
(*b*), Ina Ray Hutton and her Orchestra (*c*),
Ray Noble and his Orchestra (*d*), Bill Robinson
(*e*), The Vienna Boys Choir (*f*), Lyda Roberti (*g*),
Jack Oakie (*h*), Henry Wadsworth (*i*), The
Nicholas Brothers (*j*), Burns and Allen, Amos
'n' Andy, Charles Ruggles. *Prod* Benjamin
Glazer. *Dir* Norman Taurog. *Scr* Walter De
Leon, Francis Martin, Ralph Spence. *Ph* Leo
Tover. *Art* Hans Dreier. *Ed* Ellsworth
Hoagland. *Dances* Le Roy Prinz. *Mus* Sigmund
Krumgold, Sam Wineland. *Songs* Ralph Rainger,
Richard Whiting, Leo Robin, Dorothy Parker: I
wished on the moon (*a*); Double trouble (*g-h-i*);
Why dream (*i*); Miss Brown to you (*d-e-j*);
Through the doorway of dreams; Amargura.
Ray Noble: Why stars come out at night (*d*);
Goodnight sweetheart (*d*). Harry Revel and
Mack Gordon: It's the animal in me (*b*).

The Big Broadcast Of 1937 Paramount 1936
(*100 mins*) *with* Ray Milland, Shirley Ross (*a*),
Burns and Allen, Jack Benny, Martha Raye
(*b*), Benny Fields (*c*), Bob Burns, Benny
Goodman and his Orchestra (*d*), Leopold
Stokowski and his Symphony Orchestra (*e*),

The Band Wagon/Fred Astaire and Cyd Charisse

Larry Adler (*f*), Frank Forrest (*g*), Sam Hearn, Louis Da Pron. *Prod* Lewis E Gensler. *Dir* Mitchell Leisen. *Scr* Francis Martin, Walter De Leon from a story by Erwin Gelsey, Arthur Kober, Barry Trivers. *Ph* Theodore Sparkuhl. *Art* Hans Dreier, Robert Usher. *Ed* Stuart Heisler. *Mus* Boris Moross. *Songs* Ralph Rainger and Leo Robin: Cross patch (*d*); La bomba (*g*); Vote for Mr Rhythm (*b*); I'm talking through my heart (*a*); Hi-ho the radio (*quartet*); Here's love in your eyes (*c-d-f*); You came to my rescue (*a-g*); Night in Manhattan (*c-d*) (deleted before release). *Interpolated songs* Here comes the bride (*b-d*); Bugle call rag (*d*); Fugue in G minor (Bach) (*e*).

The Big Broadcast Of 1938 Paramount 1937 (*90 mins*) *with* Bob Hope (*a*), Shirley Ross (*b*), Dorothy Lamour (*c*), Martha Raye (*d*), Kirsten Flagstad (*e*), Leif Ericson (*f*), Grace Bradley (*g*), Tito Guizar (*h*), Shep Fields and his Rippling Rhythm (*i*), W C Fields, Ben Blue. *Prod* Harlan Thompson. *Dir* Mitchell Leisen. *Scr* Walter De Leon, Francis Martin, Ken Englund from a story by Frederick Hazlitt Brennan. *Ph* Harry Fishbeck, Gordon Jennings *Art* Hans Dreier. *Ed* Eda Warren, Chandler House. *Dances* Le Roy Prinz. *Mus* Boris Moross. *Songs* Ralph Rainger and Leo Robin: Thanks for the memory (*a-b*); You took the words right out of my heart (*c-f*); This little ripple had rhythm (*i*); Brunnhilde's battle cry – 'Die Walkure' (Wagner) (*e*); Zuni zuni; Don't tell a secret to a rose; Mama that moon is here again; The waltz lives on.

Billy Rose's Diamond Horseshoe Fox 1945 (*105 mins*) *with* Dick Haymes (*a*), Betty Grable (*b*), Phil Silvers (*c*), William Gaxton (*d*), Carmen Cavallaro (*e*), Beatrice Kay (*f*), Kenny Williams (*g*). *Prod* William Perlberg. *Dir/Scr* George Seaton. *Ph* Ernest Palmer, Fred Sersen. *Art* Lyle Wheeler, Joseph C Wright. *Sets* Thomas Little. *Ed* Robert Simpson. *Sound* E Clayton Ward, Roger Heman. *Dances* Hermes Pan. *Mus* Alfred Newman, Charles Henderson. *Songs* Harry Warren, Mack Gordon: Come to the Diamond Horseshoe (*b and chorus*); The more I see you (*a-b*); I wish I knew (*a-b*); In Acapulco (*b-e*); Play me an old-fashioned melody (*d-f*); A nickel's-worth of jive.
Title in GB: *The Diamond Horseshoe*.

Billy Rose's Jumbo Euterpe-Arwin/MGM 1962 (*125 mins*) *with* Doris Day (*a*), Stephen Boyd (*b*), Jimmy Durante (*c*), Martha Raye (*d*), Dean Jagger. *Prod* Joe Pasternak, *assoc* Martin Melcher, Roger Edens. *Dir* Charles Walters, Busby Berkeley (2nd unit). *Ph* William Daniels. *Scr* Sidney Sheldon from a story by Ben Hecht and Charles McArthur. *Art* George

W Davis, Preston Ames. *Ed* Richard W Farrell. *Sound* Wesley C Miller. *Dances* Busby Berkeley. *Mus* Georgie Stoll. *Songs* Richard Rodgers and Lorenz Hart: My romance (*a*), The circus is on parade (*a-b-c-d*); Over and over again (*a*); Little girl blue (*a*); Why can't I (*a-d*); The most beautiful girl in the world (*b-c*); This can't be love (*a*). Roger Edens: Sawdust, spangles and dreams (*a-b-c-d*).
Title in GB: *Jumbo*.

The Birth Of The Blues Paramount 1941 (*85 mins*) *with* Bing Crosby (*a*), Mary Martin (*b*), Jack Teagarden (*c*), Brian Donlevy (*d*), Ruby Elzy (*e*), Eddie 'Rochester' Anderson, Harry Barris, Carolyn Lee. *Prod* B G De Sylva, Monte Bell. *Dir* Victor Schertzinger. *Scr* Harry Tugend, Walter De Leon. *Ph* William C Mellor. *Art* Hans Dreier, Ernest Fegte. *Ed* Paul Weatherwax. *Cost* Edith Head. *Mus* Robert Emmett Dolan. *Songs* The birth of the blues (*a*); Memphis blues (*a*); My melancholy baby (*a*); Wait till the sun shines Nellie (*a-b*); By the light of the silvery moon (*a*); The waiter, the porter and the upstairs maid (*a-b-c*); Waiting at the church (*b*); St James Infirmary blues (*a*); Cuddle up a little closer (*e*); St Louis blues (*e*). *Instrumentals* Gotta go to the jailhouse; At a Georgia camp meeting; Tiger rag; Shine. *Note* Clarinet playing for Bing Crosby dubbed by Danny Polo, trumpet playing for Brian Donlevy by Poky Carriere.

Blues In The Night Warner/1st National 1941 *with* Priscilla Lane (*a*), Betty Field, Richard Whorf, Jack Carson, Elia Kazan, Lloyd Nolan, Jimmy Lunceford and his Orchestra (*b*), Will Osborne and his Orchestra (*c*). *Prod* Hal B Wallis, *assoc* Henry Blanke. *Dir* Anatole Litvak. *Scr* Robert Rossen from a play by Edwin Gilbert. *Ph* Ernie Haller. *Art* Max Parker. *Ed*

Owen Marks. *Cost* Damon Giffard. *Sound*
Everett A Brown. *Mus* Leo F Forbstein.
Songs Harold Arlen and Johnny Mercer: Blues
in the night (*b*); Says who, says you, says I
(*c*); This time the dream's on me (*a*); Hang on
to your lids, kids.

Blue Skies Paramount 1946 (*104 mins*) *with*
Bing Crosby (*a*), Fred Astaire (*b*), Billy De
Wolfe (*c*), Olga San Juan (*d*), Joan Caulfield
(*e*). *Prod* Sol C Siegel. *Dir* Stuart Heisler.
Scr Arthur Sheekman, Allan Scott. *Ph* Charles
Lang Jnr, William Snyder. *Art* Hal Pereira,
Hans Dreier. *Sets* Sam Comer, Maurice
Goodman. *Ed* Le Roy Stone. *Cost* Edith Head.
Sound Hugo Grenzbach, John Cope. *Dances*
Hermes Pan. *Mus* Robert Emmett Dolan.
Songs Irving Berlin: Blue skies (*a*); All by myself
(*a*); Heat wave (*b-d*); Getting nowhere (*a*); You
keep coming back like a song (*a-e*); I've got
my captain working for me now (*a-c*); Russian
lullaby (*a*); A serenade to an old-fashioned girl
(*e*); I'll be seeing you in C-u-b-a (*a-d*); A pretty
girl is like a melody (*quartet*); Not for all the
rice in China (*a*); How deep is the ocean (*a and
quartet*); Everybody step (*a*); A couple of song
and dance men (*a-b*); Puttin' on the Ritz (*b*);
The little things in life (*a*); You'd be surprised
(*d*); Always (*orch.*).

Born To Dance MGM 1936 *with* Virginia
Bruce (*a*), Eleanor Powell (*b*), James Stewart
(*c*), Sid Silvers (*d*), Frances Langford (*e*), Buddy
Ebsen (*f*), Una Merkel (*g*), Georges and Jalna
(*h*). *Prod* Jack Cummings. *Dir* Roy Del Ruth.
Scr Jack McGowan, B G De Sylva, Sid Silvers.
Ph Ray June. *Art* Cedric Gibbons. *Sets* Edwin
B Willis. *Ed* Blanche Sewell. *Cost* Adrian.
Sound Douglas Shearer. *Dances* Dave Gould.
Mus Alfred Newman, *assoc* Roger Edens.
Songs Cole Porter: I've got you under my skin
(*a*); Rap tap on wood (*b*); Easy to love (*b-c-e*);
Swingin' the jinx away (*b-f-e*); Entrance of
Lucy James (*a*); Love me, love my Pekinese
(*a*); Hey babe hey (*b-c-d-e-f-g*); I'm nuts about
you; Rolling home.

The Boys From Syracuse Universal 1940
(*74 mins*) *with* Allan Jones (*a*), Joe Penner (*b*),
Charles Butterworth (*c*), Rosemary Lane (*d*),
Martha Raye (*e*), Irene Hervey, Alan Mowbray.
Prod Jules Levey. *Dir* Edward A Sutherland.
Scr Leonard Spiegelgass, Charles Grayson,
Paul Gerard Smith from the Broadway show
by George Abbott and Rodgers and Hart
based on Shakespeare's 'A Comedy Of Errors'.
Ph Joseph Valentine. *Art* Jack Otterson.
Ed Milton Carruth. *Dances* Dave Gould. *Mus*
Charles Previn. *Songs* Richard Rodgers and
Lorenz Hart: Falling in love with love (*a-d*);
This can't be love (*d*); Sing for your supper

(*e*); Who are you (*a*); He and she (*b-e*); The
Greeks have no word for it (*e*).

Brigadoon MGM 1955 (*102 mins*) *with* Gene
Kelly (*a*), Cyd Charisse (*b*), Van Johnson (*c*),
Jimmy Thompson (*d*), Elaine Stewart, Barry
Jones, Hugh Laing, Eddie Quillan. *Prod* Arthur
Freed. *Dir* Vincente Minnelli. *Scr* Alan Jay
Lerner from his book for the Broadway show.
Ph Joseph Ruttenberg. *Art* Cedric Gibbons,
Preston Ames. *Sets* Edwin B Willis, Keogh
Gleason. *Ed* Albert Akst. *Cost* Irene Sharaff.
Sound Douglas Shearer. *Dances* Gene Kelly.
Mus Johnny Green. *Songs* Frederick Loewe
and Alan Jay Lerner: Brigadoon (*chorus*);
Almost like being in love (*a*); Waiting for my
dearie (*b*); Dance of the clans (*chorus*); I'll go
home with Bonnie Jean (*c-d*); Once in the
highlands/Prologue (*chorus*); There but for you
go I (*a*); Down on McConachy Square (*chorus*);
The heather on the hill (*a-b*); Come to me, bend
to me (*d*) (deleted before release).
Note Vocals for Cyd Charisse and Jimmy
Thompson dubbed by Carole Richards and
John Gustafson respectively.

The Broadway Melody MGM 1929 (*106 mins*)
with Charles King (*a*), Anita Page (*b*), Bessie
Love (*c*), Mary Doran (*d*), Jed Prouty, Kenneth
Thompson, Edward Dillon. *Prod* Lawrence
Weingarten. *Dir* Harry Beaumont. *Scr* Edmund
Goulding, Sarah Y Mason, Norman Houston,
James Gleason. *Ph* John Arnold. *Art* Cedric
Gibbons. *Ed* Sam Zimbalist. *Mus* Arthur
Lange. *Songs* Nacio Herb Brown and Arthur
Freed: The Broadway melody (*a-b-c*); The
wedding of the painted doll; You were meant
for me (*a-c*); The boy friend; Love boat; Truthful
Deacon Brown; Give my regards to Broadway
(Cohan).

The Broadway Melody Of 1936 MGM 1935
(*110 mins*) *with* Eleanor Powell (*a*), Robert
Taylor (*b*), Buddy Ebsen (*c*), Frances Langford
(*d*), June Knight (*e*), Vilma Ebsen (*f*), Jack
Benny, Sid Silvers. *Prod* John W Considine
Jnr. *Dir* Roy Del Ruth. *Scr* Moss Hart, Jack
McGowan, Harry Conn, Sid Silvers. *Ph*
Charles Rosher. *Art* Cedric Gibbons. *Ed*
Blanche Sewell. *Cost* Adrian. *Sound* Douglas
Shearer. *Dances* Dave Gould, Albertina Rasch.
Mus Alfred Newman, *assoc* Roger Edens.
Songs Nacio Herb Brown and Arthur Freed:
You are my lucky star (*a-d*); I've got a feeling
you're fooling (*a-b-d-e*); Sing before breakfast
(*a-c-f*); On a Sunday afternoon (*c-f*); Broadway
rhythm (*a-d*).

The Broadway Melody Of 1938 MGM 1937
with Eleanor Powell (*a*), George Murphy (*b*),
Sophie Tucker (*c*), Judy Garland (*d*), Buddy
Ebsen (*e*) Robert Taylor, Sid Silvers, The

Robert Mitchell Boychoir (*f*). *Prod* Jack Cummings. *Dir* Roy Del Ruth. *Scr* Jack McGowan, Sid Silvers. *Ph* William Daniels. *Art* Cedric Gibbons. *Sets* Edwin B Willis. *Ed* Blanche Sewell. *Sound* Douglas Shearer. *Dances* Dave Gould. *Mus* Georgie Stoll, *assoc* Roger Edens. *Songs* Nacio Herb Brown and Arthur Freed: I'm feeling like a million (*a-b*); Yours and mine (*a-d-f*); Everybody sing (*d*); Follow in my footsteps (*a-b-e*); Your Broadway and my Broadway (*c*); Broadway rhythm; Sun showers; Got a new pair of shoes (*a*) (deleted before release). Roger Edens: Dear Mr Gable (based on You made me love you) (*d*).

The Broadway Melody Of 1940 MGM 1940 (*102 mins*) *with* Fred Astaire (*a*), Eleanor Powell (*b*), George Murphy (*c*), Douglas McPhail (*d*), Florence Rice, Frank Morgan, Ian Hunter. *Prod* Jack Cummings. *Dir* Norman Taurog. *Scr* Leon Gordon, George Oppenheimer from a story by Jack McGowan and Dore Schary. *Ph* Oliver T Marsh, Joseph Ruttenberg. *Art* Cedric Gibbons *Sets* Edwin B Willis. *Ed* Blanche Sewell. *Cost* Dolly Tree. *Sound* Douglas Shearer. *Dances* Bobby Connolly. *Mus* Alfred Newman, *assoc* Roger Edens. *Songs* Cole Porter: I've got my eyes on you (*a*); I concentrate on you (*a-b-d*); Begin the beguine (*a-b*); Between you and me (*c*); Please don't monkey with Broadway (*a-c*); I happen to be in love.

Broadway Rhythm MGM 1943 (*115 mins*) *with* George Murphy (*a*), Ginny Simms (*b*), Charles Winninger (*c*), Gloria De Haven (*d*), Lena Horne (*e*), Nancy Walker (*f*), Eddie 'Rochester' Anderson (*g*), Hazel Scott (*h*), Tommy Dorsey and his Orchestra (*i*), The Ross Sisters (*j*), Ben Blue. *Prod* Jack Cummings. *Dir* Roy Del Ruth. *Scr* Dorothy Kingsley, Harry Clork based on the Jerome Kern-Oscar Hammerstein Broadway show 'Very Warm For May'. *Ph* Leonard Smith. *Art* Cedric Gibbons, Jack Martin Smith. *Sets* Edwin B Willis, McLean Nisbet. *Ed* Albert Akst. *Cost* Irene. *Sound* Douglas Shearer. *Dances* Robert Alton, Don Loper, Charles Walters, Jack Donohue. *Mus* Johnny Green. *Songs* Jerome Kern and Oscar Hammerstein II: All the things you are (*b*). Don Raye and Gene De Paul: Irresistible you (*b*); I love corny music (*i*); Milkman keep those bottles quiet (*f*); Who's who; Solid potato salad. Hugh Martin and Ralph Blane: Brazilian boogie (*b*). Nacio Herb Brown and Arthur Freed: Broadway rhythm (*a*). *Interpolated songs* Pretty baby (*c-d*); Amor (*b*).

Bye Bye Birdie Kohlmar-Sidney/Columbia 1963 (*112 mins*) *with* Dick Van Dyke (*a*), Janet Leigh (*b*), Ann-Margret (*c*), Paul Lynde (*d*),

Bobby Rydell (*e*), Jesse Pearson (*f*), Mary La Roche (*g*), Bryan Russell (*h*), Maureen Stapleton (*i*), Robert Paige. *Prod* Fred Kohlmar. *Dir* George Sidney. *Scr* Irving Brecher from Michael Stewart's book for the Broadway show. *Ph* Joseph F Biroc. *Art* Ted Haworth. *Ed* Charles Nelson. *Dances* Onna White. *Mus* Johnny Green. *Songs* Charles Strouse and Lee Adams: Bye bye Birdie (*c*); One last kiss (*f*); Put on a happy face (*a-b*); Honestly sincere (*f*); The telephone hour (*c-e and chorus*); A lot of livin' to do (*c-f-e*); Kids (*a-d-h-i*); Rosie (*b-c-e*); One boy (*c*); How lovely to be a woman (*c*); Hymn for a Sunday evening (*c-d-g-h*).

Cabin In The Sky MGM 1942 (*99 mins*) *with* Lena Horne (*a*), Ethel Waters (*b*), Eddie 'Rochester' Anderson (*c*), Duke Ellington Orchestra (*d*), Louis Armstrong (*e*), John W Bublett (*f*), The Hall Johnson Choir (*g*). *Prod* Arthur Freed, *assoc* Albert Lewis. *Dir* Vincente Minnelli. *Scr* Joseph Schrank from Lynn Root's musical play. *Ph* Sidney Wagner. *Art* Cedric Gibbons, Leonid Vasian. *Sets* Edwin B Willis, Hugh Hunt. *Ed* Harold F Kress. *Cost* Irene, Howard Shoup, Gile Steele. *Sound* Douglas Shearer, William Steinkamp. *Mus* Georgie Stoll, *assoc* Roger Edens. *Songs* Harold Arlen and E Y Harburg: Happiness is just a thing called Joe (*b*); Life's full of consequence (*a-c*); Li'l black sheep (*g*); Ain't it the truth. Vernon Duke, John La Touche and Ted Fetter: Cabin in the sky (*b-c*); Taking a chance on love (*b-c*); Honey in the honeycomb (*a-b*); In my old Virginia home. Duke Ellington: Going up (*d*); Things ain't what they used to be (*d*). *Interpolated song* Shine (*f*).

Calamity Jane Warner 1953 (*101 mins*) *with* Doris Day (*a*), Howard Keel (*b*), Allyn McLerie (*c*), Gale Robbins, Phil Carey, Dick Wesson. *Prod* William Jacobs. *Dir* David Butler. *Scr* James O'Hanlon. *Ph* Wilfrid M Cline. *Art* John Beckman. *Sets* G M Bernsten. *Ed* Irene Morra. *Cost* Howard Shoup. *Dances* Jack Donohue. *Sound* Stanley Jones, David Forrest. *Mus* Ray Heindorf. *Songs* Sammy Fain and Paul Francis Webster: I can do without you (*a-b*); The Deadwood stage (*a*); Secret love (*a*); A woman's touch (*c*); The Black Hills of Dakota (*a*); Higher than a hawk (*b*); 'Tis Harry I'm plannin' to marry (*c*); Just blew in from the Windy City (*a*).

Call Me Madam Fox 1953 (*117 mins*) *with* Ethel Merman (*a*), George Sanders (*b*), Donald O'Connor (*c*), Vera-Ellen (*d*), Billy De Wolfe, Walter Slezak, Helmut Dantine. *Prod* Sol C Siegel. *Dir* Walter Lang. *Scr* Arthur Sheekman from the book by Howard Lindsay and Russell Crouse for the Broadway show by Irving

Berlin. *Ph* Leon Shamroy. *Art* Lyle Wheeler, John De Cuir. *Sets* Walter M Scott. *Ed* Robert Simpson. *Cost* Irene Sharaff. *Sound* Bernard Freericks, Roger Heman. *Dances* Robert Alton. *Mus* Alfred Newman, *assoc* Ken Darby. *Songs* Irving Berlin: Welcome to Lichtenburg (*chorus*); The hostess with the mostest (*a*); Can you use any money today (*a*); Mrs Sally Adams ('*secretaries*'); Marrying for love (*b*); It's a lovely day today (*c-d*); You're just in love (*a-b-c*); The Ocarina (*d and chorus*); What chance have I with love (*c*); Something to dance about (*c-d*); The best thing for you (*a-b*); International rag (*a*).
Note Vocals for Vera-Ellen dubbed by Carole Richards.

Camelot 7 Arts/Warner 1967 *with* Vanessa Redgrave (*a*), Richard Harris (*b*), Franco Nero (*c*), David Hemmings, Lionel Jeffries, Laurence Naismith. *Prod* Jack L Warner, Joel Freeman. *Dir* Joshua Logan. *Scr* Alan Jay Lerner from his own book for the Broadway show based on T H White's 'The First And Future King'. *Ph* Richard H Kline. *Art* Edward Carrere. *Ed* Folmar Blangstead. *Cost* John Truscott. *Sound* M A Merrick, Dan Wallin. *Dances/musical staging* Buddy Schwab. *Mus* Alfred Newman, *assoc* Ken Darby. *Songs* Frederick Loewe and Alan Jay Lerner: Camelot (*b*); C'est moi (*c*); Follow me (*a*); The lusty month of May (*a*); Guenevere (*chorus*); Wedding ceremony (*chorus*); How to handle a woman (*b*); I wonder what the King is doing tonight (*b*); Then you may take me to the fair (*a*); If ever I would leave you (*c*); The simple joys of maidenhood (*a*); What do the simple folks do (*a-b*); I loved you once in silence (*a*).

Can Can Suffolk-Cummings/Fox 1960 (*131 mins*) *with* Frank Sinatra (*a*), Shirley MacLaine (*b*), Maurice Chevalier (*c*), Louis Jourdan (*d*), Juliet Prowse (*e*), Marcel Dalio. *Prod* Jack Cummings, Saul Chaplin. *Dir* Walter Lang. *Scr* Dorothy Kingsley, Charles Lederer from Abe Burrows' book for the Broadway show by Cole Porter. *Ph* William Daniels. *Art* Lyle Wheeler, Jack Martin Smith. *Sets* Walter M Scott, Paul S Fox. *Ed* Robert Simpson. *Sound* Fred Hynes, W D Flick. *Dances* Hermes Pan. *Mus* Nelson Riddle. *Songs* Cole Porter: Can Can (*b-e and chorus*); It's alright with me (*a-d*); Come along with me (*b*); Live and let live (*c-d*); You do something to me (*d*); Adam and Eve ballet (*b-e*); Let's do it (*a-b*); Montmartre (*a-c*); C'est magnifique (*a-b*); Maidens typical of France (*e and chorus*); Snake dance (*e*); Just one of those things (*c*); I love Paris (*a-c*); Apache dance (Allez-vous en) (*b*); I am in love (*background*).

Can't Help Singing Universal 1945 *with*

Deanna Durbin (*a*), Robert Paige (*b*), Akim Tamiroff, June Vincent, David Bruce, Leonid Kinskey. *Prod* Felix Jackson. *Dir* Frank Ryan. *Scr* Lewis R Foster, Frank Ryan based on 'Girl Of The Overland Trail' by Samuel J and Curtis B Warshawsky. *Ph* Woody Bredell, W Howard Greene. *Art* John B Goodman Robert Clatworthy. *Sets* Russell A Gausman, Edward R Robinson. *Ed* Ted J Kent. *Cost* Walter Plunkett. *Sound* Bernard W Brown. *Mus* Hans J Salter. *Songs* Jerome Kern and E Y Harburg: Can't help singing (*a-b*); More and more (*a*); Any moment now (*a*); Californ-i-ay (*a-b*); Elbow room (*b*); Swing your sweet heart.

Carefree RKO 1938 (*80 mins*) *with* Fred Astaire (*a*), Ginger Rogers (*b*), Ralph Bellamy, Jack Carson, Luella Gear, Clarence Kolb. *Prod* Pandro S Berman. *Dir* Mark Sandrich. *Scr* Allan Scott, Ernest Pagano from a story by Dudley Nichols and Hager Wilde. *Ph* Robert De Grasse. *Art/Sets* Van Nest Polglaze. *Ed* William Hamilton. *Dances* Hermes Pan. *Mus* Victor Baravelle. *Songs* Irving Berlin: Change partners (*a-b*); The Yam (*a-b*); Carefree (golf number) (*a*); I used to be colour-blind (*a-b*); The night is filled with music (*orchestra*).

Carmen Jones Fox 1954 (*105 mins*) *with* Dorothy Dandridge (voice of Marilyn Horne) (*a*), Harry Belafonte (voice of Laverne Hutchinson) (*b*), Olga James (*c*), Pearl Bailey (*d*), Diahann Carroll (voice of Bernice Peterson) (*e*), Joe Adams (voice of Marvin Hayes) (*f*), Brock Peters (*g*), Roy Glenn (*h*), Nick Stewart (*i*). *Prod/Dir* Otto Preminger. *Scr* Harry Kleiner from Oscar Hammerstein's book for Billy Rose's Broadway show based on Bizet's 'Carmen'. *Ph* Sam Leavitt. *Art* Edward L Ilou, John De Cuir. *Sound* Roger Heman. *Dances* Herbert Ross. *Mus* Herschel Burke Gilbert, Dmitri Tiomkin. *Songs* Oscar Hammerstein II from arias by Georges Bizet: Dat's love (*a*); You talk just like my maw (*b*); Dere's a cafe on de corner (*a*); Dis flower (*b*); Beat out dat rhythm on a drum (*d*); Stand up and fight (*f*); My Joe (*c*); Whizzin' away along de track (*quintet*); Card song; Duet and finale (*a-b*).

Carousel Fox 1956 (*128 mins*) *with* Shirley Jones (*a*), Gordon Macrae (*b*), Robert Rounseville (*c*), Barbara Ruick (*d*), Cameron Mitchell (*e*), Claramae Turner (*f*), Jacques D'Amboise (*g*), Gene Lockhart. *Prod* Henry Ephron. *Dir* Henry King. *Scr* Phoebe and Henry Ephron from the book by Oscar Hammerstein for the Broadway show based on Ferenc Molnar's 'Liliom'. *Ph* Charles G Clarke. *Art* Lyle Wheeler, Jack Martin Smith. *Sets* Walter M Scott. *Ed* William Reynolds. *Cost* Mary Wills.

Sound Bernard Freericks, H Leonard. *Dances*
Rod Alexander. *Mus* Alfred Newman, *assoc*
Ken Darby. *Songs* Richard Rodgers and Oscar
Hammerstein II: Carousel waltz (*overture*);
You're a queer one, Julie Jordan (*a-d*); Mister
Snow (*d*); If I loved you (*a-b*); When the
children are asleep (*c-d*); June is bustin' out all
over (*d-f and ensemble*); Soliloquy (*b*); Blow high,
blow low (*e*); Stonecutters cut it on stone (*e*);
What's the use of wonderin' (*a*); A real nice
clambake (*c-d-f and ensemble*); You'll never walk
alone (*a-f*); Carousel ballet (*g and 'Louise'*).

Centennial Summer Fox 1946 *with* Jeanne
Crain (*a*), Cornel Wilde (*b*), Constance Bennett
(*c*), Dorothy Gish (*d*), Walter Brennan, Linda
Darnell, William Eythe, Buddy Swan (*e*).
Prod/Dir Otto Preminger. *Scr* Michael Kanin.
Ph Ernest Palmer. *Art* Lyle Wheeler, Leland
Fuller. *Ed* Harry Reynolds. *Cost* Rene Hubert.
Sound W D Flick, Roger Heman. *Dances*
Dorothy Fox. *Mus* Alfred Newman. *Songs*
Jerome Kern, lyrics by Oscar Hammerstein II,
Leo Robin, Johnny Mercer: All through the
day (*a*); In love in vain (*a*); Up with the lark
(*a-c-d-e*); The right romance (*a*); Cinderella Sue;
Two hearts together.
Note Vocals for Jeanne Crain dubbed by
Louanne Hogan.

Chitty Chitty Bang Bang Warfield/United
Artists 1968 *with* Dick Van Dyke (*a*), Sally Ann
Howes (*b*), Lionel Jeffries (*c*), Gert Frobe (*d*),
Anna Quayle (*e*), James Robertson Justice (*f*),
Adrian Hill (*g*), Heather Ripley (*h*), Benny Hill
Robert Helpmann. *Prod* Albert R Broccoli.
Dir Ken Hughes. *Scr* Ken Hughes, Ronald
Dahl based on Ian Fleming's stories. *Ph*
Christopher Challis. *Art* Harry Pottle. *Sets*
Ken Adam. *Ed* John Shirley. *Cost* Elizabeth
Haffenden, Joan Bridge. *Sound* John Mitchell,
Fred Hynes. *Dances* Marc Breaux and Dee Dee
Wood. *Mus* Irwin Kostal. *Songs* Richard M and
Robert B Sherman: Chitty Chitty Bang Bang
(*a-b-g-h*); You two (*a-g-h*); Toot sweets (*a-b-f-g-h*
and ensemble); Hushabye Mountain (*a-b*); Me ol'
bamboo (*a and chorus*); Truly scrumptious
(*b-g-h*); A lovely lonely man (*b*); Posh (*c*); The
roses of success (*c and 'inventors'*); Chu-chi face
(*d-e*); Doll on a music box (*a-b*).

**A Connecticut Yankee At King Arthur's
Court** Paramount 1948 (*107 mins*) *with* Bing
Crosby (*a*), Rhonda Fleming (*b*), William
Bendix (*c*), Cedric Hardwick (*d*), Murvyn Vye
(*e*), Virginia Field, Henry Wilcoxon. *Prod*
Robert Fellows. *Dir* Tay Garnett. *Scr* Edmund
Beloin from the musical play based on
Mark Twain's story. *Art* Hans Dreier, Roland
Anderson. *Ed* Archie Marshek. *Cost* Edith
Head. *Mus* Victor Young. *Songs* Jimmy Van

Heusen and Johnny Burke: Once and for
always (*a-b*); Busy doing nothing (*a-c-d*); If you
stub your toe on the moon (*a*); When is
sometime (*b*); 'Twixt myself and me (*e*).
Title in GB: *A Yankee At King Arthur's Court*.

The Country Girl Paramount 1954 *with* Bing
Crosby (*a*), Grace Kelly, William Holden,
Jacqueline Fontaine (*b*), Anthony Ross, Gene
Reynolds. *Prod* William Perlberg. *Dir/Scr*
George Seaton from Clifford Odetts' play.
Ph John F Warren. *Art* Hal Pereira, Roland
Anderson. *Sets* Sam Comer, Grace Gregory.
Cost Edith Head. *Sound* Gene Merritt, John
Cope. *Dances* Robert Alton. *Mus* Joseph J
Lilley. *Songs* Harold Arlen and Ira Gershwin:
Dissertation on the state of bliss (*a-b*); The
search is through (*a*), It's mine, it's yours (*a*),
The land around us (*a*).

Cover Girl Columbia 1944 (*107 mins*) *with*
Gene Kelly (*a*), Rita Hayworth (*b*), Phil Silvers
(*c*), Lee Bowman, Otto Kruger, Eve Arden,
Jinx Falkenburg. *Prod* Arthur Schwartz, *assoc*
Norman Deming. *Dir* Charles Vidor. *Scr* Vir-
ginia Van Upp. *Ph* Allen M Davey, Rudolph
Mate. *Art* Lionel Banks, Cary Odell. *Sets*
Fay Babcock. *Ed* Viola Lawrence. *Cost* Jean
Louis. *Sound* Lambert Day. *Dances* Stanley
Donen, Gene Kelly, Seymour Felix, Jack Cole.
Mus Morris Stoloff, *assoc* Carmen Dragon, Saul
Chaplin. *Songs* Jerome Kern and Ira Gershwin:
Cover girl (*b and chorus*); Make way for tomorrow
(*a-b-c*); Long ago and far away (*a-b*); Put me to
the test (*a-b-c*); 'Alter Ego' dance (*a*); Sure
thing (*b*), Who's complaining? (*c*); The show
must go on (*chorus*). *Interpolated song* Poor John (*b*).
Note Vocals for Rita Hayworth dubbed by
Nan Wynn.

Cuban Love Song MGM 1931 (*90 mins*) *with*
Laurence Tibbett (*a*), Lupe Velez (*b*), Jimmy

Durante (*c*), Gus Arnheim and his Orchestra
(*d*), Karen Morley, Louise Fazenda, Ernest
Torrance. *Dir* W S Van Dyke. *Scr* John Lynch
from a story by C Gardiner Sullivan and Bess
Meredith. *Ph* Harold Rosson. *Art* Cedric
Gibbons. *Ed* Margaret Booth. *Cost* Adrian.
Sound Douglas Shearer. *Mus* Herbert Stothart,
Charles Maxwell. *Songs* Jimmy McHugh,
Dorothy Fields, Herbert Stothart: Cuban love
song (*a*); Tramps at sea (*a*). *Interpolated song*
The peanut vendor (*a*).

Daddy Long Legs Fox 1955 (*126 mins*) *with*
Fred Astaire (*a*), Leslie Caron (*b*), Fred Clark,
Thelma Ritter, Terry Moore, Ray Anthony
and his Orchestra (*c*). *Prod* Samuel G Engel.
Dir Jean Negulesco. *Scr* Phoebe and Henry
Ephron from the novel by Jean Webster. *Ph*
Leon Shamroy. *Art* Lyle Wheeler, John De
Cuir. *Ed* William Reynolds. *Dances* Roland
Petit, David Robel, Fred Astaire. *Mus* Alfred
Newman. *Ballet Music* Alex North. *Songs*
Johnny Mercer: Daddy long legs (*chorus*);
Something's gotta give (*a-b*); Sluefoot (*a-b-c*);
Dream (*a-b-c*); Thunderbird (*c*); How I made
the team (*a*); C-a-t spells cat (*b*); Welcome
egghead (*chorus*). *Ballets* Daddy long legs (*a*);
Guardian angel (*a-b*); Classical ballet/Hong
Kong cafe/Carnival in Rio (*b*).

Dames Warner-Vitaphone 1934 (*90 mins*) *with*
Dick Powell (*a*), Ruby Keeler (*b*), Joan
Blondell (*c*), Phil Regan (*d*), Zasu Pitts, Hugh
Herbert, Guy Kibbee. *Prod* Darryl F Zanuck.
Dir Ray Enright. *Scr* Delmer Daves, Robert
Lord. *Ph* Sid Hickox, George Barnes. *Ed*
Harold McLernon. *Cost* Orry-Kelly. *Dances*
Busby Berkeley. *Mus* Leo F Forbstein. *Songs*
Harry Warren and Al Dubin: Dames (*a*); I only
have eyes for you (*a-b*); The girl at the ironing
board (*c*). Mort Dixon and Allie Wrubel: Try

to see it my way (*a-c*). Sammy Fain and Irving
Kahal: When you were a smile on your
mother's lips.

Damn Yankees Warner 1958 (*110 mins*) *with*
Gwen Verdon (*a*), Ray Walston (*b*), Tab
Hunter (*c*), Shannon Bolin (*d*), Rae Allen (*e*),
Russ Brown (*f*), Jimmie Komack (*g*), Nathaniel
Frey (*h*), Robert Schaffer (*i*), Albert Linville
(*j*), Bob Fosse (*k*). *Prod/Dir* George Abbott
and Stanley Donen. *Scr* George Abbott from
his own book for the Broadway show based
on Douglass Wallop's novel 'The Year The
Yankees Lost The Pennant'. *Ph* Harold
Lipstein. *Art* Bert Tuttle. *Sets* John P Austin.
Ed Frank Bracht, Russell Graziano. *Sound*
Stanley Jones. *Dances* Bob Fosse, Pat Ferrier.
Mus Ray Heindorf. *Songs* Richard Adler and
Jerry Ross: Whatever Lola wants (*a*); Shoeless
Joe from Hannibal, Mo (*e and ensemble*);
There's something about an empty chair (*d*);
Two lost souls (*a-c*); A little brains, a little
talent (*a*); Goodbye old girl (*c-i*); Who's got
the pain (*a*); Six months out of every year
(*d-i*); Those were the good old days (*b*); Heart
(*f-g-h-j*).

Title in GB: *What Lola Wants*.

A Damsel In Distress RKO 1937 *with* Fred
Astaire (*a*), Joan Fontaine (*b*), George Burns,
Gracie Allen (*c*), Ray Noble and his Orchestra
(*d*), Reginald Gardiner, Mary Dean (*e*), Pearl
Amatore (*f*), Betty Rone (*g*), Jan Duggen (*h*).
Prod Pandro S Berman. *Dir* George Stevens.
Scr P G Wodehouse, S K Lauren, Ernest
Pagano from the book by P G Wodehouse.
Ph Joseph H August. *Art/Sets* Van Nest
Polglaze. *Ed* Henry Berman. *Dances* Hermes
Pan. *Mus* Alfred Newman. *Songs* George and
Ira Gershwin: A foggy day (*a*); Nice work if
you can get it (*a-e-g-h*); I can't be bothered now

*Cover Girl/Gene Kelly, Phil Silvers and Rita
Hayworth; Daddy Long Legs/Leslie Caron*

(*a*); Stiff upper lip (*c*); The jolly tar and the milkmaid (*a-e-f-g-h and chorus*); Things are looking up (*a-b*).

Dancing Lady MGM 1933 (*82 mins*) *with* Joan Crawford (*a*), Clark Gable, Franchot Tone, Fred Astaire (*b*), Nelson Eddy (*c*), Art Jarrett (*d*), Winnie Lightner (*e*), Ted Healy and The Three Stooges, Robert Benchley, Lynn Bari. *Prod* David O Selznick. *Dir* Robert Z Leonard. *Scr* Allen Rivkin, P J Wolfson from the novel by James Warner Bellah. *Ph* Oliver T Marsh. *Art* Cedric Gibbons. *Ed* Margaret Booth. *Cost* Adrian. *Sound* Douglas Shearer. *Dances* Sammy Lee. *Mus* Louis Silvers. *Songs* Nacio Herb Brown and Arthur Freed: Hold your man (*e*). Richard Rodgers and Lorenz Hart: Rhythm of the day (*c*). Jimmy McHugh and Dorothy Fields: My dancing lady (*a-d*); Hey young fella (*chorus*). Burton Lane and Harold Adamson: Everything I have is yours (*a-d*); Let's go Bavarian (*a-b*); Hey ho the gang's all here.

Deep In My Heart MGM 1954 (*132 mins*) *with* Jose Ferrer (*a*), Helen Traubel (*b*), Howard Keel (*c*), Ann Miller (*d*), William Olvis (*e*), Vic Damone (*f*), Jane Powell (*g*), Rosemary Clooney (*h*), Fred and Gene Kelly (*i*), Tony Martin (*j*), Cyd Charisse (*k*), James Mitchell (*l*), Merle Oberon. *Prod* Roger Edens. *Dir* Stanley Donen. *Scr* Leonard Spiegelgass. *Ph* George J Folsey. *Art* Cedric Gibbons, Edward Carfagno. *Sets* Edwin B Willis. *Ed* Adrienne Fazan. *Sound* Douglas Shearer. *Dances* Eugene Loring. *Mus* Adolph Deutsch. *Songs* Sigmund Romberg: Leg of mutton (*a-b*); Your land and my land (*c*); You will remember Vienna (*b*); It (*d*); Auf wiedersehn (*b*); I love to go swimmin' with women (*i*); Softly as in a morning sunrise (*b*); The road to paradise (*f*); When I grow too old to dream (*a*); Lover come back to me (*j*); Serenade (*e*); Mr and Mrs (*a-h*); Stouthearted men (*b*); Will you remember (*f-g*); One alone (*k-l*).

Biography of composer Sigmund Romberg.

The Desert Song Warner 1952 (*110 mins*) *with* Gordon Macrae (*a*), Kathryn Grayson (*b*), Allyn McLerie (*c*), Raymond Massey, Dick Wesson, Steve Cochran, Ray Collins. *Prod* Rudi Fehr. *Dir* Bruce Humberstone. *Scr* Roland Kibbee, from the Romberg operetta. *Ph* Robert Burks. *Art* Stanley Fleischer. *Sets* William L Kuehl. *Ed* William Ziegler. *Cost* Leah Rhodes, Marjorie Best. *Sound:* C A Riggs, David Forrest. *Dances* Le Roy Prinz. *Mus* Ray Heindorf. *Songs* Sigmund Romberg, Oscar Hammerstein, Otto Harbach: The desert song (*a-b*); Long live the night (*b*); The riff song (*a and chorus*); Romance (*b*); Eastern dance (*c*); One alone (*a-b*); One flower (*b*). Serge Walter

and Jack Scholl: Gay Parisienne (*b*).
Note Third version; previously filmed 1929, 1943.

Doctor Dolittle APJAC/Fox 1967 *with* Rex Harrison (*a*), Anthony Newley (*b*), Samantha Eggar (*c*), Richard Attenborough (*d*), William Dix, Geoffrey Holder, Peter Bull. *Prod* Arthur P Jacobs. *Dir* Richard Fleischer. *Scr* Leslie Bricusse based on Hugh Lofting's stories. *Ph* Robert Surtees. *Art* Ed Graves, Jack Martin Smith. *Sets* Mario Chiari, Walter M Scott, Stuart Reiss. *Ed* Samuel Beetley, Marjorie Fowler. *Cost* Ray Aghayan. *Sound* Murray Spivack, James Corcoran. *Dances* Herbert Ross. *Mus* Lionel Newman, Alexander Courage. *Songs* Leslie Bricusse: Doctor Dolittle (*b*); I've never seen anything like it (*d*); After today (*b*); When I look in your eyes (*a*); Fabulous places (*a-b-c*); My friend the doctor (*b*); Talk to the animals (*a*); Beautiful things (*b-c*); The vegetarian (*a*); At the crossroads (*c*); I think I like you (*a-c*); Like animals (*a*); Something in your smile (*a*); Where are the words (*b*).

Down Argentine Way Fox 1940 (*94 mins*) *with* Betty Grable (*a*), Don Ameche (*b*), Carmen Miranda (*c*), Charlotte Greenwood (*d*), Six Hits and a Miss (*e*), The Nicholas Brothers (*f*). *Prod* Darryl F Zanuck, *assoc* Harry Joe Brown. *Dir* Irving Cummings. *Scr* Karl Tunberg, Darrel Ware from a story by Rian James and Ralph Spence. *Ph* Ray Rennahan, Leon Shamroy. *Art* Richard Day, Joseph C Wright. *Dances* Nick Castle, Geneva Sawyer. *Mus* Emil Newman. *Songs* Harry Warren and Mack Gordon: Two dreams met (*a-b-e*); Sing to your Senorita (*d*); Down Argentina way; Nenita; Doin' the conga. Jimmy McHugh and Al Dubin: South American way (*c*). *Interpolated song* Mama yo quiero (*c*).

Dubarry Was A Lady MGM 1943 (*101 mins*) *with* Gene Kelly (*a*), Lucille Ball (*b*), Red Skelton (*c*), Virginia O'Brien (*d*), Zero Mostel (*e*), Tommy Dorsey and his Orchestra (*f*) with Dick Haymes, Jo Stafford and the Pied Pipers (*g*), Rags Ragland (*h*). *Prod* Arthur Freed. *Dir* Roy Del Ruth. *Scr* Irving Brecher from the book by B G De Sylva and Herbert Fields for the Cole Porter Broadway show. *Ph* Karl Freund. *Art* Cedric Gibbons. *Sets* Edwin B Willis, Henry Grace. *Ed* Blanche Sewell. *Cost* Irene. *Sound* James K Brock. *Dances* Charles Walters. *Mus* Georgie Stoll, *assoc* Roger Edens. *Songs* Cole Porter: Friendship (*a-b-c-d-e-f-h*); Katie went to Haiti (*f-g*); Do I love you (*a*). Burton Lane and Ralph Freed: Dubarry was a lady. Freed, Lane and E Y Harburg: Salome. Freed, Lane and Lew Brown Madame, I love your crepes suzettes. Freed, Brown and Roger Edens: I love an Esquire

girl. Roger Edens: Ladies of the bath.
Note Vocals for Lucille Ball dubbed by Martha Mears.

Easter Parade MGM 1948 (*103 mins*) *with* Fred Astaire (*a*), Judy Garland (*b*), Peter Lawford (*c*), Ann Miller (*d*), Dick Beavers (*e*), Jules Munshin, Clinton Sundberg. *Prod* Arthur Freed. *Dir* Charles Walters. *Scr* Sidney Sheldon, Frances Goodrich, Albert Hackett. *Ph* Harry Stradling. *Art* Cedric Gibbons, Jack Martin Smith. *Sets* Edwin B Willis, Arthur Krams. *Ed* Albert Akst. *Cost* Irene, Valles. *Sound* Douglas Shearer. *Dances* Robert Alton. *Mus* Johnny Green, *assoc* Roger Edens. *Songs* Irving Berlin: Easter parade (*a-b*); Drum crazy (*a*); The girl on the magazine cover (*e*); A couple of swells (*a-b*); Happy Easter (*a*); Shakin' the blues away (*d*); It only happens when I dance with you (*a-b-d*); Steppin' out with my baby (*a*); I wish I was in Michigan (*b*); A feller with an umbrella (*b-c*); Better luck next time (*b*); Vaudeville medley – I love a piano/Ragtime violin/Snooky ookums/When the midnight choo-choo leaves for Alabam (*a-b*); Mr Monotony (*b*) (deleted before release).

The Eddie Cantor Story Warner 1953 (*116 mins*) *with* Keefe Braselle, Marilyn Erskine, Will Rogers Jnr, Aline McMahon, Alex Gerry, Arthur Franz, Greta Granstedt. *Prod* Sidney Skolsky. *Dir* Alfred E Green. *Scr* Jerome Weidman, Ted Sherdeman, Sidney Skolsky. *Ph* Edwin Du Par. *Art* Charles H Clarke. *Sets* William Wallace. *Ed* William Ziegler. *Cost* Howard Shoup, Marjorie Best. *Sound* C A Riggs, David Forrest. *Dances* Le Roy Prinz. *Mus* Ray Heindorf. *Songs* (all sung by Eddie Cantor, dubbing for Braselle): Josephine please no lean on da bell; Now's the time to fall in love; When I'm the President; If you knew Susie; Ida; Pretty baby; You must have been a beautiful baby; Yes sir, that's my baby; Makin' whoopee; Margie; Row row row; Bye bye blackbird; One hour with you; Ma!; How ya gonna keep 'em down on the farm.

Every Night At Eight Paramount 1935 (*80 mins*) *with* Alice Faye (*a*), George Raft (*b*), Frances Langford (*c*), Ted Fio Rito and his Orchestra (*d*), Patsy Kelly (*e*), The Radio Rogues (*f*), Harry Barris (*g*). *Prod* Walter Wanger. *Dir* Raoul Walsh. *Scr* Gene Towne, Graham Baker. *Ph* James Van Trees. *Art* Hans Dreier. *Ed* W. Don Hayes. *Mus* Sam Wineland. *Songs* Jimmy McHugh and Dorothy Fields: I'm in the mood for love (*c*); Take it easy (*a-c-e*); Speaking confidentially (*a-c-e*); I feel a song coming on (*a-c-e-g*); Every night at eight. Ted Fio Rito and Joe Young: Then you've never been blue (*c*).

The Farmer Takes A Wife Fox 1952 *with* Betty Grable (*a*), Dale Robertson (*b*), Thelma Ritter (*c*), John Carroll (*d*), Eddie Foy Jnr (*e*), Charlotte Austin, Merry Anders, Kathleen Crowley. *Prod* Frank P Rosenberg. *Dir* Henry Levin. *Scr* Walter Bullock, Sally Benson, Joseph Fields from Walter D Edmonds' novel 'Rome Haul'. *Ph* Arthur E Arling. *Art* Lyle Wheeler, Addison Hehr. *Sets* Claude Carpenter, Joseph C Wright. *Ed* Louis Loeffler. *Cost* Billy Travilla. *Sound* Roger Heman, Eugene Grossman. *Dances* Jack Cole. *Mus* Lionel Newman. *Songs* Harold Arlen and Dorothy Fields: On the Erie Canal (*b and chorus*); We're doin' it for the natives in Jamaica (*c-e and chorus*); We're in business (*a-b and chorus*); Today I love everybody (*a*); Something real special (*a-b*); When I close my door (*a*); Can you spell Schenectady (*e*); With the sun warm upon me (*a-b*). Remake of the 1935 film.

Finian's Rainbow Seven Arts/Warner 1968 *with* Fred Astaire (*a*), Petula Clark (*b*), Tommy Steele (*c*), Don Francks (*d*), Barbara Hancock (*e*), Brenda Arnau (*f*), Keenan Wynn (*g*), Avon Long (*h*), Jester Hairston (*i*), Roy Glenn (*j*). *Prod* Joseph Landon, Joel Freeman. *Dir* Francis Ford Coppola. *Scr* Fred Saidy, E Y Harburg from their book for the Broadway show. *Ph* Philip Lathrop. *Art* Hilyard M Brown. *Sets* William L Kuehl, Philip Abramson. *Ed* Melvin Shapiro. *Cost* Dorothy Jeakins. *Dances* Hermes Pan. *Mus* Ray Heindorf *assoc* Ken Darby. *Songs* Burton Lane and E Y Harburg: Look to the rainbow (*a-b-d*); This time of the year (*chorus*); If this isn't love (*a-b-d-e and chorus*); Old devil moon (*b-d*); Necessity* (*f and chorus*); How are things in Glocca Morra (*a-b-c-d-e and chorus*); Something sort of grandish (*b-c*); When the idle poor become the idle rich (*a-b and chorus*); That great come-and-get-it day (*b-d and chorus*); The begat (*g-h-i-j*); When I'm not near the girl I love (*c-e*); Woody's here (*chorus*); Rain dance ballet int. Old devil moon (*e*).
Note Necessity* was pre-recorded and issued on Soundtrack album but never filmed.

The Five Pennies Dena/Paramount 1959 (*117 mins*) *with* Danny Kaye (*a*), Louis Armstrong (*b*), Barbara Bel Geddes (*c*), Susan Gordon (*d*), Harry Guardino (*e*), Bob Crosby (*f*), Ray Anthony, Tuesday Weld. *Prod* Jack Rose, *assoc* Sylvia Fine. *Dir* Melville Shavelson. *Scr* Rose and Shavelson from a story by Robert Smith. *Ph* Daniel L Fapp, John P Fulton. *Art* Hal Pereira, Tambi Larsen. *Sets* Sam Comer, Grace Gregory. *Ed* Frank P Keller. *Cost* Edith Head. *Sound* Hugo Grenzbach, John

Wilkinson. *Mus* Leith Stevens. *Songs* The five pennies (*a-b*); After you've gone (*b*); Bill Bailey (*a-b*); Goodnight, sleep tight (*a-b-c-d*); When the saints go marching in (*a-b*); Battle hymn of the republic (*a-b*); The music goes round and round (*a-d*); Lullaby in ragtime (*a-c*); Carnival of Venice (*a*); Jingle bells (*a*); Paradise (*f*); Follow the leader (*a-c-e*); My blue heaven (*f*).
Instrumentals by 'The Five Pennies' The wail of the winds; Indiana; Washington and Lee swing; Running wild; Out of nowhere.
Biography of bandleader Red Nichols, who dubbed the trumpet playing for Danny Kaye. Vocals for Barbara Bel Geddes dubbed by Eileen Wilson.

The Fleet's In Paramount 1942 (*93 mins*) *with* William Holden, Dorothy Lamour (*a*), Eddie Bracken (*b*), Betty Hutton (*c*), Jimmy Dorsey and his Orchestra (*d*) with Bob Eberle (*e*) and Helen O'Connell (*f*), Gil Lamb (*g*), Betty Jane Rhodes (*h*), Cass Daley (*i*). *Prod* Paul Jones. *Dir* Victor Schertzinger. *Scr* Walter de Leon, Sid Silvers from a story by Monte Brice and J Walter Ruben. *Ph* William C Mellor. *Art* Hans Dreier. *Ed* Paul Weatherwax. *Cost* Edith Head. *Dances* Jack Donohue. *Mus* Victor Young. *Songs* Victor Schertzinger and Johnny Mercer: The fleet's in (*h*); I remember you (*d-e*); Tangerine (*d-e-f*); Build a better mousetrap (*c-d*); Tomorrow

you belong to Uncle Sam (*i*); Not mine (*a-b-c-d-e*); When you hear the time signal (*a*); Arthur Murray taught me dancing in a hurry (*c*).

Flirtation Walk 1st National 1934 (*97 mins*) *with* Dick Powell (*a*), Ruby Keeler (*b*), Ross Alexander, Pat O'Brien, John Eldredge, John Arledge, Henry O'Neill, Tyrone Power. *Prod* Robert Lord. *Dir* Frank Borzage. *Scr* Delmer Daves, Lou Edelman. *Ph* Sol Polito, George Barnes. *Art* Jack Okey. *Ed* William Holmes. *Cost* Orry-Kelly. *Dances* Bobby Connolly. *Mus* Leo F Forbstein. *Songs* Mort Dixon and Allie Wrubel: Flirtation walk (*a-b*); Mr and Mrs is the name (*a-b*); No house, no wife, no moustache (*a*), I see two lovers (*a*); Smoking in the dark; When do we eat?

The Flower Drum Song Hunter-Fields/ Universal 1961 (*133 mins*) *with* Nancy Kwan (*a*), Juanita Hall (*b*), James Shigeta (*c*), Reiko Sato (*d*), Miyoshi Umeki (*e*), Victor Sen Yung (*f*), Benson Fong (*g*), Jack Soo (*h*), Kam Tong (*i*), Patrick Adiarte (*j*). *Prod* Ross Hunter. *Dir* Henry Koster. *Scr* Joseph Fields from Oscar Hammerstein's book for the Broadway show based on the novel by C Y Lee. *Ph* Russell Metty. *Art* Alexander Golitzen, Joseph C Wright. *Sets* Howard Bristol. *Ed* Milton Carruth. *Cost* Irene Sharaff. *Sound* Waldon O Watson, Joe Lapis. *Dances* Hermes Pan. *Mus*

Flirtation Walk/Dick Powell and Ruby Keeler

Alfred Newman, *assoc* Ken Darby. *Songs*
Richard Rodgers and Oscar Hammerstein II:
Fan Tan Fanny (*a and chorus*); Grant Avenue
(*a and chorus*); I enjoy being a girl (*a*); You are
beautiful (*c*); Chop suey (*b-c-j and chorus*); I am
going to like it here (*e*); The other generation
(*b-g-i*); Love look away and dream ballet (*d*);
Gliding through my memoree (*a-f and chorus*);
Sunday – comedy ballet (*a-b*); Don't marry me
(*e-b*); A hundred million miracles (*e-i and chorus*).
Flying Down To Rio RKO 1933 (*89 mins*)
with Gene Raymond (*a*), Dolores Del Rio (*b*),
Fred Astaire (*c*), Ginger Rogers (*d*), Raul
Roulien (*e*), Etta Moten (*f*), Blanche Frederici.
Prod/Scr Lou Brock. *Dir* Thornton Freeland.
Ph J J Faulkner. *Art/Sets* Van Nest Polglaze.
Ed Jack Kitchin. *Dances* Dave Gould, Hermes
Pan. *Mus* Max Steiner. *Songs* Vincent Youmans,
Gus Kahn, Edward Eliscu: Flying down to
Rio (*c and chorus*); Orchids in the moonlight
(*a-b-c-e*); The Carioca (*c-d-f*); Music makes me
(*d*).
Follow The Fleet RKO 1935 (*110 mins*) *with*
Fred Astaire (*a*), Ginger Rogers (*b*), Harriet
Hilliard (*c*), Randolph Scott, Lucille Ball, Betty
Grable, Tony Martin. *Prod* Pandro S Berman.
Dir Mark Sandrich. *Scr* Dwight Taylor from
the play 'Shore Leave' by Hubert Osborne and
Allan Scott. *Ph* David Abel. *Art/Sets* Van
Nest Polglaze, Carroll Clark. *Ed* Henry
Berman. *Cost* Bernard Newman. *Sound* John
Tribby. *Dances* Hermes Pan. *Mus* Max Steiner.
Songs Irving Berlin: Let yourself go (*a-b*); Let's
face the music and dance (*a-b*); Get thee behind
me Satan (*c*); I'd rather lead a band (*a*); I'm
putting all my eggs in one basket (*a-b*); But
where are you (*c*); We saw the sea (*a and chorus*).
Footlight Parade Warner 1933 (*100 mins*)
with Dick Powell (*a*), James Cagney (*b*), Joan

Blondell (*c*), Ruby Keeler (*d*), Frank McHugh
(*e*), Claire Dodd (*f*), Dorothy Lamour. *Prod* Darryl
F Zanuck. *Dir* Lloyd Bacon. *Scr* Manuel Seff,
James Seymour. *Ph* George Barnes. *Ed* George
Amy. *Cost* Orry-Kelly. *Dances* Busby Berkeley.
Mus Leo F Forbstein. *Songs* Harry Warren and
Al Dubin: Shanghai Lil (*b-d and chorus*);
Honeymoon Hotel (*a-d*). Sammy Fain and
Irving Kahal: By a waterfall (*a-d and chorus*);
Sitting on a backyard fence (*d*); Ah, the moon
is here (*a-e and chorus*).
For Me And My Gal MGM 1942 (*104 mins*)
with Judy Garland (*a*), Gene Kelly (*b*), George
Murphy (*c*), Marta Eggerth (*d*), Ben Blue,
Richard Quine, Horace (later Stephen) McNally.
Prod Arthur Freed. *Dir* Busby Berkeley. *Scr*.
Richard Sherman, Sid Silvers, Fred Finklehoffe
from a story by Howard Emmett Rogers. *Ph*
William Daniels. *Art* Cedric Gibbons. *Sets*
Edwin B. Willis. *Ed* Ben Lewis. *Cost* Irene.
Sound Douglas Shearer. *Dances* Bobby Connolly.
Mus Georgie Stoll, *assoc* Roger Edens. *Songs*
For me and my gal (*a-b*); After you've gone
(*a*); Ballin' the jack (*a-b*); Till we meet again
(*a*); Oh you beautiful doll (*a*); When you wore
a tulip (*a-b*); Where do we go from here;
What are you going to do to help the boys;
How ya gonna keep 'em down on the farm;
Smiles (*a*).
Forty-Second Street Warner/Vitaphone 1933
(*85 mins*) *with* Dick Powell (*a*), Ruby Keeler
(*b*), Bebe Daniels (*c*), Ginger Rogers (*d*), Una
Merkel (*e*), Toby Wing (*f*), Clarence Nordstrom
(*g*), Warner Baxter. *Prod* Darryl F Zanuck. *Dir*
Lloyd Bacon. *Scr* Rian James, James Seymour
from a story by Bradford Ropes. *Ph* Sol Polito.
Ed Thomas Pratt. *Cost* Orry-Kelly. *Dances*
Busby Berkeley. *Mus* Leo F Forbstein. *Songs*
Harry Warren and Al Dubin: Forty-Second

*Follow the Fleet/Fred Astaire, Ginger Rogers
and Lucille Ball*

Street (*a-b and chorus*); You're getting to be a
habit with me (*a-c*); Shuffle off to Buffalo
(*b-d-e-g*); Young and healthy (*a-f and chorus*).
Fox Follies Of 1929 Fox 1929 *with* Sue Carol
(*a*), Lola Lane (*b*), Sharon Lynn (*c*), Dixie Lee
(*d*), David Rollins (*e*), Stepin Fetchit, Melva
Cornell, Paula Langlen, David Percy. *Dir/Scr*
David Butler, William K Wells. *Ph* Charles
Van Enger. *Ed* Ralph Dietrich. *Dances* Marcel
Silver, Fanchon. *Songs* Con Conrad, Sidney
Mitchell and Archie Gottler: The breakaway
(*a-e*); Why can't I be like you; Pearl of old
Japan; Walking with Susie; That's your baby;
Legs; Look what you've done to me; Big City
blues.
Title in GB: *Movietone Follies Of 1929.*
Funny Face Paramount 1956 (*103 mins*) *with*
Fred Astaire (*a*), Audrey Hepburn (*b*), Kay
Thompson (*c*), Robert Flemyng, Michel
Auclair, Suzy Parker. *Prod* Roger Edens. *Dir*
Stanley Donen. *Scr* Leonard Gershe, based on
the Gershwin Broadway show. *Ph* Ray June,
John P Fulton. *Art* George W Davis, Hal
Pereira. *Ed* Frank Bracht. *Cost* Edith Head.
Dances Eugene Loring, Fred Astaire. *Mus*
Adolph Deutsch, *assoc* Roger Edens. *Songs*
George and Ira Gershwin: Funny face (*a-b*);
How long has this been going on (*b*); Clap yo'
hands (*a-c*); He loves and she loves (*a-b*); 'S
wonderful (*a-b*); Let's kiss and make up
(courtyard dance) (*a*); Cellar dance (Basal
Metabolism) – How long has this been going
on/Funny Face (*a*). Roger Edens and Leonard
Gershe: Think pink (*c*); Bonjour Paris (*a-b-c*);
On how to be lovely (*b-c*).
Funny Girl Rastar/Columbia 1968 (*147 mins*)
with Barbra Streisand (*a*), Omar Sharif (*b*),
Walter Pidgeon, Kay Medford (*c*), Mae Questel
(*d*), Tommy Rall (*e*), Anne Francis, Lee Allen.

Prod Ray Stark. *Dir/Ass. Prod* William Wyler.
Scr Isobel Lennart from her book for the
Broadway show. *Ph* Harry Stradling. *Art*
Robert Luthardt. *Sets* William Kiernan. *Ed*
Maury Winetrobe, William Sands. *Cost* Irene
Sharaff. *Sound* Charles J Rice. *Dances* Herbert
Ross. *Prodn Design* Gene Callahan. *Mus* Walter
Scharf. *Songs* Jule Styne and Bob Merrill:
Funny girl (*a*); People (*a*); You are woman, I
am man (*a-b*); Don't rain on my parade (*a*);
If a girl isn't pretty (*c-d*); His love makes me
beautiful (Bridal scene) (*a and Ziegfeld girls*);
I'm the greatest star (*a*); A temporary arrange-
ment (*b*). *Interpolated songs* Second hand rose
(*a*); My man (*a*); Swan ballet burlesque (*a-e and
chorus*); I'd rather be blue over you (*a*); Roller
skate rag (*a and chorus*); Sadie Sadie (*a*).
Biography of Fanny Brice.
The Gang's All Here Fox 1943 (*103 mins*)
with Alice Faye (*a*), Carmen Miranda (*b*), Benny
Goodman and his Orchestra (*c*), Phil Baker
(*d*), Sheila Ryan (*e*), Tony De Marco (*f*),
Charlotte Greenwood (*g*), Edward Everett
Horton. *Prod* William Le Baron. *Dir/Dances*
Busby Berkeley. *Scr* Walter Bullock. *Ph* Edward
Cronjager. *Art* James Basevi, Joseph C Wright
Sets Paul S Fox. *Ed* Ray Curtiss. *Sound* George
Leverett, Roger Heman. *Mus* Alfred Newman,
Charles Henderson. *Songs* Harry Warren and
Leo Robin: No love, no nothing (*a-c-e-f*); The
lady in the tutti-frutti hat (*b*); Paducah (*b-c*);
Minnie's in the money (*c*); A journey to a star
(*a-b-c-d-e-f-g*); You discover you're in New
York (*b-d*). David Raksin: Polka dot ballet
(*a and chorus*). *Interpolated music* Brazil (*d*); Let's
dance (*c*); Soft winds (*c*).
Title in GB: *The Girls He Left Behind.*
The Gay Divorce(e) RKO 1934 (*107 mins*)
with Fred Astaire (*a*), Ginger Rogers (*b*),

Edward Everett Horton (*c*), Erik Rhodes (*d*), Alice Brady, Eric Blore, Betty Grable (*e*), Lillian Miles (*f*). *Prod* Pandro S Berman. *Dir* Mark Sandrich. *Scr* George Marion Jnr, Edward Kaufman, Dorothy Yost from the Cole Porter Broadway show based on a story by Dwight Taylor. *Ph* David Abel. *Art/Sets* Van Nest Polglaze. *Ed* William Hamilton. *Cost* Bernard Newman. *Dances* Hermes Pan, Dave Gould. *Mus* Max Steiner. *Songs* Cole Porter: Night and day (*a-b*). Con Conrad and Herb Magidson: The Continental (*a-b-d-f*); Needle in a haystack (*a*). Harry Revel and Mack Gordon: Let's knock knees (*c-e*); Don't let it bother you (*a*).

Gay Purr-ee UPA/Warner 1962 (*86 mins*) Cartoon feature with voices of: Judy Garland (*a*), Robert Goulet (*b*), Hermione Gingold (*c*), Red Buttons (*d*), Paul Frees (*e*), Morey Amsterdam, Mel Blanc. *Prod* Henry G Saperstein, Lee Orgel. *Dir* Abe Levitow. *Scr* Dorothy and Chuck Jones. *Ph* Roy Hutchcroft, Dan Miller, Jack Stevens, Duane Keegan. *Art* Victor Haboush. *Ed* Ted Baker, Earl Bennett, Sam Horta. *Mus* Mort Lindsey, *assoc* Joseph J. Lilley. *Songs* Harold Arlen and E Y Harburg: Mewsette (*b*); Little drops of rain (*a-b*); Money cat (*c*); Take my hand Paree (*a*); Paris is a lonely town (*a*); Bubbles (*b-d*); Roses red, violets blue (*a*); The horse won't talk (*e*); Portraits of Newsette (*orch.*); Mewsette finale (*a-b*).

Gentlemen Marry Brunettes United Artists 1955 (*97 mins*) *with* Jane Russell (*a*), Jeanne Crain (*b*), Alan Young (*c*), Scott Brady (*d*), Rudy Vallee (*e*), Johnny Desmond (voice only over credits) (*f*). *Prod* Robert Bassett, Richard Sale, Bob Waterfield. *Dir* Richard Sale. *Scr* Richard Sale, Mary Loos from Anita Loos' book 'But Gentlemen Marry Brunettes'. *Ph* Desmond Dickinson. *Art* Paul Sheriff. *Ed* G Turney-Smith. *Sound* John Cox. *Dances* Jack Cole. *Mus* Robert Farnon. *Songs* Richard Rodgers and Lorenz Hart: Have you met Miss Jones (*a-b-c-d-e*); My funny valentine (*b-c*); I've got five dollars (*a-d*). *Other songs* Gentlemen marry brunettes (*f*); You're driving me crazy (*a-b*); Miss Annabelle Lee (*orch.*); I wanna be loved by you (*a-b-e*); Daddy (*a-b*); Ain't misbehavin' (*a-b-c*).

Note Vocals for Jeanne Crain dubbed by Anita Ellis, for Scott Brady by Robert Farnon.

Gentlemen Prefer Blondes Fox 1953 (*91 mins*) *with* Jane Russell (*a*), Marilyn Monroe (*b*), Charles Coburn, Elliott Reid, Tommy Noonan, George Winslow, Marcel Dalio. *Prod* Sol C Siegel. *Dir* Howard Hawks. *Scr* Charles Lederer from the book by Anita Loos and Joseph Fields for the Broadway musical based

on Miss Loos' story. *Ph* Harry J Wild. *Art* Lyle Wheeler, Joseph C Wright. *Sets* Claude Carpenter. *Ed* Hugh S Fowler. *Cost* Billy Travilla. *Sound* Roger Heman, E Clifton Ward. *Dances* Jack Cole. *Mus* Lionel Newman. *Songs* Jule Styne and Leo Robin: Bye bye baby (*a-b* and chorus); A little girl from Little Rock (*a-b*); Diamonds are a girl's best friend (*a-b* and chorus). Hoagy Carmichael and Harold Adamson: Ain't there anyone here for love (*a*); When love goes wrong (*a-b*). *Background music* How blue the night.

George White's Scandals Fox 1934 (*79 mins*) *with* Rudy Vallee (*a*), Jimmy Durante (*b*), Cliff Edwards (*c*), Dixie Dunbar (*d*), Alice Faye (*e*), Gertrude Michael, George White. *Prod* Winfield R Sheehan. *Dir* George White, Thornton Freeland, Harry Lachman. *Scr* George White, Jack Yellen. *Ph* Lee Garmes, George Schneiderman. *Ed* Paul Weatherwax. *Sound* A L Van Kirbach, George Leverett. *Dances* Jack Donohue. *Mus* Uncredited. *Songs* Ray Henderson, Jack Yellen and Irving Caesar: Oh you nasty man (*e*), My dog loves your dog (*a*); Hold my hand; Following in mother's footsteps; So·nice; Sweet and simple; Every day is father's day with baby; Six women. Based on George White's Broadway revue.

Gigi MGM 1958 (*116 mins*) *with* Leslie Caron (*a*), Maurice Chevalier (*b*), Louis Jourdan (*c*), Hermione Gingold (*d*), Isabel Jeans, Eva Gabor, Jacques Bergerac. *Prod* Arthur Freed. *Dir* Vincente Minnelli. *Scr* Alan Jay Lerner from the novel by Colette. *Ph* Joseph Ruttenberg. *Art* William A Horning, Preston Ames. *Sets* Henry Grace, Keogh Gleason. *Ed* Adrienne Fazan. *Cost(Prodn Design)* Cecil Beaton. *Sound* Wesley C Miller. *Mus* André Previn. *Songs* Frederick Loewe and Alan Jay Lerner: Gigi (*c*); Thank Heaven for little girls (*b* and chorus); It's a bore (*b-c*); The Parisians (*a*); She is not thinking of me (Waltz at Maxim's) (*c*); The night they invented champagne (*a-c-d*); I remember it well (*b-d*); Say a prayer for me tonight (*a*); I'm glad I'm not young anymore (*b*). *Note* Vocals for Leslie Caron dubbed by Betty Wand.

Girl Crazy MGM 1943 (*99 mins*) *with* Judy Garland (*a*), Mickey Rooney (*b*), June Allyson (*c*), Nancy Walker (*d*), Tommy Dorsey and his Orchestra (*e*), Rags Ragland, Don Taylor. *Prod* Arthur Freed. *Dir* Norman Taurog. *Scr* Fred Finklehoffe, from the book of the Gershwin Broadway show by Guy Bolton and Jack McGowan. *Ph* William Daniels, Robert Planck. *Art* Cedric Gibbons. *Sets* Edwin B Willis, Mac Alper. *Ed* Albert Akst. *Cost* Irene. *Sound* William Steinkamp. *Dances* Busby

Berkeley, Charles Walters, Jack Donohue. *Mus* Georgie Stoll, *assoc* Roger Edens. *Songs* George and Ira Gershwin: But not for me (*a*); Bidin' my time (*a and quartet*); Could you use me (*a-b*); Embraceable you (*a*); Treat me rough (*b-c*); I got rhythm (*a-b-e*); Bronco busters; Cactus time in Arizona; Fascinating rhythm (*e*); Sam and Delilah (*a*).
Note Second version, previously filmed 1932, also remade 1966.

Give A Girl A Break MGM 1953 (*82 mins*) *with* Debbie Reynolds (*a*), Marge and Gower Champion (*b*), Bob Fosse (*c*), Helen Wood, Kurt Kasznar, Richard Anderson. *Prod* Jack Cummings. *Dir* Stanley Donen. *Scr* Albert Hackett, Frances Goodrich. *Ph* William C Mellor. *Art* Cedric Gibbons, Paul Groesse. *Ed* Adrienne Fazan. *Sound* Douglas Shearer. *Dances* Gower Champion. *Mus* André Previn, *assoc* Saul Chaplin. *Songs* Burton Lane and Ira Gershwin: Give a girl a break; In our united state (*a-c*); It happens every time; Nothing is impossible; Applause, applause (*a-b*).

The Glenn Miller Story Universal 1954 (*116 mins*) *with* James Stewart, June Allyson, Louis Armstrong All Stars (*a*), Gene Krupa (*b*), Frances Langford (*c*), The Modernaires (*d*), Henry Morgan (*e*), Ben Pollack (*f*), Charles Drake, George Tobias. *Prod* Aaron Rosenberg, *assoc* William Goetz. *Dir* Anthony Mann. *Scr* Oscar Brodney, Valentine Davies. *Ph* William Daniels. *Sound* Leslie J Carey. *Dances* Kenny Williams. *Mus* Joseph Gershenson, *assoc* Henry Mancini. *Featured music* Chattanooga choo-choo (*c-d*); Basin Street blues (*a-b*); Bidin' my time (*quartet*); I'm looking at the world through rose-coloured glasses (*e-f*); Everybody loves my baby (*f*). *Played by 'Glenn Miller Orchestra'*: Moonlight serenade; Tuxedo junction; Little

brown jug; Adios; St Louis blues march; In the mood; Elmer's tune; String of pearls; Pennsylvania 6-5000; Stairway to the stars; American patrol; Over the rainbow; I know why. *By studio orchestra*: I dreamt I dwelt in marble halls; So little time ('Glenn Miller Story' theme by Henry Mancini).
Note Trombone playing for James Stewart (as Glenn Miller) was dubbed by Joe Yukl, and piano for Henry Morgan by Lyman Gandee.

Glorifying The American Girl Paramount 1929 *with* Eddie Cantor (*a*), Rudy Vallee (*b*), Dan Healy (*c*), Mary Eaton (*d*), Helen Morgan (*e*), Kaye Renard, Edward Crandall. *Dir* Millard Webb. *Scr* J P McEvoy, Millard Webb. *Songs* Walter Donaldson: Changes; Sam the old accordion man; At sundown. Irving Berlin: Blue skies (*a*). Others: I'm just a vagabond lover (*b*); What wouldn't I do for that man (*e*); There must be someone waiting; 'Pageant of lovers' scene.

Going My Way Paramount 1944 (*130 mins*) *with* Bing Crosby (*a*), Barry Fitzgerald, Rise Stevens (*b*), Frank McHugh (*c*), The Robert Mitchell Boychoir (*d*), Jean Heather (*e*), Gene Lockhart. *Exec Prod* B G De Sylva. *Prod/Dir* Leo McCarey. *Scr* Frank Butler, Frank Cavett from a story by Leo McCarey. *Ph* Lionel Lindon, Gordon Jennings. *Art* Hans Dreier, William Flannery. *Sets* Steve Seymour. *Ed* Leroy Stone. *Cost* Edith Head. *Sound* Gene Merritt, John Cope. *Mus* Robert Emmett Dolan. *Songs* Jimmy Van Heusen and Johnny Burke: Going my way (*a-b*); Swinging on a star (*a-d*); The day after forever (*a-e*). *Interpolated songs* Silent night (*a-d*); Ave Maria (Schubert) (*a-b-d*); Too-ra-loo-ra-loo-ral (That's an Irish lullaby) (*a*); 'Carmen'-Habanera (Bizet) (*b*).

The Golddiggers In Paris Warner 1938

(*97 mins*) *with* Rudy Vallee (*a*), Rosemary Lane
(*b*), Hugh Herbert (*c*), Gloria Dickson (*d*),
Carole Landis (*e*), Allen Jenkins (*f*), Mabel
Todd (*g*), Eddie 'Rochester' Anderson (*h*).
Prod Hal B Wallis, *assoc* Samuel Bischoff. *Dir*
Ray Enright. *Scr* Earl Baldwin, Warren Duff
from a story by Jerry Wald, Richard Macaulay
and Maurice Leo. *Ph* Sol Polito, George Barnes.
Art Robert Haas. *Ed* George Amy. *Dances*
Busby Berkeley. *Mus* Leo F Forbstein. *Songs*
Harry Warren and Al Dubin: The Latin
quarter; I wanna go back to Bali (*a*); A
stranger in Paree (*a-b-c-d-f-g*); Put that down in
writing; Waltz of the flowers. Harry Warren
and Johnny Mercer: My adventure; Day
dreaming all night long (*a-b*).
Title in GB: *The Gay Impostors*.

The Golddiggers Of Broadway Warner 1929
with William Bakewell (*a*), Winnie Lightner
(*b*), Nancy Welford (*c*), Ann Pennington (*d*),
Nick Lucas (*e*), Lilyan Tashman (*f*), Conway
Tearle. *Dir* Roy Del Ruth. *Scr* Avery Hopwood,
Robert Lord from Hopwood's play. *Ed* William
Holmes. *Dances* Larry Ceballos. *Songs* Joe
Burke and Al Dubin: Painting the clouds with
sunshine (*d-e*); In a kitchenette; What will I do
without you; Tiptoe through the tulips; Go
to bed; And they still fall in love.

The Golddiggers Of 1933 1st National-
Vitaphone 1933 (*96 mins*) *with* Dick Powell (*a*),
Joan Blondell (*b*), Ruby Keeler (*c*), Ginger
Rogers (*d*), Etta Moten (*e*), Warren William,
Ned Sparks. *Prod* Darryl F Zanuck. *Dir* Mervyn
Le Roy. *Scr* Erwin Gelsey, James Seymour from
a story by Avery Hopwood. *Ph* Sol Polito.
Ed George Amy. *Cost* Orry-Kelly. *Dances*
Busby Berkeley. *Mus* Leo F Forbstein. *Songs*
Harry Warren and Al Dubin: Pettin' in the
park (*a-c and chorus*); Remember my forgotten
man (*b-e and chorus*); The shadow waltz (*a-c
and chorus*); We're in the money (*d and chorus*);
I've got to sing a torch song (*a*).

The Golddiggers Of 1935 1st National-
Vitaphone 1935 (*95 mins*) *with* Dick Powell (*a*),
Wini Shaw (*b*), Adolph Menjou, Gloria Stuart,
Alice Brady, Hugh Herbert, Glenda Farrell,
Dorothy Dare. *Prod* Robert Lord. *Dir/Dances*
Busby Berkeley. *Scr* Peter Milne, Robert Lord,
Manuel Seff. *Ph* George Barnes. *Art* Anton
Grot. *Ed* George Amy. *Cost* Orry-Kelly. *Mus*
Leo F Forbstein. *Songs* Harry Warren and Al
Dubin: The Lullaby of Broadway (*a-b*); The
words are in my heart (*a and chorus*); I'm going
shopping with you (*a*).

The Golddiggers Of 1937 1st National 1937
(*100 mins*) *with* Dick Powell (*a*), Joan Blondell
(*b*), Lee Dixon (*c*), Glenda Farrell, Jane Wyman.
Prod Hal B Wallis, *assoc* Earl Baldwin.

Dir Lloyd Bacon. *Scr* Warren Duff from a
story by Richard Maibaum, Michael Wallach
and George Haight. *Ph* Arthur Edeson. *Ed*
Thomas Richards. *Dances* Busby Berkeley.
Mus Leo F Forbstein. *Songs* Harry Warren and
Al Dubin: With plenty of money and you (*a*),
All's fair in love and war (*a-b and chorus*). Harold
Arlen and E Y Harburg: Speaking of the
weather (*a*); Let's put our heads together (*a-b*);
The life insurance song.

The Goldwyn Follies Goldwyn/United
Artists 1938 (*120 mins*) *with* Andrea Leeds (*a*),
Vera Zorina (*b*), Kenny Baker (*c*), Ella Logan
(*d*), Adolph Menjou (*e*), Bobby Clark (*f*), The
Ritz Brothers (*g*), Phil Baker (*h*), Edgar
Bergen (*i*), The Metropolitan Opera Ballet (*j*).
Prod Samuel Goldwyn, *assoc* George Haight.
Dir George Marshall. *Scr* Ben Hecht. *Ph* Gregg
Toland. *Art* Richard Day. *Ed* Sherman Todd.
Dances George Balanchine. *Mus* Alfred Newman.
Songs Lyrics by Ira Gershwin, music by (1)
George Gershwin: I was doing alright (*d*);
Love is here to stay (*c*); Love walked in (*c*).
(2) Vernon Duke: I love to rhyme (*h-i*).
(3) Kurt Weill: Spring again (*c*); I'm not com-
plaining. Ray Golden and Sid Kuller: Here
pussy pussy (*g*). Water-lily ballet (*b-j*).

The Great Ziegfeld MGM 1936 (*180 mins*)
with William Powell, Luise Rainer, Myrna Loy,
Fanny Brice (*a*), Virginia Bruce (*b*), Ray
Bolger (*c*), Dennis Morgan (*d*), Harriet Hoctor
(*e*), Frank Morgan, Buddy Doyle. *Prod* Hunt
Stromberg. *Dir* Robert Z Leonard. *Scr* William
Anthony McGuire. *Ph* Oliver T Marsh. *Art*
Cedric Gibbons. *Sets* Edwin B Willis. *Ed*
William S Gray. *Cost* Adrian. *Sound* Douglas
Shearer. *Dances* Seymour Felix. *Mus* Arthur
Lange. *Songs* Walter Donaldson and Harold
Adamson: You (*chorus*); It's been so long; You
never looked so beautiful; Follies girl; Queen
of the jungle (*e*); You gotta pull strings. Irving
Berlin: A pretty girl is like a melody (*d*).

Guys And Dolls Goldwyn/MGM 1955 (*150
mins*) *with* Frank Sinatra (*a*), Marlon Brando (*b*),
Jean Simmons (*c*), Vivian Blaine (*d*), Stubby
Kaye (*e*), Sheldon Leonard, Robert Keith, The
Goldwyn Girls. *Prod* Samuel Goldwyn. *Dir/Scr*
Joseph L Mankiewicz, adapted from the book
by Jo Swerling and Abe Burrows for the
Broadway show based on Damon Runyon's
stories. *Ph* Harry Stradling. *Art* Joseph C
Wright, Oliver Smith. *Ed* Daniel Mandell.
Sound Fred Lau. *Dances* Michael Kidd. *Mus*
Jay Blackton. *Songs* Frank Loesser: Guys and
dolls (*chorus*): Fugue for tinhorns (*e and quartet*);
The oldest established permanent floating crap
game in New York (*chorus*); I'll know (*b-c*); A
bushel and a peck (*d*); Adelaide's lament (*d*);

If I were a bell (*c*); I've never been in love
before (*b*); Take back your mink (*d and chorus*);
Sit down, you're rocking the boat (*e and chorus*);
Luck be a lady (*b*); Sue me (*a-c*); A woman in
love (*b-c*); Adelaide (*a*); Pet me poppa (*d*).
Gypsy Warner 1962 (*149 mins*) *with* Rosalind
Russell (*a*), Natalie Wood (*b*), Karl Malden
(*c*), Ann Jillian (*d*), Paul Wallace (*e*), Roxanne
Arlen (*f*), Betty Bruce (*g*), Faith Dane (*h*).
Prod/Dir Mervyn Le Roy. *Scr* Leonard
Spiegelgass from the play by Arthur Laurents
based on the life story of Gypsy Rose Lee.
Ph Harry Stradling. *Art* John Beckman. *Sets*
Ralph S. Hurst. *Ed* Philip W Anderson. *Cost*
Orry-Kelly. *Sound* M A Merrick, Dolph
Thomas. *Dances* Robert Tucker. *Mus* Frank
Perkins. *Songs* Jule Styne and Stephen Sondheim: Small world (*a-c*); Rose's turn (*a*); Some
people (*a*); You'll never get away from me
(*a-c*); Let me entertain you (*b*); Everything's
coming up roses (*a*); If mama was married
(*b-d*); Baby June and her newsboys (*d*); All I
need is the girl (*b-e*); Dainty June and her
farmboys (*d*); Together wherever we go (*a-b*);
Little lamb (*b*); You gotta have a gimmick
(*f-g-h*); Mr Goldstone I love you (*a*);
Toreadorable (*a*).
Note Vocals for Rosalind Russell partially
dubbed by Lisa Kirk.
Hans Christian Andersen Goldwyn/RKO 1952
(*120 mins*) *with* Danny Kaye (*a*), Zizi Jeanmaire
(*b*), Farley Granger (*c*), Jane Wyman (*d*), Joey
Walsh (*e*), Roland Petit (*f*), Erik Bruhn (*g*).
Prod Samuel Goldwyn. *Dir* Charles Vidor.
Scr Moss Hart from a story by Myles Connolly.
Ph Harry Stradling. *Art* Richard Day. *Sets*
Howard Bristol. *Ed* Daniel Mandell. *Sound*
Fred Lau. *Dances* Roland Petit. *Mus* Walter
Scharf. *Songs* Frank Loesser (ballet music by
Jerome Moross): I'm Hans Christian Andersen
(*a*); Wonderful Copenhagen (*a-e*); Inch worm
(*a*); Little Mermaid ballet (*b-f-g*); Thumbelina
(*a*); Anywhere I wander (*a*); No two people
(*a-d*); The ugly duckling (*a*); The King's new
clothes (*a*).
Story based on the life of Andersen, with
sequences created around his fairy tales.
The Happiest Millionaire Disney/Buena Vista
1967 (*141 mins*) *with* Tommy Steele (*a*), Fred
MacMurray (*b*), Lesley Ann Warren (*c*), John
Davidson (*d*), Greer Garson (*e*), Geraldine
Page (*f*), Gladys Cooper (*g*), Hermione
Baddeley (*h*), Eddie Hodges (*i*), Paul Petersen
(*j*), Joyce Bulifant (*k*). *Prod* Walt Disney, Bill
Anderson. *Dir* Norman Tokar. *Scr* A J
Carothers from the book and Broadway play
by Cordelia Drexel Biddle and Kyle Crichton,
based on the life of Anthony J Drexel Biddle.

Ph Edward Colman. *Art* Carroll Clark, John
B Mansbridge. *Ed* Cotton Warburton. *Sound*
Robert O Cook. *Dances* Marc Breaux and Dee
Dee Wood. *Mus* Jack Elliott. *Songs* Richard M
and Robert B Sherman: Fortuosity (*a*); What's
wrong with that (*b-c*); Watch your footwork
(*i-j*); Valentine candy (*c*); Strengthen the
dwelling (*b and chorus*); I'll always be Irish
(*a-b-c-h*); Bye-yum-pum-pum (*c-k*); I believe in
this country (*b*); Detroit (*c-d*); There are those
(*a-f-g*); When a man has a daughter (*b*); Let's
have a drink on it (*a-d*); Are we dancing (*c-d*).
Interpolated song La belle fille d'amore (*b*).
The Harvey Girls MGM 1945 (*101 mins*)
with Judy Garland (*a*), Kenny Baker (*b*), Cyd
Charisse (*c*), Virginia O'Brien (*d*), Marjorie Main
(*e*), Ray Bolger (*f*), John Hodiak, Angela
Lansbury (*g*). *Prod* Arthur Freed, *assoc* Roger
Edens. *Dir* George Sidney. *Scr* Edmund
Beloin, Nathaniel Curtis from a story by
Samuel Hopkins Adams, Eleanore Griffin and
William Rankin. *Ph* George J Folsey, Warren
Newcom. *Art* Cedric Gibbons, William
Ferrari. *Sets* Edwin B Willis. *Ed* Albert Akst.
Cost Irene. *Sound* Douglas Shearer. *Dances*
Charles Walters, Robert Alton. *Mus* Lennie
Hayton. *Songs* Harry Warren and Johnny
Mercer: On the Atcheson, Topeka and the
Santa Fe (*ensemble*); In the valley (*a*); Wait
and see (*b-c-g*); The train must be fed (*ensemble*);
Swing your partner round and round (*a-e and
ensemble*); It's a great big world (*a-c-d*); The wild,
wild West (*d*); Oh you kid (*g*); Speciality dance
Interp. On the Atcheson, Topeka (*f*).
Note Vocals for Cyd Charisse dubbed by Betty
Russell.
The Helen Morgan Story Warner 1957
(*118 mins*) *with* Ann Blyth, Paul Newman,
Richard Carlson, Rudy Vallee, The De Castro
Sisters, Jimmy McHugh, Walter Winchell,
Alan King, Gene Evans. *Prod* Martin Rackin.
Dir Michael Curtiz. *Scr* Oscar Paul, Dean
Riesner, Stephen Longstreet, Nelson Gidding.
Ph Ted McCord. *Art* John Beckman. *Sets*
Howard Bristol. *Ed* Frank Bracht. *Cost*
Howard Shoup. *Sound* Francis J Scheid, Dolph
Thomas. *Dances* Le Roy Prinz. *Mus* Ray
Heindorf. *Songs* (Sung by Gogi Grant, dubbing
for Ann Blyth in the title role): Bill; Why was I
born; I can't give you anything but love; If
you were the only girl in the world; Avalon;
Do do do; Breezin' along with the breeze;
Someone to watch over me; The one I love;
Body and soul; Can't help lovin' that man;
Something to remember you by; Speak to me
of love; On the sunny side of the street; More
than you know; April in Paris; Love nest; The
man I love; Just a memory; Deep night; Don't

ever leave me; I've got a crush on you; I'll get by; My melancholy baby.

Title in GB: *Both Ends Of The Candle*.

Hello Dolly! Fox 1969 (*148 mins*) with Barbra Streisand (*a*), Walter Matthau (*b*), Michael Crawford (*c*), Louis Armstrong (*d*), Marianne McAndrew (*e*), Danny Lockin (*f*), E J Peaker (*g*), Tommy Tune (*h*), Joyce Ames (*i*). *Prod* Ernest Lehman, Roger Edens. *Dir* Gene Kelly. *Scr* Ernest Lehman from Michael Stewart's book for the Broadway show based on Thornton Wilder's 'The Matchmaker'. *Ph* Harry Stradling. *Art* Jack Martin Smith, Herman Blumenthal. *Sets* Walter M Scott. *Prodn Design* John de Cuir. *Ed* William Reynolds. *Cost* Irene Sharaff. *Sound* Jack Solomon, Murray Spivack. *Dances* Michael Kidd. *Mus* Lionel Newman, Lennie Hayton. *Songs* Jerry Herman: Hello Dolly! (*a-b-d and chorus*); Just leave everything to me (*a*); Put on your Sunday clothes (*a-c-f and chorus*); Dancing (*a-c-e-f-g-h-i and chorus*); So long dearie (*a*); It takes a woman (*a-b-c-f and chorus*); Ribbons down my back (*e*); Before the parade passes by (*a and chorus*); Elegance (*c-e-f-g*); Love is only love (*a*); It only takes a moment (*c-e and chorus*).

Hello, Frisco, Hello Fox 1943 (*98 mins*) *with* Alice Faye (*a*), John Payne (*b*), Jack Oakie (*c*), June Havoc (*d*), Lynn Bari, Laird Cregar. *Prod* Milton Sperling, William Perlberg. *Dir* Bruce Humberstone. *Scr* Robert Ellis, Helen Logan, Richard Macaulay. *Ph* Charles Clarke, Allen Davey. *Art* James Basevi, Boris Leven. *Sets* Paul S Fox, Thomas Little. *Ed* Barbara McLean. *Cost* Helen Rose. *Sound* Joseph E. Aiken, Roger Heman. *Dances* Val Raset. *Mus* Emil Newman, Charles Henderson. *Songs* Harry Warren and Mack Gordon: You'll never know (*a-b*). *Interpolated songs* Has anybody here seen Kelly? (*a-c and chorus*); Hello, Frisco, hello (*a-b-c-d*); Ragtime cowboy Joe (*a-c-d*); Bedelia (*chorus*); By the light of the silvery moon (*a*); A friend from your old home town (*c-d*); It's tulip time in Holland (*skaters*); When you wore a tulip (*skaters*); I've got a girl in every port (*c*); Lindy (*a-b-c-d*); Sweet cider time (*a and chorus*); They always pick on me (*a*); Doin' the Grizzly Bear (*a-c-d*); San Francisco (*chorus*); Hello ma baby (*chorus*); Every inch a sailor (*c-d*).

Here Come The Waves Paramount 1945 *with* Bing Crosby (*a*), Betty Hutton (*b*), Sonny Tufts (*c*), Ann Doran, Noel Neill, Harry Barris. *Prod/Dir* Mark Sandrich. *Scr* Allan Scott, Ken Englund, Zion Meyers. *Ph* Charles Lang. *Art* Hans Dreier, Roland Anderson. *Sets* Milt Gross. *Ed* Ellsworth Hoagland. *Cost* Edith Head. *Dances* Danny Dare. *Mus* Robert Emmett Dolan. *Songs* Harold Arlen and

Johnny Mercer: Here come the Waves (*title music*); Accentuate the positive (*a-c*); I promise you (*a-b*); That old black magic (*a*); Let's take the long way home (*a-b*); There's a fella waitin' in Poughkeepsie (*b*).

Higher And Higher RKO 1943 (*90 mins*) *with* Frank Sinatra (*a*), Jack Haley (*b*), Michele Morgan (*c*), Marcy Maguire (*d*), Mel Tormé (*e*), Leon Errol (*f*), Barbara Hale (*g*), Victor Borge (*h*), Paul and Grace Hartman (*i*). *Prod* Charles Koerner. *Dir* Tim Whelan. *Scr* Jay Dratler, Ralph Spence from a story by Joshua Logan and Gladys Hurlbut based on the Rodgers and Hart Broadway show. *Ph* Robert De Grasse. *Art* Albert S D'Agostino, Jack Okey. *Sets* Darrell Silvera, Claude Carpenter. *Ed* Gene Milford. *Sound* Jean L Speak, James Stewart. *Dances* Ernst Matray. *Mus* Constantin Bakaleinikoff. *Songs* Jimmy McHugh and Harold Adamson: Higher and higher (*b-c-d-e-i*); I couldn't sleep a wink last night (*a*); I saw you first (*a-d*); A lovely way to spend an evening (*a*); Minuet in boogie (*a-b-c-d-e-f-g-h-i*); Today I'm a debutante (*c*); The music stopped (*a*); It's a most important affair (*b-c-d-e-f-i*); You're on your own. Richard Rodgers and Lorenz Hart: Disgustingly rich (*b-c-d-e-i*).

High Society Siegel/MGM 1956 (*107 mins*) *with* Bing Crosby (*a*), Frank Sinatra (*b*), Grace Kelly (*c*), Celeste Holm (*d*), Louis Armstrong and the All Stars (*e*); John Lund, Louis Calhern, Sidney Blackmer, Margalo Gilmore. *Prod* Sol C Siegel. *Dir* Charles Walters. *Scr* John Patrick from Philip Barry's 'The Philadelphia Story'. *Ph* Paul C Vogel. *Art* Cedric Gibbons, Hans Peters. *Sets* Edwin B Willis, Richard Pefferle. *Ed* Ralph Winters. *Cost* Helen Rose. *Sound* Wesley C Miller. *Mus* Johnny Green, *assoc* Saul Chaplin. *Songs* Cole Porter: High Society calypso (*e*); Little one (*a*); Who wants to be a millionaire (*b-d*); True love (*a-c*); You're sensational (*b*); I love you Samantha (*a-e*); Now you has jazz (*a-e*); Well did you evah (*a-b*); Mind if I make love to you (*b*). *Background music* Easy to love; I've got my eyes on you.

High, Wide And Handsome Paramount 1937 (*111 mins*) *with* Irene Dunne (*a*), Randolph Scott, Dorothy Lamour (*b*), William Frawley (*c*), Akim Tamiroff, Elizabeth Patterson, Ben Blue. *Prod* Arthur Hornblow Jnr. *Dir* Rouben Mamoulian. *Scr* Oscar Hammerstein II based on 'The Black Gold Rush'. *Ph* Victor Milner, Gordon Jennings. *Art* Hans Dreier, John Goodman. *Cost* Travis Banton. *Sound* Don Johnson. *Ed* Archie Marshek. *Dances* Le Roy Prinz. *Mus* Boris Moross. *Songs* Jerome Kern and Oscar Hammerstein II: High, wide and handsome

(*a*); Can I forget you (*a*); The things I want
(*b*); The folks who live on a hill (*a*); Will you
marry me tomorrow Maria (*c and chorus*);
Allegheny Al (*a-b*).

Hit The Deck MGM 1955 (*112 mins*) *with*
Tony Martin (*a*), Jane Powell (*b*), Russ
Tamblyn (*c*), Kay Armen (*d*), Vic Damone
(*e*), Debbie Reynolds (*f*), Ann Miller (*g*),
Walter Pidgeon, Gene Raymond. *Prod* Joe
Pasternak. *Dir* Roy Rowland. *Scr* Sonya Levien,
William Ludwig from Herbert Fields' book
for the Broadway show based on Hubert
Osborne's 'Shore Leave'. *Ph* George J Folsey.
Art Cedric Gibbons, Paul Groesse. *Sets* Edwin
B Willis. *Ed* John McSweeney Jnr. *Sound*
Douglas Shearer. *Dances* Hermes Pan. *Mus*
Georgie Stoll. *Songs* Vincent Youmans, Leo
Robin, Clifford Grey and Sidney Clare: More
than you know (*a*); Keepin' myself for you (*a*);
Sometimes I'm happy (*b-e*); A kiss or two (*f*);
Lucky bird (*b*); Join the navy/Loo loo (*f*); Why
oh why (*a-b-c-e-f-g*); I know that you know
(*e*); The lady from the Bayou (*g*); Hallelujah
(*a-c-e*). *Interpolated song* Ciribiribin (*a-b-c-d-e-f-g*).

Holiday Inn Paramount 1942 (*101 mins*) *with*
Bing Crosby (*a*), Fred Astaire (*b*) Marjorie
Reynolds (*c*), Virginia Dale (*d*), Louise Beavers
(*e*), Walter Abel. *Prod/Dir* Mark Sandrich. *Scr*
Claude Binyon. *Ph* David Abel. *Art* Hans
Dreier, Roland Anderson. *Ed* Ellsworth
Hoagland. *Cost* Edith Head. *Dances* Bernard
(Babe) Pearce. *Mus* Robert Emmett Dolan.
Songs Irving Berlin: Come to Holiday Inn
(*a-c*); Happy holiday (*a-c*); I'll capture your
heart (*a-b-c-d*); Abraham (*a-c-e*); Lazy (*a*); White
Christmas (*a-c*); Easter parade (*a*); Be careful
it's my heart (*a-b-c*); I've got plenty to be
thankful for (*a*); I can't tell a lie (*b-c*); Let's
start the New Year right (*a*); You're easy to
dance with (*b-d*); Song of freedom (*a*); Say it
with firecrackers (*b*).

Hollywood Hotel 1st National 1937 (*109 mins*)
with Dick Powell (*a*), Rosemary Lane (*b*), Lola
Lane (*c*), Frances Langford (*d*), Benny Good-
man and his Orchestra (*e*), Johnny 'Scat' Davis
(*f*), Raymond Paige and his Orchestra (*g*), Ted
Healy (*h*), Mabel Todd (*i*), Jerry Cooper (*j*),
Carole Landis. *Prod* Hal B Wallis, *assoc* Samuel
Bischoff. *Dir/Dances* Busby Berkeley. *Scr* Jerry
Wald, Maurice Leo, Richard Macaulay. *Ph*
Charles Rosher, George Barnes. *Art* Robert
Haas. *Mus* Leo F Forbstein. *Songs* Richard
Whiting and Johnny Mercer: Let that be a
lesson to you (*a-b-e-f-h-i*); Sing you son of a
gun (*a*); I'm like a fish out of water (*a-b-h-i*);
I've hitched my wagon to a star (*a-g*); I've got a
heartful of music (*e*); Silhouetted in the moon-
light (*d-j*); Hooray for Hollywood (*d-e-f*); Can't

teach my old heart new tricks (*a-g*) (deleted
before release). *Interpolated songs* Dark eyes (*g*);
Sing sing sing (*e*); California here I come (*e*).
Background music Have you got any castles, baby;
Bob White.

Hollywood Revue MGM 1929 *with* Cliff
Edwards (*a*), Marion Davies (*b*), Joan Crawford
(*c*), Buster Keaton (*d*), George K Arthur (*e*),
Nacio Herb Brown (*f*), Jack Benny (*g*), Charles
King (*h*), The Brox Sisters (*i*), Gus Edwards (*j*)
John Gilbert (*k*), Marie Dressler (*l*), Norma
Shearer (*m*), Albertina Rasch Ballet (*n*), Laurel
and Hardy. *Prod* Harry Rapf. *Dir* Charles F
Reisner. *Scr* Al Boasberg, Robert Hopkins. *Ph*
John Arnold, Irving G Reis, Maximilian Fabian.
Art Cedric Gibbons, Richard Day. *Ed* William
Gray. *Cost* David Cox. *Dances* Sammy Lee,
Albertina Rasch. *Mus* Arthur Lange. *Songs* Nacio
Herb Brown and Arthur Freed: Singin' in the rain
(*a-b-c-d-e-f-i*); You were meant for me; Tommy
Atkins on parade. Louis Alter: Got a feeling
for you (*c*). Fred Fisher: Tableau of jewels (*l*);
Bones and tambourines; Strike up the band.
Gus Edwards: Medley of his songs (*j*). Andy
Rice and Martin Broones: I'm the queen. Ray
Klages and Jesse Greer: Lowdown rhythm.
Other scenes Underwater ballet (*l-n*); 'Romeo and
Juliet' – balcony scene (*k-m*); Salome dance (*d*);
Oh, what a man (*b*).

**How To Succeed In Business Without
Really Trying** Mirisch/United Artists 1966 (*121
mins*) *with* Robert Morse (*a*), Rudy Vallee (*b*),
Anthony Teague (*c*), Kay Reynolds (*d*), Michele
Lee (*e*), Maureen Arthur (*f*), Sammy Smith (*g*),
Ruth Kobart (*h*), John Myhers (*i*), Murray
Matheson. *Prod/Dir/Scr* David Swift from the
Broadway show by Abe Burrows, Jack
Weinstock and Willie Gilbert, based on
Shepherd Mead's novel. *Ph* Burnett Guffey.

*High, Wide and Handsome/Irene Dunne and
Randolph Scott*

Art Robert Boyle. *Sets* Edward G Boyle. *Ed* Ralph Winters, Allan Jacobs. *Cost* Micheline. *Sound* Robert Martin. *Dances* Dale Moreda based on original stage choreography by Bob Fosse. *Mus* Nelson Riddle. *Songs* Frank Loesser How to succeed in business (*a*); I believe in you (*a-e*); Coffee break (*c-d*); The company way (*a-c-g*); A secretary is not a toy (*i*); Been a long day (*a-d-e*); Finch's frolic (*a*); Grand old ivy (*a-b*); The brotherhood of man (*a-b-b*); Paris original (*orch.*).

I Dood It MGM 1943 (*102 mins*) *with* Eleanor Powell (*a*), Red Skelton (*b*), John Hodiak, Lena Horne (*c*), Jimmy Dorsey and his Orchestra (*d*), Hazel Scott (*e*), Richard Ainley. *Prod* Jack Cummings. *Dir* Vincente Minnelli. *Scr* Sig Herzig, Fred Saidy. *Ph* Ray June. *Art* Cedric Gibbons, Jack Martin Smith. *Sets* Edwin B Willis, Helen Conway. *Ed* Robert J Kern. *Cost* Irene Sharaff, Gile Steele. *Sound* Douglas Shearer. *Dances* Bobby Connolly. *Mus* Georgie Stoll, *assoc* Kay Thompson. *Songs* Gene De Paul and Don Raye: Star eyes (*d*); So long Sarah Jane. Kay Thompson, Richard Meyers and Leo Robin: Jericho (*c-e and chorus*). *Interpolated songs* Swinging the jinx away (*a*); One o'clock jump (*d*); Taking a chance on love (*c*); Hola e pae.
Title in GB: *By Hook Or By Crook.*

I Live For Love Warner 1935 (*63 mins*) *with* Dolores Del Rio (*a*), Everett Marshall (*b*), Allen Jenkins (*c*), Eddie Conrad (*d*), Shaw and Lee (*e*), Guy Kibbee, Berton Churchill. *Prod* Bryan Foy. *Dir/Dances* Busby Berkeley. *Scr* Jerry Wald, Julius Epstein, Robert Andrews. *Ph* George Barnes. *Art* Esdras Hartley. *Ed* Terry Morse. *Cost* Orry-Kelly. *Mus* Leo F Forbstein. *Songs* Mort Dixon and Allie Wrubel: Mine alone (*b*); Silver wings (*b*); Shaving song (*b-c-d-e*); I live for love (*a*); I wanna play house with you.

I'll See You In My Dreams Warner 1951 (*113 mins*) *with* Doris Day (*a*), Danny Thomas (*b*), Frank Lovejoy (*c*), Patrice Wymore (*d*), James Gleason, Jim Backus. *Prod* Louis F Edelman. *Dir* Michael Curtiz. *Scr* Melville Shavelson, Jack Rose. *Ph* Ted McCord. *Art* Douglas Bacon. *Sets* George James Hopkins. *Ed* Owen Marks. *Cost* Leah Rhodes, Marjorie Best. *Sound* Oliver Garretson, David Forrest. *Dances* Le Roy Prinz. *Mus* Ray Heindorf. *Songs* Gus Kahn: I'll see you in my dreams (*a*); I'm nobody's sweetheart now (*a*); Makin' whoopee (*a-b*); It had to be you (*b*); I wish I had a girl (*a-b*); Love me or leave me (*d*); Pretty baby (*b*); My buddy (*a*); Carolina in the morning (*d*); Yes sir, that's my baby (*a-b*); The one I love (*a*); Swingin' down the lane (*a-b*);

Memories (*unidentified tenor as John McCormack*). *Background music* The Carioca; Your eyes have told me so; Shine on harvest moon.
Biography of song-writer Gus Kahn.

In Caliente 1st National 1935 (*84 mins*) *with* Dolores Del Rio (*a*), Phil Regan (*b*), Winifred Shaw (*c*), Tony and Sally De Marco (*d*), Judy Canova (*e*), Edward Everett Horton (*f*), George Humbert (*g*), Pat O'Brien, Glenda Farrell, Leo Carrillo. *Dir* Lloyd Bacon. *Scr* Jerry Wald, Julius Epstein, story and adaptation by Ralph Block and Warren Duff. *Ph* Sol Polito, George Barnes. *Art* Robert M Haas. *Ed* Jimmy Gibbons. *Cost* Orry-Kelly. *Dances* Busby Berkeley. *Mus* Leo F Forbstein. *Songs* Mort Dixon and Allie Wrubel: In Caliente; The lady in red (*c-e-f-g*); Muchacha (*a-b and chorus*). Harry Warren and Al Dubin; To call you my own.

It Happened In Brooklyn MGM 1946 (*105 mins*) *with* Frank Sinatra (*a*), Kathryn Grayson (*b*), Jimmy Durante (*c*), Peter Lawford (*d*), Billy Roy (*e*), Gloria Grahame, Marcy Maguire. *Prod* Jack Cummings. *Dir* Richard Whorf. *Scr* Isobel Lennart from a story by John McGowan. *Ph* Robert Planck. *Art* Cedric Gibbons, Leonid Vasian. *Sets* Edwin B Willis. *Ed* Blanche Sewell. *Cost* Irene. *Sound* Douglas Shearer. *Dances* Jack Donohue. *Mus* Johnny Green. *Songs* Jule Styne and Sammy Cahn: Time after time (*a-b*); The Brooklyn Bridge (*a*); I believe (*a-c-e*); Whose baby are you (*d*); It's the same old dream (*a*); The song's gotta come from the heart (*a-c*). *Interpolated arias* 'Lakme' – Bell song (Delibes) (*b*); 'Don Giovanni' – La ci darem la mano (Mozart) (*a-b*).
Note Piano playing for Frank Sinatra dubbed by André Previn.

It's A Great Feeling Warner 1949 (*85 mins*) *with* Doris Day (*a*), Jack Carson (*b*), Dennis Morgan (*c*). Guests incl. Ronald Reagan, Danny Kaye, Jane Wyman, Errol Flynn, Gary Cooper, Joan Crawford. *Prod* Alex Gottlieb. *Dir* David Butler. *Scr* Melville Shavelson, Jack Rose from a story by I A L Diamond. *Ph* Wilfrid M Cline. *Art* Stanley Fleischer. *Ed* Irene Morra. *Dances* Le Roy Prinz. *Mus* Ray Heindorf. *Songs* Jule Styne and Sammy Cahn: It's a great feeling (*a*); Give me a song with a beautiful melody (*b*); At the Cafe Rendezvous (*a*); That was a big fat lie (*a-b*); Blame my absentminded heart (*a*); Fiddle-de-dee; There's nothing rougher than love.

It's Always Fair Weather MGM 1955 (*102 mins*) *with* Gene Kelly (*a*), Dan Dailey (*b*), Michael Kidd (*c*), Dolores Gray (*d*), Cyd Charisse (*e*), Lou Lubin (*f*), David Burns (*g*), Jay C Flippen. *Prod* Arthur Freed. *Dir/Dances*

Stanley Donen and Gene Kelly. *Scr/Lyr* Betty
Comden and Adolph Green. *Ph* Robert
Bronner. *Art* Cedric Gibbons, Arthur Loner-
gan. *Sets* Edwin B Willis, Hugh Hunt. *Ed*
Adrienne Fazan. *Cost* Helen Rose. *Sound*
Wesley C Miller. *Mus/Songs* André Previn:
Music is better than words (*d*); March march
(*a-b-c and chorus*); I like myself (*a*); Once upon a
time (*a-b-c*); Thanks a lot but no thanks (*d and
chorus*); The time for parting (*a-b-c-g*); Situation-
wise (*b*); Why are we here (Blue Danube)
(*a-b-c*); Baby you knock me out (*e and chorus*);
Stillman's gym (*f and chorus*).

The Jazz Singer Warner-Vitaphone 1927 (*90
mins*) *with* Al Jolson (*a*), Eugenie Besserer, May
McAvoy, Warner Oland, Myrna Loy, William
Demarest. *Dir* Alan Crosland. *Ph* Hal Mohr.
Scr Al Cohn from Samson Raphaelson's play.
Mus Al Goodman, Louis Silvers. *Songs* My
mammy; Kol Nidrei; Dirty hands, dirty face;
Mother I still have you; Blue skies; Toot toot
tootsie goodbye (all (*a*)).
The first part-talkie, premiered on 23 October
1927.

The Jazz Singer Warner 1952 (*107 mins*) *with*
Danny Thomas (*a*), Peggy Lee (*b*), Mildred
Dunnock (*c*), Eduard Franz. *Prod* Louis F
Edelman. *Dir* Michael Curtiz. *Scr* Frank Davis,
Leonard Stern, Lewis Meltzer from Samson
Raphaelson's play. *Ph* Carl Guthrie. *Art* Leo
F Kuter. *Sets* George James Hopkins. *Ed* Alan
Crosland Jnr. *Cost* Howard Shoup. *Sound*
C A Riggs, David Forrest. *Dances* Le Roy
Prinz. *Mus* Ray Heindorf. *Songs* Sammy Fain
and Jerry Seelen: I hear the music now (*a-b*);
O moon (*a*); Living the life I love (*a*); What
are New Yorkers made of (*a-b and chorus*).
Interpolated songs Kol Nidrei (*a and choir*); This
is a very special day (*a-b*); Hush-a-bye (*a-c*);
Lover (*b*); Just one of those things (*b*). *Montage*
I'm looking over a four leaf clover, Breezing
along with the breeze, If I could be with you,
The birth of the blues (*a*).
Remake of the 1927 film.

Jolson Sings Again Columbia 1949 (*96 mins*)
with Larry Parks (*a*), Barbara Hale, Ludwig
Donath, William Demarest, Myron McCormick
Bill Goodwin, Tamara Shayne, Morris Stoloff.
Prod Sidney Buchman. *Dir* Henry Levin.
Scr Sidney Buchman. *Ph* William Snyder. *Art*
Walter Holscher. *Sets* William Kiernan. *Ed*
William Lyons. *Cost* Jean Louis. *Sound* George
Cooper. *Mus* Morris Stoloff, *assoc* Saul Chaplin.
Songs (all (*a*)) After you've gone; Chinatown my
Chinatown; Give my regards to Broadway;
I only have eyes for you; I'm just wild about
Harry; I'm looking over a four leaf clover;
Is it true what they say about Dixie; Rockabye

your baby with a Dixie melody; Carolina in
the morning; Toot toot tootsie goodbye;
Pretty baby; Back in your own backyard;
When the red red robin; Sonny boy; For me
and my gal; California here I come; Baby face.
Montage of clips from *The Jolson Story*: April
showers, You made me love you, Ma blushin'
Rosie, Let me sing and I'm happy, The Spaniard
that blighted my life, About a quarter to nine
(*w. Evelyn Keyes*); The anniversary song;
Swanee; My mammy. *Background music* Learn
to croon (*Bing Crosby record*); I'll take romance;
It's a blue world.
Sequel to *The Jolson Story*. Biography of Al
Jolson, who dubbed vocals for Larry Parks.

The Jolson Story Columbia 1946 (*128 mins*)
with Larry Parks (*a*), Evelyn Keyes (*b*), Scotty
Beckett (*c*), The Robert Mitchell Boychoir (*d*),
Ludwig Donath (*e*), William Demarest, Bill
Goodwin, Tamara Shayne. *Prod* Sydney
Skolsky, *assoc* Gordon Griffith. *Dir* Alfred E
Green. *Scr* Sidney Buchman, Stephen Long-
street. *Ph* Joseph Walker. *Art* Stephen Gooson,
Walter Holscher. *Sets* William Kiernan, Louis
Diage. *Ed* William Lyons. *Cost* Jean Louis.
Sound Hugh McDowell. *Dances* Joseph Lewis,
Jack Cole. *Mus* Morris Stoloff, *assoc* Saul
Chaplin. *Songs* On the banks of the Wabash
(*c*); Swanee (*a*); Ave Maria (Schubert) (*c-d*);
April showers (*a*); Latin from Manhattan (*a-b*);
When you were sweet sixteen (*c*); The anni-
versary song (*a-e*); Waiting for the Robert E Lee
(*a*); The Spaniard who blighted my life (*a and
chorus*); American patrol (*c*); *Montage* The lullaby
of Broadway, Forty-Second Street, We're in the
money (*b*); California here I come (*a-b*);
Rockabye your baby with a Dixie melody (*a*);
Let me sing and I'm happy (*a*); Goodbye my
bluebell (*c*); Liza (*a-b*); My mammy (*a*); Ma

The Jazz Singer/Al Jolson and May McAvoy

blushin' Rosie (*a*); ᴅʏ the light of the silvery
moon (*c*); There's a rainbow round my shoulder
(*a*); Eli Eli (*c-e*); Avalon (*a*); I want a girl
(*a and minstrels*); I'm sitting on top of the world
(*a*); About a quarter to nine (*a-b*); Toot toot
tootsie goodbye (*a*); After the ball (*c*); Every
little movement (*orch.*); You made me love you
(*a*); Blue skies (*a*) (deleted before release).
Biography of Al Jolson, who dubbed vocals
for Larry Parks.

The Kid From Spain Goldwyn/United
Artists 1932 (*90 mins*) *with* Eddie Cantor (*a*),
Lyda Roberti (*b*), Robert Young, Noah Beery,
Ruth Hall, The Goldwyn Girls, Frances Dean
(Betty Grable), *Prod* Samuel Goldwyn. *Dir*
Leo McCarey. *Scr* William Anthony McGuire,
Bert Kalmar, Harry Ruby. *Ph* Gregg Toland.
Ed Stuart Heisler. *Sound* Vinton Vernon.
Dances Busby Berkeley. *Mus* Alfred Newman.
Songs Bert Kalmar and Harry Ruby, with
Harry Akst*: In the moonlight (*b*); Look what
you've done (*a*); What a perfect combination (*a*)*
The King And I Fox 1956 (*133 mins*) *with*
Deborah Kerr (*a*), Yul Brynner (*b*), Rita
Moreno (*c*), Terry Saunders (*d*), Carlos Rivas
(*e*), Alan Mowbray. *Prod* Charles Brackett. *Dir*
Walter Lang. *Scr* Ernest Lehman from the
book by Oscar Hammerstein II for the
Broadway show based on Anna Leonowens'

book. *Ph* Leon Shamroy. *Art* Lyle Wheeler,
John De Cuir. *Sets* Walter M Scott, Paul S Fox
Ed Robert Simpson. *Cost* Irene Sharaff. *Dances*
Jerome Robbins. *Mus* Alfred Newman, *assoc*
Ken Darby. *Songs* Richard Rodgers and Oscar
Hammerstein II: I whistle a happy tune (*a*);
My lord and master (*c*); Hello young lovers (*a*);
The march of the Siamese children (*children*);
Shall I tell you what I think of you (*a*); A
puzzlement (*b*); We kiss in a shadow (*c-e*); I
have dreamed (*c-e*); Something wonderful (*d*);
Shall we dance (*a-b*); The small house of Uncle
Thomas (*children's ballet*); The song of the
King (*b*); Getting to know you (*a and children*).
Note Vocals for Deborah Kerr partially dubbed
by Marni Nixon.
The King Of Burlesque Fox 1935 (*83 mins*)
with Alice Faye (*a*), Jack Oakie (*b*), Fats Waller
(*c*), Dixie Dunbar (*d*), Kenny Baker (*e*), Shaw
and Lee (*f*), Warner Baxter (*g*). *Prod* Kenneth
Magowan. *Dir* Sidney Lanfield. *Scr* James
Seymour, Harry Tugend, Gene Markey from a
story by Vina Delmar. *Ph* J Peverell Marley.
Ed Ralph Dietrich. *Dances* Sammy Lee. *Songs*
Jimmy McHugh and Ted Koehler: I'm shooting
high (*a-b-f-g*); Lovely lady (*e*); Whose big baby
are you (*a*); Spreadin' rhythm around (*a*); Too
good to be true (*c*); I've got my fingers crossed
(*a-c-d*). Jack Yellen and Lew Pollack: I love to

The King of Jazz

ride the horses. Trad: Oh Susannah (*c*).

The King Of Jazz Universal 1930 (*105 mins*) *with* Paul Whiteman and his Orchestra (*a*), Jeanette Loff (*b*), Stanley Smith (*c*), The Rhythm Boys (Bing Crosby, Harry Barris and Al Rinker) (*d*), The Brox Sisters (*e*), John Boles (*f*), The Markert Dancers (*g*), The Sisters G. (*h*), Al Norman (*i*), Charles Irwin (*j*), George Chiles (*k*), Jeanie Lang (*l*), Don Rose (*m*), Marion Statler (*n*), Jacques Cartier (*o*), Tommy Atkins Sextet (*p*), Nell O'Day (*q*), The Hollywood Beauties (*r*), Johnny Fulton (*s*). *Prod* Carl Laemmle Jnr. *Dir* John Murray Anderson. *Scr* Harry Ruskin, John Murray Anderson. *Ph* Ray Rennahan, Hal Mohr, Jerome Ash. *Art/Cost* Herman Rosse. *Ed* Robert Carlisle. *Sound* C Roy Hunter. *Dances* Russell E Markert. *Mus* Paul Whiteman, Ferde Grofe. *Songs* Milton Ager and Jack Yellen: Music has charms (*a-d*); My bridal veil (*a-b-c-g-l-r*); A bench in the park (*a-b-c-d-e-g-h-r*); The song of the dawn (*f*); Happy feet (*a-d-g-h-i-j*); I like to do things for you (*a-d*). Mabel Wayne and Billy Rose: It happened in Monterey (*b-f-g-h-k-s*); My ragamuffin Romeo (*k-l-m-n*). Billy Moll and Harry Barris: So the blackbirds and the bluebirds got together (*d*). *Interpolated items* Rhapsody in blue (*a-g-h-o-r*); 'Melting Pot' medley (*a-b-g-k-p-q-r*).

Kismet MGM 1955 (*113 mins*) *with* Howard Keel (*a*), Ann Blyth (*b*), Dolores Gray (*c*), Vic Damone (*d*), Monty Woolley, Sebastian Cabot. *Prod* Arthur Freed. *Dir* Vincente Minnelli. *Scr* Charles Lederer, Luther Davis from the Broadway show based on the book by Edward Knoblock. *Ph* Joseph Ruttenberg. *Art* Cedric Gibbons, Preston Ames. *Sets* Edwin B Willis, Keogh Gleason. *Ed* Adrienne Fazan. *Cost* Tony Duquette. *Sound* Douglas Shearer. *Dances* Jack Cole. *Mus* André Previn, Jeff Alexander. *Songs* Robert Wright and Chet Forrest based on themes by Borodin: Fate (*a*); Bored (*c*); Gesticulate (*a*); And this is my beloved (*a-b-d*); The olive tree (*a*); Not since Nineveh (*c*); Rahadlakum (*a-c*); The sands of time (*a*); A stranger in paradise (*b-d*); Baubles, bangles and beads (*b*); Night of my nights (*d*); Rhymes have I (*a*).

Kiss Me Kate MGM 1953 (*109 mins*) *with* Howard Keel (*a*), Kathryn Grayson (*b*), Ann Miller (*c*), Tommy Rall (*d*), Bobby Van (*e*), Bob Fosse (*f*), Keenan Wynn (*g*), James Whitmore (*h*), Carol Haney (*i*). *Prod* Jack Cummings. *Dir* George Sidney. *Scr* Dorothy Kingsley from Sam and Bella Spewack's book for the Cole Porter Broadway show based on 'The Taming Of The Shrew'. *Ph* Charles Rosher. *Art* Cedric Gibbons, Urie McLeary.

Ed Ralph E Winters. *Sound* Douglas Shearer. *Dances* Hermes Pan. *Mus* André Previn, *assoc* Saul Chaplin. *Songs* Cole Porter: Too darn hot (*c*); So in love (*a-b*); I hate men (*b*); Wunderbar (*a-b*); Tom, Dick and Harry (*c-d-e-f*); Were thine that special face (*a*); We open in Venice (*a-b-c-d*); I've come to wive it wealthily in Padua (*a*); Why can't you behave (*c-d*); Where is the life that late I led (*a*); From this moment on (*c-d-e-f-i*); Always true to you in my fashion (*c-d*); Brush up your Shakespeare (*g-h*).

Note This was the first (and only) musical to be filmed in 3-D, although it was eventually released 'flat'.

Lady Be Good MGM 1941 (*111 mins*) *with* Eleanor Powell (*a*), Robert Young (*b*), Ann Sothern (*c*), Red Skelton (*c*), Dan Dailey (*d*), Virginia O'Brien (*e*), Phil Silvers (*f*), Jimmy Dorsey and his Orchestra (*g*), John Carroll (*h*), Lionel Barrymore, Reginald Owen. *Prod* Arthur Freed. *Dir* Norman Z McLeod. *Scr* Jack McGowan, Kay Van Riper, John McClain, based very loosely on the 1924 Gershwin show. *Ph* George J Folsey, Oliver T Marsh. *Art* Cedric Gibbons. *Sets* Edwin B Willis. *Ed* Frederick Y Smith. *Cost* Dolly Tree. *Sound* Douglas Shearer. *Dances* Busby Berkeley. *Mus* Georgie Stoll, *assoc* Roger Edens. *Songs* George and Ira Gershwin: Oh lady be good; Hang on to me; Fascinating rhythm. Jerome Kern and Oscar Hammerstein II: The last time I saw Paris (*b*). Roger Edens and Arthur Freed: You'll never know; Your words and my music.

Lady In The Dark Paramount 1943 (*100 mins*) *with* Ginger Rogers (*a*), Ray Milland, Warner Baxter, Jon Hall, Barry Sullivan, Mischa Auer, Don Loper, Richard Blumenthal. *Prod* B G De Sylva, Richard Blumenthal. *Dir* Mitchell Leisen. *Scr* Frances Goodrich, Albert Hackett from the Broadway play by Moss Hart. *Ph* Ray Rennahan, Gordon Jennings. *Art* Hans Dreier. *Sets* Raoul Pene Du Bois, Ray Moyer. *Ed* Alma Macorie. *Cost* Edith Head. *Sound* Earl Hayman, Walter Oberst. *Mus* Robert Emmett Dolan. *Songs* Kurt Weill and Ira Gershwin: The saga of Jenny (*a*); My ship (*a*); One life to live (*a*); Girl of the moment (*chorus*); It looks like Liza; This is new (*a*). Jimmy Van Heusen and Johnny Burke: Suddenly it's spring (*a*).

Les Girls MGM 1957 (*114 mins*) *with* Gene Kelly (*a*), Mitzi Gaynor (*b*), Taina Elg (*c*), Kay Kendall (*d*), Jacques Bergerac, Leslie Phillips, Patrick MacNee. *Prod* Sol C Siegel, *assoc* Saul Chaplin. *Dir* George Cukor. *Scr* John Patrick from a story by Vera Caspary. *Ph* Robert Surtees. *Art* William A Horning, Gene Allen. *Sets* Edwin B Willis. *Ed* Ferris Webster. *Cost* Orry-Kelly. *Sound* Wesley C Miller. *Dances* Jack

Cole. *Mus* Adolph Deutsch. *Songs* Cole Porter: Les Girls (*a-b-c-d*); You're just too too (*a-d*); Ca c'est l'amour (*b-c*); Ladies in waiting (*b-c-d*); Why am I so gone on you (*a*); Drinking song.

Look For The Silver Lining Warner 1949 (*100 mins*) *with* June Haver (*a*), Ray Bolger (*b*), Gordon Macrae (*c*), Lee and Lyn Wilde (*d*), Walter Catlett, Charles Ruggles (*e*), Rosemary De Camp (*f*). *Prod* William Jacobs. *Dir* David Butler. *Scr* Phoebe and Henry Ephron from a story by Bert Kalmar and Harry Ruby. *Ph* J Peverell Marley. *Art* John Hughes. *Ed* Irene Morra. *Cost* Milo Anderson. *Sound* David Forrest. *Dances* Le Roy Prinz. *Mus* Ray Heindorf. *Songs* Jerome Kern: Look for the silver lining (*a-c*); Who (*b*); Sunny (*a*); Whippoor-will (*a*); Wild rose (*a*). Others: A kiss in the dark (*a-c*); Time on my hands (*a-c*); The night before Christmas (*a-b*).
Biography of Marilyn Miller.

Love And Hisses Fox 1937 *with* Walter Winchell, Ben Bernie and his Orchestra (*a*), Bert Lahr (*b*), Simone Simon (*c*), Joan Davis (*d*), The Peters Sisters (*e*), The Brewster Twins (*f*), Ruth Terry (*g*), Dick Baldwin (*h*), Douglas Fowley, Charles Barnet. *Prod* Darryl F Zanuck, *assoc* Kenneth MacGowan. *Dir* Sidney Lanfield. *Scr* Art Arthur, Curtis Kenyon. *Ph* Robert Planck. *Art* Bernard Herzbrun, Thomas Little, Mark Lee Kirk. *Ed* Robert Simpson. *Dances* Nick Castle, Geneva Sawyer. *Mus* Louis Silvers. *Songs* Harry Revel and Mack Gordon: Broadway's gone Hawaiian (*a-e-g*); Sweet someone (*c*); Be a good sport (*f*); Darling, je vous aime beaucoup (*c*); I wanna be in Winchell's column (*h*); Lost in your eyes.

Lovely To Look At MGM 1952 (*112 mins*) *with* Howard Keel (*a*), Kathryn Grayson (*b*), Marge and Gower Champion (*c*), Red Skelton (*d*), Ann Miller (*e*), Kurt Kasznar, Zsa Zsa Gabor. *Prod* Arthur Freed. *Dir* Mervyn Le Roy, Vincente Minnelli. *Scr* Harry Ruby, George Wells from Otto Harbach's book for the Broadway show 'Roberta' based on Alice Duer Miller's novel 'Gowns By Roberta'. *Ph* George J Folsey. *Art* Cedric Gibbons, Gabriel Scognamiglio. *Sets* Edwin B Willis. *Ed* John McSweeney Jnr. *Sound* Douglas Shearer. *Cost* Adrian. *Dances* Hermes Pan. *Mus* Carmen Dragon, *assoc* Saul Chaplin. *Songs* Jerome Kern, Otto Harbach, Dorothy Fields, Jimmy McHugh: Lovely to look at (*a and chorus*); Lafayette (*a-c-d*); Smoke gets in your eyes (*b*); I won't dance (*c*); You're devastating (*a-b*); Yesterdays (*b*); The most exciting night (*a*); The touch of your hand (*a-b*); Hard to handle (*e*); Opening night (*chorus*).
Remake of the 1934 film *Roberta*.

Love Me Or Leave Me MGM 1955 (*122 mins*) *with* Doris Day (*a*), James Cagney, Cameron Mitchell, Claude Stroud (*b*), Robert Keith, Tom Tully. *Prod* Joe Pasternak. *Dir* Charles Vidor. *Scr* Daniel Fuchs, Isobel Lennart. *Ph* Arthur E Arling. *Art* Cedric Gibbons, Urie McLeary. *Sets* Edwin B Willis. *Ed* Ralph E Winters. *Sound* Douglas Shearer. *Dances* Alex Romero. *Mus* Georgie Stoll, Percy Faith. *Songs* Nicholas Brodszky and Sammy Cahn: I'll never stop loving you (*a*); Never look back (*a*). *Interpolated songs* I'm sitting on top of the world (*b*); (remainder by (*a*)) Love me or leave me; It all depends on you; You made me love you; Everybody loves my baby; Stay on the right side, sister; Sam the old accordion man; Mean to me; Shakin' the blues away; Ten cents a dance; At sundown. *Background music* Wang wang blues; Sugar blues; Five foot two, eyes of blue; June night; I cried for you; My blue heaven; What can I say after I say I'm sorry.
Biography of singer Ruth Etting.

Love Me Tonight Paramount 1932 (*62 mins*) *with* Maurice Chevalier (*a*), Jeanette MacDonald (*b*), Charles Ruggles (*c*), Charles Butterworth (*d*), Myrna Loy (*e*), C Aubrey Smith (*f*), Elizabeth Patterson (*g*), Ethel Griffies (*h*), Blanche Frederici (*i*), Joseph Cawthorne (*j*), Robert Greig (*k*), Marion Byron (*l*), Cecile Cunningham (*m*), Tyler Brook (*n*), Edgar Norton (*o*), Rita Owin (*p*), Rolf Sedan (*q*), Gabby Hayes (*r*), George Humbert (*s*), Bert Roach (*t*). *Prod/Dir* Rouben Mamoulian. *Scr* Samuel Hoffenstein, Waldemar Young, George Marion Jnr from the play by Paul Armont and Leopold Marshand. *Ph* Victor Milner. *Songs* Richard Rodgers and Lorenz Hart; Love me tonight (*a-b*); That's the song of Paree (*a-l-r*); Isn't it romantic? (*a-b-n-q-t*); Lover (*b*); Mimi (*a-c-f-h*); A woman needs something like that (*b-j*); Poor Apache (*a*); The son of a gun is nothing but a tailor (*b-e-f-g-h-i-k-m-o-p-s*).

The Love Parade Paramount 1929 (*111 mins*) *with* Maurice Chevalier (*a*), Jeanette MacDonald (*b*), Lilian Roth (*c*), Lupino Lane, Virginia Bruce, Eugene Pallette, Ben Turpin. *Prod* Jesse L Lasky. *Dir* Ernst Lubitsch. *Scr* Ernest Vajda, Guy Bolton from the play 'The Prince Consort' by Leon Xanrof and Jules Cheneel. *Ph* Victor Milner. *Art* Hans Dreier. *Ed* Merrill White. *Songs* Victor Schertzinger and Clifford Grey: Dream lover (*b*); Nobody's using it now (*a*); My love parade (*a*); Anything to please the Queen; Paris stay the same (*a*); March of the Grenadiers (*b*); Let's be common.

Make Mine Music Disney/RKO 1946 (*74 mins*) *Voices and talents of* Nelson Eddy (*a*),

Dinah Shore (*b*), The Pied Pipers (*c*), Jerry
Colonna (*d*), Andy Russell (*e*), Sterling Hollo-
way (*f*), Benny Goodman and his Orchestra/
Quartet (*g*), The Andrews Sisters (*h*), The
King's Men (*i*), Riabouchinska and Lichine
(*j*), The Ken Darby Chorus (*k*). *Prod* Walt
Disney, *assoc* Joe Grant. *Dir* Jack Kinney,
Clyde Geronimi, Josh Meador, Hamilton
Luske, Bob Cormack. *Scr* (16 writers). *Process
Effects* Ub Iwerks. *Art* Mary Blair, Elmer
Plummer, John Hench. *Sound* S O Slyfield,
Robert O Cook. *Mus* Charles Wolcott, Oliver
Wallace, Edward Plumb, Ken Darby. *Songs*
Charles Wolcott, Ray Gilbert, Allie Wrubel,
Eliot Daniel, Bobby Worth, Ken Darby: Make
mine music (*chorus*); Willie the whale (*a*); Two
silhouettes (*b-j*); The Martins and the Coys (*i*);
Blue bayou (*k*); Without you (*e*); Johnny
Fedora (*h*); Casey at the bat (*d*). *Interpolated
music* Peter and the Wolf (Prokoviev) (*f*); All
the cats join in (Wilder-Sauter-Gilbert) (*c-g*);
After you've gone (Creamer-Layton) (*g*
(*quartet*)).
Full-length cartoon feature, made 1944 but not
released until 1946. In later years various
sequences were extracted and issued as shorts.
Mammy Warner-Vitaphone 1930 *with* Al Jolson
(*a*), Lowell Sherman (*b*), Hobart Bosworth (*c*),
Louise Dresser, Lee Moran, Stanley Fields.
Dir Michael Curtiz. *Scr* L G Rigby, Joseph
Jackson. *Ph* Barney McGill. *Sound* George
Groves. *Mus* Leo F Forbstein. *Songs* Irving
Berlin: Here we are (*minstrel chorus*); Let me
sing and I'm happy (*a*); To my mammy (*a*); In
the morning (*minstrel chorus*); The call of the
South (interp. Swanee river) (*a*); Knights of
the road (*a-b-c*); Looking at you (across the
breakfast table) (*a*). *Interpolated songs* Yes we
have no bananas (operatic adaptation by Irving
Berlin) (*a and minstrel chorus*); Who paid the
rent for Mrs Rip Van Winkle (*a*); Why do
they all take the night boat to Albany (*a*).
Mary Poppins Disney/Buena Vista 1964 (*139
mins*) *with* Julie Andrews (*a*), Dick Van Dyke
(*b*), Glynis Johns (*c*), David Tomlinson (*d*), Ed
Wynn (*e*), Karen Dotrice (*f*), Matthew Garber
(*g*), Hermione Baddeley, Elsa Lanchester,
Arthur Treacher, Marcy Maguire. *Prod* Walt
Disney, *assoc* Bill Walsh. *Dir* Robert Stevenson.
Scr Bill Walsh, Don Da Gradi based on the
books by P L Travers. *Ph* Edward Colman.
Visual Effects Peter Ellenshaw. *Art* Carroll
Clark, William H Tuntke. *Sets* Tony Walton,
Emile Kuri, Hal Gausman. *Ed* Cotton War-
burton. *Cost* Tony Walton, Bill Thomas.
Sound Robert O Cook, Dean Thomas. *Dances*
Marc Breaux and Dee Dee Wood. *Mus* Irwin
Kostal. *Songs* Richard M and Robert B Sher-

man: The perfect nanny (*f-g*); Sister Suffragette
(*c*); The life I lead (*d*); A spoonful of sugar
(*a*); Chim chim cheree (*a-b-f-g*); Jolly holiday
(*a-b*); Stay awake (*a*); I love to laugh (*a-b-e*);
Supercalifragilisticexpialidocious (*a-b*); Feed
the birds (*a*); Fidelity Fiduciary Bank (*b-d*);
Step in time (*a-b and 'sweeps'*); A man has
dreams (*b-d*); Let's go fly a kite (*c-d*).
Meet Me In Las Vegas MGM 1956 (*112
mins*) *with* Dan Dailey (*a*), Cyd Charisse (*b*),
Lena Horne (*c*), Frankie Laine (*d*), The Four
Aces (*e*), Sammy Davis Jnr (voice only) (*f*),
John Brasia (*g*), *Prod* Joe Pasternak, *Dir* Roy
Rowland. *Scr* Isobel Lennart. *Ph* Robert
Bronner. *Art* Cedric Gibbons, Urie McLeary.
Dances Hermes Pan, Eugene Loring.
Sets Edwin B Willis. *Ed* Albert Akst. *Cost*
Helen Rose. *Sound* Wesley C Miller. *Mus*
Georgie Stoll, Johnny Green. *Songs* Nicholas
Brodszky and Sammy Cahn: The gal with the
yaller shoes (*a*); If you can dream (*c*); My lucky
charm (*a*); Frankie and Johnnie ballet (*a-b-f-g*).
Sleeping Beauty ballet (*b*).
Title in GB: *Viva Las Vegas.*
Meet Me In St Louis MGM 1944 (*113 mins*)
with Judy Garland (*a*), Leon Ames (*b*), Margaret
O'Brien (*c*), Mary Astor (*d*), Lucille Bremer
(*e*), Henry Daniels Jnr (*f*), Tom Drake (*g*),
Harry Davenport (*h*). *Prod* Arthur Freed. *Dir*
Vincente Minnelli. *Scr* Fred Finklehoffe, Irving
Brecher from the novel by Sally Benson. *Ph*
George Folsey, Henri Jaffa. *Art* Cedric Gibbons,
Lemuel Ayres, Jack Martin Smith. *Sets* Edwin
B Willis, Paul Holdchinsky. *Ed* Albert Akst. *Cost*
Irene Sharaff. *Sound* Douglas Shearer. *Dances*
Charles Walters. *Mus* Georgie Stoll, *assoc* Roger
Edens. *Songs* Hugh Martin and Ralph Blane: The
trolley song (*a and chorus*); The boy next door (*a*);
Over the banisters (*a*); Skip to my Lou (*a-e-f-g
and ensemble*); Have yourself a merry little
Christmas (*a-c*). *Interpolated songs* Meet me in
St Louis (*a-c-e-f-h*); You and I (*b-d*); Under the
bamboo tree (*a-c*).
Note Vocal for Leon Ames dubbed by Arthur
Freed.
Merry Andrew MGM 1958 (*100 mins*) *with*
Danny Kaye (*a*), Pier Angeli (*b*), Robert Coote
(*c*), Rex Evans (*d*), Salvatore Baccaloni (*e*),
Tommy Rall (*f*), Noel Purcell (*g*), Walter
Kingsford (*h*), Patricia Cutts. *Prod* Sol C Siegel,
assoc Saul Chaplin. *Dir/Dances* Michael Kidd.
Scr Isobel Lennart, I A L Diamond from a
story by Paul Gallico. *Ph* Robert Surtees. *Art*
William A Horning, Gene Allen. *Sets* Henry
Grace, Richard Pefferle. *Ed* Harold F Kress.
Cost Walter Plunkett. *Sound* Wesley C Miller.
Mus Nelson Riddle. *Songs* Saul Chaplin and
Johnny Mercer: The pipes of Pan (*a and*

boys); Chin up, stout fellow (a-c-d); Everything
is ticketty-boo (a); You can't always have
what you want (a-b); The square of the
hypotenuse (a and boys); Salud (a-b-e-f and
ensemble) parodied as Here's cheers (c-d-g-b).

The Merry Widow MGM 1934 (99 mins) with
Maurice Chevalier (a), Jeanette MacDonald (b),
Edward Everett Horton (c), Una Merkel,
George Barbier, Donald Meek, Sterling
Holloway, Shirley Ross. Prod Irving Thalberg.
Dir Ernst Lubitsch. Scr Samson Raphaelson,
Ernest Vajda from the libretto of the Lehar
operetta by Victor Leon and Leo Stein. Ph
Oliver T Marsh. Art Cedric Gibbons, Frederick
Hope. Sets Edwin B Willis. Ed Frances March.
Cost Adrian. Sound Douglas Shearer. Mus
Herbert Stothart. Songs Franz Lehar, with new
lyrics by Lorenz Hart and Gus Kahn: The
Merry Widow waltz (a-b); Vilia (b); I'm going
to Maxim's (a-c); Girls girls girls (a).

The Merry Widow MGM 1952 (105 mins)
with Lana Turner (a), Fernando Lamas (b),
Richard Haydn (c), Una Merkel, Thomas
Gomez, John Abbott. Prod Joe Pasternak. Dir
Curtis Bernhardt. Scr Sonya Levien, William
Ludwig from the libretto of the Lehar operetta
by Victor Leon and Leo Stein. Ph Robert
Surtees. Art Cedric Gibbons, Paul Groesse.
Sets Edwin B Willis. Ed Conrad A Nervig.
Sound Douglas Shearer. Dances Jack Cole.
Mus Jay Blackton. Songs Franz Lehar, with new
lyrics by Paul Francis Webster: The Merry
Widow waltz (a-b); Vilia (b); Girls girls girls
(b); I'm going to Maxim's (b-c); Night (b);
Can Can (chorus).
Note Vocals for Lana Turner dubbed by Trudy
Erwin.

Mississippi Paramount 1935 with Bing Crosby
(a), Joan Bennett, W C Fields, Gail Patrick,
Queenie Smith, Ann Sheridan. Prod Arthur
Hornblow Jnr. Dir Edward A Sutherland.
Scr Francis Martin, Jack Cunningham from
Booth Tarkington's 'The Fighting Coward'.
Ph Charles Lang. Art Hans Dreier, Bernard
Herzbrun. Ed Chandler House. Sound Gene
Merritt. Mus Sigmund Krumgold, Maurice
Lawrence. Songs Richard Rodgers and Lorenz
Hart: Soon; Down by the river; It's easy to
remember; Roll Mississippi; You are my heart.
Interpolated song Swanee river (all (a)).

Mr Music Paramount 1950 (113 mins) with
Bing Crosby (a), Dorothy Kirsten (b), Peggy
Lee (c), Groucho Marx (d), The Merry Macs
(e), Marge and Gower Champion (f), Nancy
Olson, Charles Coburn, Richard Haydn, Robert
Stack. Prod Robert L Welch. Dir Richard
Haydn. Scr Arthur Sheekman from Samson
Raphaelson's play 'Accent On Youth'. Ph
George Barnes. Art Hans Dreier, Earl Hedrick.
Ed Duane Harrison. Cost Edith Head. Mus
Joseph J Lilley. Songs Jimmy Van Heusen and
Johnny Burke: Mr Music (chorus); Accidents
will happen (a-b); And you'll be home (a); Life
is so peculiar (a-c-d-e-f); High on the list (a);
Milady (uncredited duettists); Wouldn't it be
funny (a); Wasn't I there (a).

Monte Carlo Paramount 1930 (90 mins) with
Jeanette MacDonald (a), Jack Buchanan (b),
Claude Allister (c), Zasu Pitts, Tyler Brooke,
Donald Novis, Lionel Belmore. Prod Adolph
Zukor. Dir Ernst Lubitsch. Scr Ernest Vajda
from 'Monsieur Beaucaire' by Booth Tarking-
ton and Evelyn Sutherland, and 'The Blue
Coast' by Hans Mueller. Ph Victor Milner.
Mus W Franke Harling. Songs Richard Whiting,
W Franke Harling and Leo Robin: Beyond
the blue horizon (a-b and chorus); Give me a
moment, please (a-b); She'll love me and like
it (c); Always in all ways (a-b); Trimmin' the
women; Whatever it is, it's grand; Day of days.

Moon Over Miami Fox 1941 (91 mins) with
Betty Grable (a), Robert Cummings (b), Don
Ameche (c), Charlotte Greenwood (d), Jack
Haley (e), Carole Landis (f), The Condos
Brothers (g), Jack Cole (h). Prod Harry Joe
Brown. Dir Walter Lang. Scr Vincent Lawrence,
Brown Tolmes from a story by Stephen Powys.
Ph Leon Shamroy, Allen M Davey, J Peverell
Marley. Art Wiard Ihnen, Richard Day.
Dances Hermes Pan. Mus Alfred Newman.
Songs Ralph Rainger and Leo Robin: You
started something (a-b-c); Kindergarten conga
(a and chorus); Oh me, oh Mi-ami (a-d-f); Miami
here I am (chorus); Is that good (d-e); Solitary
Seminole; Loveliness and love; Hooray for
today; I've got you all to myself. Joe Burke
and Edgar Leslie: Moon over Miami (orch.).

Mother Wore Tights Fox 1947 *with* Betty
Grable (*a*), Dan Dailey (*b*), Vanessa Brown,
Mona Freeman, Robert Arthur, Lee Patrick,
Senor Wences. *Prod/Scr* Lamar Trotti. *Dir*
Walter Lang. *Ph* Harry Jackson. *Art* Richard
Day, Joseph C Wright. *Sets* Thomas Little.
Ed J Watson Webb Jnr. *Cost* Orry-Kelly.
Sound Eugene Grossman, Roger Heman.
Dances Seymour Felix, Kenny Williams. *Mus*
Alfred Newman. *Songs* Josef Myrow and Mack
Gordon: Kokomo, Indiana (*a-b*); You do (*a-b*);
There's nothing like a song; Rolling down to
Bowling Green; This is my favourite city; Fare
thee well, dear Alma Mater.

The Music Man Warner 1962 (*151 mins*) *with*
Robert Preston (*a*), Shirley Jones (*b*), Buddy
Hackett (*c*), Ronny Howard (*d*), Pert Kelton
(*e*), Hermione Gingold (*f*), The Buffalo Bills
(*g*). *Prod/Dir* Morton Da Costa. *Scr* Marion
Hargrove from the Broadway show by Meredith
Willson and Franklin Lacey. *Ph* Robert Burks.
Art Paul Groesse. *Sets* George James Hopkins.
Ed William Ziegler. *Cost* Dorothy Jeakins.
Sound M A Merrick, Dolph Thomas. *Dances*
Onna White, Tom Panko. *Mus* Ray Heindorf.
Songs Meredith Willson: Rock Island/Iowa
stubborn (*chorus*); Ya got trouble (*a and chorus*);
The piano lesson (*b-e*); If you don't mind
(*b-e*); Goodnight my someone (*a-b*); 76
trombones (*a*); Sincere (*g*); A sadder but wiser
girl (*a-c*); Pick a little (*f and women*); Marian the
librarian (*a*); Being in love (*b*); Gary, Indiana
(*a-d*); The Wells Fargo wagon (*d*); Lida
Rose/Will I ever tell you (*b-g*); Shipoopi (*a-b-g
and chorus*); Till there was you (*a-b*).

My Fair Lady Warner 1964 (*170 mins*) *with*
Audrey Hepburn (*a*), Rex Harrison (*b*), Stanley
Holloway (*c*), Wilfred Hyde-White (*d*), Jeremy
Brett (*e*), Gladys Cooper, Theodore Bikel,
Mona Washbourne. *Prod* Jack L Warner. *Dir*
George Cukor. *Scr* Alan Jay Lerner from his
book for the Broadway show based on George
Bernard Shaw's 'Pygmalion'. *Ph* Harry
Stradling. *Art* Cecil Beaton, Gene Allen. *Sets*
George James Hopkins. *Ed* William Ziegler.
Cost Cecil Beaton. *Sound* Francis J Scheid,
Murray Spivack. *Dances* Hermes Pan. *Mus* André
Previn. *Songs* Frederick Loewe and Alan Jay
Lerner: Get me to the church on time (*c*); Why
can't the English (*b*); Wouldn't it be luverly (*a*);
With a little bit of luck (*c*); A hymn to him (*b-d*);
An ordinary man (*b*); You did it (*b-d*); Just you
wait (*a*); The rain in Spain (*a-b-d*); The Ascot
gavotte (*chorus*); I could have danced all night (*a*);
Show me (*a-e*); The Embassy waltz (*orch.*); I've
grown accustomed to her face (*b*); On the
street where you live (*e*); The servants' chorus
(*chorus*); Without you (*a*).

Note Vocals for Audrey Hepburn dubbed by
Marni Nixon:

Naughty But Nice Warner 1939 (*90 mins*)
with Dick Powell (*a*), Ann Sheridan (*b*), Gale
Page (*c*), Ronald Reagan, Helen Broderick,
Zasu Pitts, Jerry Colonna. *Prod* Hal B Wallis,
assoc Sam Bischoff. *Dir* Ray Enright. *Scr* Jerry
Wald, Richard Macaulay. *Ph* Arthur L Todd.
Art Max Parker. *Ed* Thomas Richards. *Cost*
Howard Shoup. *Sound* Francis J Scheid, David
Forrest. *Mus* Leo F Forbstein. *Songs* Harry
Warren and Johnny Mercer: Hooray for
spinach (*b*); In a moment of weakness (*a*); I'm
happy about the whole thing (*a-c*); I don't
believe in signs (*a*); Corn pickin'.

Naughty Marietta MGM 1935 (*80 mins*) *with*
Nelson Eddy (*a*), Jeanette MacDonald (*b*),
Elsa Lanchester, Frank Morgan, Akim
Tamiroff, Cecilia Parker. *Prod* Hunt Stromberg.
Dir W S Van Dyke. *Scr* Albert Hackett,
Frances Goodrich, John Lee Mahin from the
operetta by Victor Herbert and Rida Johnson
Young. *Ph* William Daniels. *Art* Cedric
Gibbons. *Cost* Adrian. *Ed* Blanche Sewell. *Sound*
William Steinkamp. *Mus* Herbert Stothart. *Songs*
Victor Herbert and Rida Johnson Young, new
lyrics by Gus Kahn: Naughty Marietta (*chorus*);
Tramp tramp tramp (*a*); 'Neath the Southern
moon (*a*); It never can be love; Italian street
song (*b*); I'm falling in love with someone (*a*);
Loves of New Orleans; Ah, sweet mystery of life
(*a-b*); Live for today; Dance of the marionettes:
The first of the Nelson Eddy–Jeanette Mac-
Donald series.

New Orleans United Artists 1947 (*89 mins*)
with Louis Armstrong and the All Stars (*a*),
Billie Holiday (*b*), Woody Herman and his
Orchestra (*c*), Meade Lux Lewis (*d*), Dorothy
Patrick (*e*), Arturo de Cordova, Richard

*Naughty Marietta/Nelson Eddy and Jeanette
MacDonald*

Hageman. *Prod* Jules Levey, *assoc* Herbert J Biberman. *Dir* Arthur Lubin. *Scr* Elliot Paul, Dick Irving Hyland from a story by Elliot Paul and Herbert J Biberman. *Ph* Lucien Andriot. *Art* Rudi Feld. *Ed* Bernard W Burton. *Sound* Roy Raguse, Roy Meadows. *Mus* Nathaniel Finston. *Songs* Louis Alter and Eddie De Lange: Do you know what it means to miss New Orleans (*a-b-c-e*); Endie (*a-c*); The blues are brewin' (*a-b-c*). *Interpolated music* West End blues (*a*); Buddy Bolden's blues (*a*); Honky tonk train blues (*d*); Dippermouth blues (*a*); New Orleans stomp (*c*); Where the blues were born in New Orleans (*a*); Shimme-sha-wabble (*a*); Basin Street blues (*a*); Farewell to Storyville (*a-b*); Mahogany Hall stomp (*a*); Maryland my Maryland (*a*).

Night And Day Warner/1st National 1945 (*128 mins*) with Cary Grant (*a*), Alexis Smith (*b*), Jane Wyman (*c*), Ginny Simms (*d*), Monty Woolley (*e*), Carlos Ramirez (*f*), Mary Martin (*g*), George Zoritch and Milada Mladova (*h*), Adam and Jayne Gatano (*i*). *Prod* Hal B Wallis, *assoc* Arthur Schwartz. *Dir* Michael Curtiz. *Scr* Charles Hoffman, Leo Townsend, William Bowers. *Ph* William V Skall, J Peverell Marley, Robert Burks. *Art* John Hughes. *Sets* Armor Marlowe. *Ed* David Weisbart. *Sound* Everett A Brown, David Forrest. *Dances* Le Roy Prinz. *Mus* Leo F Forbstein, *assoc* Ray Heindorf. *Songs* Cole Porter: Night and day (*a-b*); Begin the beguine (*f-h*); What is this thing called love (*d*); Miss Otis regrets (*a-e*); Just one of those things (*d*); Do I love you (*d*); An old-fashioned garden (*a*); My heart belongs to Daddy (*g*); I get a kick out of you (*d*); You do something to me (*c*); You're the top (*a-d*); I've got you under my skin (*d*); In the still of the night; Let's do it; Easy to love. Biography of composer Cole Porter.

Oklahoma Magna/Todd AO 1955 (*145 mins*) *with* Gordon Macrae (*a*), Shirley Jones (*b*), Gloria Grahame (*c*), Gene Nelson (*d*), Charlotte Greenwood (*e*), James Whitmore (*f*), Rod Steiger (*g*), Bambi Lynn (*h*), James Mitchell (*i*), Jay C Flippen (*j*), Marc Platt (*k*). *Prod* Arthur Hornblow Jnr. *Dir* Fred Zinneman. *Scr* Sonya Levien, William Ludwig from Oscar Hammerstein's book for the Theatre Guild production based on Lynn Riggs' play 'Green Grow The Lilacs'. *Ph* Robert Surtees. *Art* Joseph C Wright. *Ed* Gene Ruggiero. *Sound* Fred Hynes. *Dances* Agnes De Mille. *Mus* Jay Blackton, Adolph Deutsch. *Songs* Richard Rodgers and Oscar Hammerstein II: Oklahoma! (*a-b-d-e-f-j and chorus*); Oh what a beautiful morning (*a*); The surrey with the fringe on top (*a-b-e*); Kansas City (*d-e*); I cain't say no (*c*); Many a new day (*b*); People will say we're in love (*a-b*); Pore Jud (*a-g*); Out of my dreams (*b*) and ballet sequence (*h-i*); The farmer and the cowman (*a-c-d-e-f-j*); All er nothin' (*c-d*).

Oliver. Romulus/Columbia 1968 (*146 mins*) *with* Ron Moody (*a*), Shani Wallis (*b*), Harry Secombe (*c*), Oliver Reed (*d*), Mark Lester (*e*), Jack Wild (*f*), Peggy Mount (*g*). *Prod* John Woolf. *Dir* Sir Carol Reed. *Scr* Vernon Harris from Lionel Bart's musical play based on Charles Dickens's 'Oliver Twist'. *Ph* Oswald Morris. *Art* John Box, Terence Marsh. *Sets* Vernon Dixon, Ken Muggleston. *Ed* Ralph Kemplen. *Cost* Phyllis Dalton. *Sound* Robert Jones, Buster Ambler. *Dances* Onna White. Tom Panko. *Mus* Johnny Green. *Songs* Lionel Bart: Oliver! (*c-g*); As long as he needs me (*b*); Consider yourself (*e-f and boys*); You've got to pick a pocket or two (*a and boys*); Who will buy (*e*); Food, glorious food (*e and boys*); Be back soon (*a and boys*); Where is love (*e*); I'd do anything (*a-b-e-f and boys*); Boy for sale (*c*); Reviewing the situation (*a*); It's a fine life (*b and chorus*); Oom-pah-pah (*b and chorus*); Finale – Where is love/Consider yourself (*ensemble*).

On A Clear Day You Can See Forever Paramount 1969 (*129 mins*) with Barbra Streisand (*a*), Yves Montand (*b*), Bob Newhart, Jack Nicholson, Simon Oakland, Mabel Albertson, Larry Blyden. *Prod* Alan Jay Lerner, Howard Koch. *Dir* Vincente Minnelli. *Scr* Alan Jay Lerner from his book for the Broadway show. *Ph* Harry Stradling. *Art* John De Cuir. *Ed* David Bretherton, Flo Williamson. *Cost* Cecil Beaton. *Mus* Nelson Riddle. *Songs* Burton Lane and Alan Jay Lerner: On a clear day (*a-b*); What did I have that I don't have (*a*); He isn't you (*a*); Melinda (*b*); Go to sleep (*a*); Hurry, it's lovely up here (*a*); Come back to me (*b*); Love with all the trimmings (*a*).

One Hour With You/Jeanette MacDonald and Maurice Chevalier

One Hour With You Paramount 1932
(*80 mins*) *with* Jeanette MacDonald (*a*), Maurice
Chevalier (*b*), Genevieve Tobin (*c*), Charles
Ruggles (*d*), Donald Novis (*e*), Roland Young,
Josephine Dunn. *Prod/Dir* Ernst Lubitsch.
Part-Dir George Cukor. *Scr* Samson Raphaelson
from Lothar Schmidt's screenplay for Lubitsch's
silent film 'The Marriage Circle'. *Ph* Victor
Milner. *Art* Hans Dreier. *Mus* W Franke
Harling. *Songs* Oscar Straus, Richard Whiting
and Leo Robin: One hour with you (*a-b-c-d-e*);
Mitzi (*b*); Now I ask you (*b*); We will always
be sweethearts (*a-b*); What would you do (*b*);
Three times a day.

100 Men And A Girl Universal 1936 *with*
Deanna Durbin (*a*), Leopold Stokowski (*b*),
Adolphe Menjou, Mischa Auer, Alice Brady,
Eugene Pallette. *Prod* Charles R. Rogers, Joe
Pasternak. *Dir* Henry Koster. *Scr* Bruce
Manning, Charles Kenyon, Hans Kraly. *Ph*
Joseph Valentine. *Ed* Bernard W Burton.
Mus Charles Previn. *Songs* Frederick Hollander
and Sam Coslow: It's raining sunbeams (*a*).
R J Robyn and T Railey: A heart that's free
(*a*). *Interpolated music* Hungarian Rhapsody
No. 2 (Liszt) (*b*); Alleluia (Mozart) (*a*);
'Lohengrin' – Prelude Act 3 (Wagner) (*b*); 'La
Traviata' – Brindisi (Verdi) (*a*); Symphony
No. 5 – finale (Tchaikovsky) (*b*).

One Night In The Tropics Universal 1940
with Allan Jones (*a*), Nancy Kelly (*b*), Abbott
and Costello, Robert Cummings, Leo Carrillo,
Peggy Moran, Mary Boland. *Prod* Leonard
Spiegelgass. *Dir* Edward A Sutherland. *Scr*
Gertrude Purcell, Charles Grayson from Earl
Derr Biggs's 'Love Insurance'. *Ph* Joseph
Valentine. *Art* Jack Otterson. *Ed* Milton
Carruth. *Dances* Larry Ceballos. *Mus* Charles
Previn. *Songs* Jerome Kern, Oscar Hammer-

stein II, Otto Harbach and Dorothy Fields:
Remind me (*a*); Your dream is the same as my
dream (*a-b*); You and your kiss (*a*); Back in my
shell.

One Night Of Love Columbia 1934 (*80 mins*)
with Grace Moore (*a*), Tullio Carminati (*b*),
Lyle Talbot, Mona Barrie, Jessie Ralph. *Prod*
Everett Riskin. *Dir* Victor Schertzinger. *Scr*
S K Lauren from a story by Dorothy Speare
and Charles Beahan. *Ph* Joseph Walker. *Ed*
Gene Milford. *Cost* Jean Louis. *Sound* Paul
Neal. *Mus* Morris Stoloff, Louis Silvers. *Songs*
Victor Schertzinger and Gus Kahn: One night
of love (*a*). *Interpolated songs* Ciribiribin (*a*);
'Madame Butterfly' – One fine day (Puccini)
(*a*); 'Carmen' – excerpt (Bizet) (*a*).

One Touch Of Venus Universal 1948 (*81
mins*) *with* Dick Haymes (*a*), Ava Gardner (*b*),
Robert Walker (*c*), Olga San Juan (*d*), Eve
Arden (*e*), Tom Conway. *Prod* Lester Cowan.
Dir William A Seiter. *Scr* Harry Kurnitz,
Frank Tashlin from the Broadway show by
S J Perelman, Ogden Nash and Kurt Weill,
suggested by the novel 'The Tinted Venus'.
Ph Frank Planer, David S Horsley. *Art* Bernard
Herzbrun, Emerich Nicholson. *Sets* Al Fields,
Russell A Gausman. *Ed* Otto Ludwig. *Cost*
Orry-Kelly. *Sound* Leslie Carey, Joe Lapis.
Dances Billy Daniels. *Mus* Leo Arnaud. *Songs*
Kurt Weill, lyrics by (1) Ogden Nash: Speak
low (*a-b*); Falling in love with you; The trouble
with women. (2) Ann Ronell: Don't look now,
but my heart is showing (*a-b-c-d*); My week
(*a-b*) (deleted before release). Lyrics by 1 & 2:
That's him (*b-d-e*).
Note Vocals for Ava Gardner dubbed by
Eileen Wilson.

On The Avenue Fox 1937 *with* Dick Powell
(*a*), Alice Faye (*b*), The Ritz Brothers (*c*), Joan

Davis (d), E E Clive (e), Madeleine Carroll, Walter Catlett. *Prod* Gene Markey. *Dir* Roy Del Ruth. *Scr* Gene Markey, William Consel-man. *Ph* Lucien Andriot. *Ed* Allen McNeil. *Dances* Seymour Felix. *Mus* Arthur Lange. *Songs* Irving Berlin: I've got my love to keep me warm (a-b-e); Slumming on Park Avenue (b-c and chorus); You're laughing at me (a); The girl on the Police Gazette (a-b and chorus); This year's kisses (b); He ain't got rhythm (b-c and chorus). Songs written and recorded for the film but deleted before release: On the avenue; Swing sister; On the steps of Grant's Tomb.

On The Town MGM 1949 (*98 mins*) *with* Gene Kelly (a), Frank Sinatra (b), Jules Munshin (c), Ann Miller (d), Betty Garrett (e), Vera-Ellen (f), Alice Pearce (g). *Prod* Arthur Freed. *Dir/Dances* Stanley Donen, Gene Kelly. *Scr* Betty Comden and Adolph Green from their book for the Broadway musical based on Leonard Bernstein's ballet 'Fancy Free'. *Ph* Harold Rosson. *Art* Cedric Gibbons, Jack Martin Smith. *Sets* Edwin B Willis. *Ed* Ralph E Winters. *Sound* Douglas Shearer. *Mus* Lennie Hayton, *assoc* Roger Edens. *Songs* Edens, Bernstein, Comden and Green: On the town (a-b-c-d-e-f); New York, New York (a-b-c); You're awful (b-e); You can count on me (a-b-c-d-e-g); I feel like I'm not out of bed yet (*uncredited bass*); Prehistoric man (a-b-c-d-e); Come up to my place (b-e); Main Street (a-f); 'Miss Turnstiles' ballet (f); 'A Day in New York' ballet (a-f).

On Your Toes 1st National 1939 (*94 mins*) *with* Vera Zorina (a), Eddie Albert (b), Alan Hale, Frank McHugh, James Gleason, Donald O'Connor, Gloria Dickson. *Prod* Hal B Wallis, Robert Lord. *Dir* Ray Enright. *Scr* Jerry Wald, Richard Macaulay from the Broadway show by

George Abbott and Rodgers and Hart. *Ph* James Wong Howe, Sol Polito. *Art* Robert Haas. *Ed* Clarence Kolster. *Cost* Orry-Kelly. *Sound* O S Garretson, David Forrest. *Dances* George Balanchine. *Mus* Leo F Forbstein. *Ballet Music* Richard Rodgers: Slaughter on 10th Avenue (a-b); Princess Zenobia ballet (a). *Songs by Rodgers and Lorenz Hart as background music* On your toes; Quiet night; There's a small hotel.

Orchestra Wives Fox 1942 (*98 mins*) *with* Glenn Miller and his Orchestra (a), The Nicholas Brothers (b), Lynn Bari (c), Ray Eberle (d), Marion Hutton (e), Tex Beneke and The Modernaires (f), Cesar Romero, George Montgomery, Ann Rutherford, Carole Landis, Jackie Gleason. *Prod* William Le Baron. *Dir* Archie Mayo. *Scr* Karl Tunberg, Darrell Ware from a story by James Prindle. *Ph* Lucien Ballard. *Art* Richard Day, Joseph C Wright. *Ed* Robert Bischoff. *Sound* Roger Heman. *Dances* Nick Castle. *Mus* Alfred Newman. *Songs* Harry Warren and Mack Gordon: People like you and me (a-e-f); Serenade in blue (a-c); Kalamazoo (a-b-e-f); At last (a-c-d); That's sabotage (a-e); Chattanooga choo-choo (a). *Interpolated instrumentals* Moonlight serenade; Boom shot; Bugle call rag; American patrol; Moonlight sonata (all (a)).

Note Vocals for Lynn Bari dubbed by Pat Friday, trumpet for George Montgomery by Steve Lipkins, piano for Cesar Romero by Chummy McGregor, and bass for Jackie Gleason by Doc Goldberg.

Paint Your Wagon Paramount 1969 (*164 mins*) *with* Jean Seberg (a), Clint Eastwood (b), Lee Marvin (c), Ray Walston (d), Harve Presnell (e), Alan Dexter (f), The Nitty Gritty Dirt Band (g). *Prod* Alan Jay Lerner, *assoc* Tom Shaw. *Dir* Joshua Logan. *Scr* Paddy Chayefsky and Alan Jay Lerner from Lerner's book for the Broadway show. *Ph* William Fraker. *Art* Carl Braunger. *Sets* James I Berkey. *Cost/Prodn Design* John Truscott. *Ed* Robert Jones. *Sound* William Randall, Fred Hynes. *Dances* Jack Baker. *Mus* Nelson Riddle, *assoc* Roger Wagner, Joseph J Lilley. *Songs* Frederick Loewe and Alan Jay Lerner: I talk to the trees (b); They call the wind Maria (e and chorus); I'm on my way (chorus); I still see Elisa (b); Hand me down that can o' beans (c-g and chorus); Whoop-ti-ay (Shivaree) (chorus); Wanderin' star (c); There's a coach comin' in (e and chorus). André Previn and Lerner: Gold fever (b and chorus); The first thing you know (c); A million miles away behind the door (a); The gospel of No Name City (f); Best things (b-c and chorus).

The Pajama Game Warner 1957 (*101 mins*) *with* Doris Day (a), John Raitt (b), Eddie Foy

Jnr (*c*), Carol Haney (*d*), Reta Shaw (*e*), Buzz
Miller (*f*), Kenneth Leroy (*g*). *Prod/Dir* George
Abbott and Stanley Donen. *Scr* George Abbott
and Richard Bissell from their book for the
Broadway show based on Bissell's novel
'7½ Cents'. *Ph* Harry Stradling. *Art* Malcolm
Bert. *Sets* William Kuehl. *Ed* William Ziegler.
Cost William and Jean Eckart. *Sound* M A
Merrick, Dolph Thomas. *Dances* Bob Fosse.
Mus Ray Heindorf. *Songs* Richard Adler and
Jerry Ross: The pajama game (*chorus*); Racing
with the clock (*c and chorus*); I'm not at all in
love (*a*); I'll never be jealous again (*c-e*); Hey
there (*b*); Once-a-year day (*d and ensemble*); Small
talk (*a*); There once was a man (*a-b*); Steam
heat (*d-f-g*); Hernando's hideaway (*d*); 7½ cents
(*a and chorus*).

Pal Joey Essex-Sidney/Columbia 1957 (*111
mins*) *with* Frank Sinatra (*a*), Rita Hayworth
(*b*), Kim Novak (*c*), Bobby Sherwood, Barbara
Nichols, Hank Henry, Elizabeth Patterson
Prod Fred Kohlmar. *Dir* George Sidney. *Scr*
Dorothy Kingsley from John O'Hara's book
for the Broadway show based on his own
stories. *Ph* Harold Lipstein. *Art* Walter
Holscher. *Sets* William Kiernan, Louis Diage.
Ed Viola Lawrence, Jerome Thoms. *Cost* Jean
Louis. *Dances* Hermes Pan. *Mus* Morris Stoloff,
assoc Nelson Riddle. *Songs* Richard Rodgers and
Lorenz Hart: That terrific rainbow (*c and chorus*);
I didn't know what time it was (*a*); Zip (*b*);
There's a small hotel (*a*); I could write a book
(*a-c*); The lady is a tramp (*a*); My funny
valentine (*c*); Bewitched (*a-b*); What do I care
for a dame (*a*). *Source music* Plant you now,
dig you later; You mustn't kick it around; Do
it the hard way; Great big town.
Note Vocals for Rita Hayworth dubbed by Jo
Ann Greer, and for Kim Novak by Trudy Erwin.

Palmy Days Goldwyn/United Artists 1931 (*77
mins*) *with* Eddie Cantor (*a*), Charlotte Greenwood
(*b*), George Raft, Barbara Weeks, Walter
Catlett, Harry Woods, The Goldwyn Girls (*c*).
Frances Dean (Betty Grable). *Prod* Samuel
Goldwyn. *Dir* Edward A Sutherland. *Scr* Eddie
Cantor, Morris Ryskind, David Freedman,
Keene Thompson. *Ph* Gregg Toland. *Sound*
Vinton Vernon. *Dances* Busby Berkeley. *Mus*
Alfred Newman. *Songs* Ballard McDonald and
Con Conrad: Bend down sister (*b-c*). Cliff
Friend: My baby said yes, yes (*a*). Eddie Cantor,
Benny Davis and Harry Akst: There's nothing
too good for my baby (*a*).

Paramount On Parade Paramount 1930 (*102
mins*) *with* Maurice Chevalier (*a*), Buddy Rogers,
(*b*), Lillian Roth (*c*), Nancy Carroll (*d*), Dennis
King (*e*), Helen Kane (*f*), Virginia Bruce (*g*),
Abe Lyman and his Orchestra (*h*), Jack Oakie,
Ruth Chatterton, William Powell, Richard
Arlen, Gary Cooper, Fay Wray, Jean Arthur,
Mitzi Green, Philip Holmes, Kay Francis,
Mary Brian, Clara Bow (*i*), Evelyn Brent, Clive
Brook, George Bancroft. *Dir* Victor Schert-
zinger, Dorothy Arzner, Frank Tuttle, Rowland
Lee, Ernst Lubitsch, Lothar Mendes, Edward
A Sutherland, Edmund Goulding, Otto
Brower, Victor Heerman. *Ph* Harry Fischbeck,
Victor Milner. *Songs* Elsie Janis and Jack King:
The time to fall in love (*b-c*); I'm true to the
navy (*i*); Paramount on parade. L Wolfe Gilbert
and Abel Baer: Dancing to save your soul (*d*);
Drink to the girl of my dreams; I'm in training
for you. Others: Sweeping the clouds away
(*a*); Nichavo (Nothing matters) (*e*); All I want
is just one girl (*a*); The origin of the Apache
dance (*a*); Come back to Sorrento; My marine.

Pepe Posa/Columbia 1960 (*195 mins*) *with*
Cantinflas, Maurice Chevalier (*a*), Bing Crosby

(b), Bobby Darin (c), Shirley Jones (d), Sammy Davis Jnr (e), Judy Garland (voice only) (f), André Previn (g), Matt Mattox (h), Michael Callan (i), Dan Dailey, Jimmy Durante, Frank Sinatra, Janet Leigh, Tony Curtis, Debbie Reynolds, Cesar Romero, Edward G Robinson, Jack Lemmon, Greer Garson, Ernie Kovacks, Kim Novak, Donna Reed. *Prod/Dir* George Sidney. *Scr* Dorothy Kingsley, Claude Binyon from a story by Leonard Spiegelgass and Sonya Levien based on a play by L Bush-Fekete. *Ph* Joseph MacDonald. *Art* Ted Haworth. *Ed* Viola Lawrence, Al Clark. *Cost* Jean Louis. *Dances* Eugene Loring. *Mus* Johnny Green. *Songs* André Previn and Dore Langdon: That's how it went alright (c-d-h-i); The faraway part of town (f); The rumble (g); Suzy's theme (orch.). *Interpolated songs* Lovely day (d); Pepe (d); Mimi (a); September song (a); Hooray for Hollywood (e). Extracts from: Let's fall in love, Pennies from heaven, South of the border (b).

Pete Kelly's Blues Mark VII/Warner 1955 (95 mins) with Jack Webb, Peggy Lee (a), Ella Fitzgerald (b), 'Pete Kelly's Big 7' (Matty Matlock's Dixielanders) (c), Teddy Buckner (d), Janet Leigh, Edmond O'Brien, Lee Marvin, Martin Milner, Jayne Mansfield, The Tuxedo Band (e). *Prod/Dir* Jack Webb. *Scr* Richard L. Breen. *Ph* Hal Rosson. *Prodn Design* Harper Goff. *Art* Feild Gray. *Sets* John Sturtevant. *Ed* Robert M Leeds. *Cost* Howard Shoup. *Sound* Leslie Hewitt, Dolph Thomas. *Mus* Ray Heindorf. *Songs* Ray Heindorf and Sammy Cahn: Pete Kelly's blues (b). Arthur Hamilton: Sing me a rainbow (a); He needs me (a). *Interpolated songs and instrumentals* Gonna meet my sweetie now (c); Somebody loves me (a-c); Oh didn't he ramble (d-e); Sugar (a-c); I never knew (c); Hard hearted Hannah (b); Breezin' along with the breeze (c); Bye bye blackbird (c) What can I say after I say I'm sorry (c). *Note* Trumpet playing for Jack Webb dubbed by Dick Cathcart.

The Pirate MGM 1947 (102 mins) with Judy Garland (a), Gene Kelly (b), Walter Slezak, Gladys Cooper, The Nicholas Brothers (c), Reginald Owen, George Zucco. *Prod* Arthur Freed. *Dir* Vincente Minnelli. *Scr* Frances Goodrich, Albert Hackett from the play by S N Behrman. *Ph* Harry Stradling. *Art* Cedric Gibbons, Jack Martin Smith. *Sets* Edwin B Willis, Arthur Krams. *Ed* Blanche Sewell. *Cost* Irene, Tom Keogh, Karinska. *Sound* Douglas Shearer. *Dances* Gene Kelly, Robert Alton. *Mus* Lennie Hayton. *Songs* Cole Porter Be a clown (a-b-c); Love of my life (a); The Pirate ballet (b); You can do no wrong (a);

Niña (b); Mack the Black (a-b); Voodoo (c).

Porgy And Bess Goldwyn/Columbia 1959 (138 mins) with Sidney Poitier (voice of Robert McFerrin) (a), Dorothy Dandridge (voice of Adele Addison) (b), Sammy Davis Jnr (c), Pearl Bailey (d), Brock Peters (e), Ruth Attaway (voice of Inez Matthews) (f), voice of Loulie Jean Norman (g), Leslie Scott (h). *Prod* Samuel Goldwyn. *Dir* Otto Preminger. *Scr* N Richard Nash from the libretto by DuBose Heyward and Ira Gershwin for the opera based on Heyward's novel 'Porgy'. *Ph* Leon Shamroy. *Art* Serge Krizman, Joseph C Wright. *Sets* Oliver Smith. *Ed* Daniel Mandell. *Dances* Hermes Pan. *Mus* André Previn, *assoc* Ken Darby. *Songs* George and Ira Gershwin and DuBose Heyward: Summertime (g); A woman is a sometime thing (b); Gone gone gone (chorus); Porgy's prayer (a); My man's gone now (f); I got plenty o' nuttin' (a); Bess you is my woman (a-b); Catfish row (chorus); It ain't necessarily so (c); I ain't got no shame (c); What you want wid Bess (b); I loves you Porgy (b); A red-headed woman (e); There's a boat that's leavin' soon for New York (c); Where is my Bess (a); I'm on my way (a); I can't sit down (chorus); Clara Clara (chorus); Street cries.

Ready, Willing And Able Warner 1937 with Ruby Keeler (a), Lee Dixon (b), Ross Alexander (c), Winifred Shaw (d), Carol Hughes, Jane Wyman, Allen Jenkins, Shaw and Lee. *Prod* Hal B Wallis, *assoc* Samuel Bischoff. *Dir* Ray Enright. *Scr* Sig Herzig, Jerry Wald, Warren Duff. *Ph* Sol Polito. *Art* Carl Jules Weyl. *Ed* Doug Gould. *Cost* Howard Shoup. *Dances* Bobby Connolly. *Mus* Leo F Forbstein. *Songs* Richard Whiting and Johnny Mercer: When a blues singer falls in love (d); Too marvellous for words (a-b-c-d); Handy with your feet (a); Sentimental and melancholy (d); Little old house (b); Just a quiet evening (c); The world is my apple (b).

Red Garters Paramount 1954 (91 mins) with Guy Mitchell (a), Rosemary Clooney (b), Jack Carson (c), Gene Barry (d), Buddy Ebsen (e), Cass Daley (f). *Prod* Pat Duggan. *Dir* George Marshall. *Scr* Michael Fessier. *Ph* Arthur E Arling. *Art* Hal Pereira, Roland Anderson. *Ed* Arthur Schmidt. *Dances* Nick Castle. *Mus* Joseph J Lilley. *Songs* Jay Livingstone and Ray Evans: Red garters (b); Meet a happy guy (a); A dime and a dollar (a); Man and woman (a-b); A brave man (b); Good intentions (b); It was a great love story (b); Ladykiller (a-b-d-e-f).

Rhapsody In Blue Warner/1st National 1945 (139 mins) with Robert Alda (a), Joan Leslie (b), Oscar Levant (c), Paul Whiteman (d), Hazel Scott (e), Al Jolson (f), Anne Brown (g),

Tom Patricola (*h*), Alexis Smith, George White, Charles Coburn, Julie Bishop, Herbert Rudley, Johnny Downs, Rosemary De Camp, Albert Basserman, Morris Carnovsky. *Prod* Jesse L Lasky. *Dir* Irving Rapper. *Scr* Sonya Levien. Howard Koch, Elliott Paul. *Ph* Sol Polito. Ernest Haller. *Art* John Hughes, Anton Grot. *Sets* Fred M MacLean. *Ed* Folmer Blangsted. *Cost* Milo Anderson. *Sound* David Forrest, Stanley Jones. *Dances* Le Roy Prinz. *Mus* Leo F Forbstein, *assoc* Ray Heindorf. *Songs* George and Ira Gershwin: Rhapsody in blue (*a-d*); An American in Paris (*ballet sequence*); Somebody loves me (*c-h*); Fascinating rhythm (*e*); Summertime (*g*); Delicious (*b*); The man I love (*e*); Piano Concerto in F (*a-c*); Liza (*a-c*); Love walked in (*a and uncredited tenor*); Mine (*a-c*); Someone to watch over me; 'S wonderful; Bidin' my time; I got plenty o' nuttin'; I got rhythm; It ain't necessarily so; Lady be good; Clap yo' hands; Do it again; I'll build a stairway to Paradise (many of these used as background music or in montage form). Lyrics by Irving Caesar: Swanee (*f*). B G De Sylva: Blue Monday blues (*negro cast*). Caesar and De Sylva: Yankee doodle blues (*b*). Biography of George Gershwin. *Note* Piano playing for Robert Alda (as Gershwin) dubbed by Ray Turner, and vocals for Joan Leslie by Louanne Hogan.

Rio Rita Radio 1929 *with* John Boles (*a*), Bebe Daniels (*b*), Bert Wheeler (*c*), Robert Woolsey (*d*), Dorothy Lee (*e*), Don Alvarado (*f*), Sam Nelson. *Prod* William Le Baron. *Dir* Luther Reed. *Scr* Guy Bolton, Fred Thompson, Russell Mack and Luther Reed from the Broadway show by Harry Tierney and Joseph McCarthy. *Ph* Robert Kurrle, Lloyd Knetchell. *Ed* William Hamilton. *Dances* Pearl Eaton. *Mus* Victor Baravelle, *assoc* Max Steiner. *Songs* Harry Tierney and Joseph McCarthy: Rio Rita (*a-b*); Sweetheart we need each other (*b*); The song of the rangers (*a and chorus*); Following the sun around (*a*); The best little lover in town (*d*); You're always in my arms (*b*); If you're in love you'll waltz; Kinkajou.

Rio Rita MGM 1942 *with* John Carroll (*a*), Ann Sothern (*b*), Kathryn Grayson (*c*), Abbott and Costello, Barry Nelson. *Prod* Pandro S Berman. *Dir* S Sylvan Simon. *Scr* Richard Connell, Gladys Lehman from the original Broadway show, as above. *Ph* George J Folsey. *Art* Cedric Gibbons. *Sets* Edwin B Willis. *Ed* Ben Lewis. *Sound* Douglas Shearer. *Mus* Herbert Stothart. *Songs* Harry Tierney and Joseph McCarthy: Rio Rita (*a-c*); The song of the rangers (*a*). Harold Arlen and E Y Harburg: Long before you came along (*a-c*). *Interpolated*

songs Brazilian dance; Ora o conga; Most unusual weather (*c*).
Remake of the 1929 film.

The Road To Hong Kong Melnor/United Artists 1962 (*91 mins*) *with* Bing Crosby (*a*), Bob Hope (*b*), Joan Collins (*c*), Dorothy Lamour (*d*), Robert Morley. Guest spots: Peter Sellers, Frank Sinatra, Dean Martin, Jerry Colonna, Dave King. *Prod/Scr* Melvin Frank. *Dir/Scr* Norman Panama. *Ph* Jack Hildyard. *Ed* Alan Osbiston, John Smith. *Sound* John Cox. *Dances* Jack Baker, Sheila Myers. *Mus* Robert Farnon. *Songs* Jimmy Van Heusen and Sammy Cahn: The road to Hong Kong (*a-b*); Team work (*a-b-c*); Warmer than a whisper (*d*); Let's not be sensible (*a-c*).
The last of the Roads travelled by this team; a journey that had begun 22 years previously with . . .

The Road To Singapore Paramount 1940 (*84 mins*) *with* Bing Crosby (*a*), Bob Hope (*b*), Dorothy Lamour (*c*), Jerry Colonna, Charles Coburn, Anthony Quinn. *Prod* Harlan Thompson. *Dir* Victor Schertzinger. *Scr* Frank Butler, Don Hartman. *Ph* William C Mellor. *Art* Hans Dreier, Richard Odell. *Ed* Paul Weatherwax. *Cost* Edith Head. *Dances* Le Roy Prinz. *Mus* Victor Young. *Songs* Lyrics by Johnny Burke, music by (1) James V Monaco: Sweet potato piper (*a-b-c*); Too romantic (*a-c*); Kaigoon (*chorus*). (2) Victor Schertzinger: The moon and the willow tree (*c*); Captain Custard (*a-b*).

Roberta RKO 1934 (*85 mins*) *with* Fred Astaire (*a*), Ginger Rogers (*b*), Irene Dunne (*c*), Helen Westley (*d*), Randolph Scott, Lucille Ball, Claire Dodd. *Prod* Pandro S Berman. *Dir* William A Seiter. *Scr* Jane Murfin, Sam Mintz, Allan Scott from Otto Harbach's book for the

Broadway show based on Alice Duer Miller's novel 'Gowns By Roberta'. *Ph* Edward Cronjager. *Art* Van Nest Polglaze, Carroll Clark. *Ed* William Hamilton. *Cost* Bernard Newman. *Sound* John Tribby. *Dances* Hermes Pan, Fred Astaire. *Mus* Max Steiner. *Songs* Jerome Kern, Otto Harbach, Dorothy Fields and Jimmy McHugh: I won't dance (*a-b*); You're devastating; Yesterdays (*c-d*); The touch of your hand; I'll be hard to handle (*b*); Smoke gets in your eyes (*a-c*); Lovely to look at (*a-c*); Let's begin (*a*); An armful of trouble; Fashion show; Don't ask me not to sing.

Robin And The Seven Hoods P.C./Warner 1964 (*103 mins*) *with* Frank Sinatra (*a*), Bing Crosby (*b*), Dean Martin (*c*), Sammy Davis Jnr (*d*), Peter Falk (*e*), Barbara Rush, Victor Buono, Edward G Robinson. *Prod* Frank Sinatra, Howard W Koch, William Daniels. *Dir* Gordon Douglas. *Scr* David R Schwartz. *Ph* William Daniels. *Art* Le Roy Deane. *Sets* Raphael Bretton. *Ed* Sam O'Steen. *Cost* Don Feld. *Sound* Everett Hughes, Vinton Vernon. *Mus* Nelson Riddle. *Songs* Jimmy Van Heusen and Sammy Cahn: My kind of town (*a*); All for one (*e*); Don't be a do-badder (*b*); Style (*a-b-c*); Any man who loves his mother (*c*); Mr Booze (*a-b-c-d*); I like to lead when I dance (*a*); Bang! Bang! (*d*); Charlotte couldn't Charleston (*chorus*); Give praise (*chorus*).

Romance On The High Seas Warner 1948 (*99 mins*) *with* Doris Day (*a*), Jack Carson (*b*). Oscar Levant (*c*), The Page Cavanagh Trio (*d*), Janis Paige, Don Defore, S Z Sakall, Sir Lancelot (*e*). *Prod* Alex Gottlieb. *Dir* Michael Curtiz. *Scr* Julius J and Philip G Epstein. *Ph* Elwood Bredell, David Curtiz. *Art* Anton Grot. *Sets* Howard Winterbottom. *Ed* Rudi Fehr. *Cost* Milo Anderson. *Sound* Everett Brown, David Forrest. *Dances* Busby Berkeley. *Mus* Ray Heindorf. *Songs* Jule Styne and Sammy Cahn: It's magic (*a*); Put 'em in a box, tie 'em with a ribbon (*a-d*); I'm in love (*a*); The tourist trade (*b-e*); It's you or no-one (*a*); Run, run, run.

Title in GB: *It's Magic*. Reverted to original American title on TV.

Roman Scandals Goldwyn/United Artists 1933 *with* Eddie Cantor (*a*), Ruth Etting (*b*), Edward Arnold, Phillips Holmes, Gloria Stuart, The Goldwyn Girls (incl. Lucille Ball, Betty Grable, Paulette Goddard). *Prod* Samuel Goldwyn. *Dir* Frank Tuttle. *Scr* William Anthony McGuire, George Oppenheimer from the play by George S Kaufman and Robert Sherwood. *Ph* Gregg Toland. *Ed* Stuart Heisler. *Dances* Busby Berkeley. *Mus* Alfred Newman. *Songs* Harry Warren and Al Dubin:

Keep young and beautiful (*a*); When we build a little home (*a*); No more love (*b*); Rome wasn't built in a day (*a*); Put a tax on love (*a*).

Rosalie MGM 1937 (*118 mins*) *with* Nelson Eddy (*a*), Eleanor Powell (*b*), Frank Morgan (*c*), Ray Bolger (*d*), Ilona Massey (*e*), Reginald Owen, Edna May Oliver, Jerry Colonna. *Prod* William Anthony McGuire. *Dir* W S Van Dyke. *Scr* William Anthony McGuire based on the play by him and Guy Bolton. *Ph* Oliver T Marsh. *Art* Cedric Gibbons. *Sets* Edwin B Willis. *Ed* Blanche Sewell. *Sound* Douglas Shearer. *Dances* Albertina Rasch. *Mus* Herbert Stothart. *Songs* Cole Porter: Rosalie (*a-b*); Close (*a*); In the still of the night (*a-b*); I've a strange new rhythm in my heart (*b*); Who knows (*a*); To love or not to love; Why should I care; Spring love is in the air; It's all over but the shouting; National anthem; I know it's not meant for me. *Interpolated song* The caissons go rolling along (*a*).

Rose Marie MGM 1935 (*110 mins*) *with* Nelson Eddy (*a*), Jeanette MacDonald (*b*), Allan Jones (*c*), James Stewart, Alan Mowbray, David Niven, Una O'Connor. *Prod* Hunt Stromberg. *Dir* W S Van Dyke. *Scr* Frances Goodrich, Albert Hackett, Alice Duer Miller from the book by Otto Harbach and Oscar Hammerstein II for the Rudolph Friml operetta. *Ph* William Daniels. *Art* Cedric Gibbons. *Ed* Blanche Sewell. *Cost* Adrian. *Sound* Douglas Shearer. *Mus* Herbert Stothart. *Songs* Rudolf Friml, Otto Harbach, Oscar Hammerstein II, with new music and lyrics by Herbert Stothart and Gus Kahn: Rose Marie (*a*); The Indian love call (*a-b*); The song of the Mounties (*a and chorus*); The door of my dreams (*b*); Totem tom tom; Just for you; Lak Jeem; Pardon me madame.

Rose Marie MGM 1954 (*115 mins*) *with* Howard Keel (*a*), Ann Blyth (*b*), Fernando Lamas (*c*), Bert Lahr (*d*), Marjorie Main (*e*), Ray Collins. *Prod/Dir* Mervyn Le Roy. *Co-Dir/Chor* Busby Berkeley. *Scr* Ronald Millar from the original as above *Ph* Paul C Vogel. *Art* Cedric Gibbons, Merrill Pye. *Sets* Edwin B Willis. *Ed* Harold F Kress. *Sound* Douglas Shearer. *Mus* Georgie Stoll. *Songs* Rudolf Friml, Otto Harbach and Oscar Hammerstein II: Rose Marie (*a*); The Indian love call (*b-c*); The song of the Mounties (*a*); The door of my dreams (*b*); Totem tom tom. Rudolf Friml and Paul Francis Webster: The right place for a girl (*a-c*); Free to be free (*b*); Love and kisses (*d-e*); I love the love (*b-c*). Georgie Stoll: I'm a Mountie who never got his man (*d*). Remake of the 1935 film.

Rose Of Washington Square Fox 1939

(*86 mins*) *with* Alice Faye (*a*), Tyrone Power
(*b*), Al Jolson (*c*), Louis Prima and his Band
(*d*), William Frawley, Joyce Compton, Moroni
Olsen. *Prod* Darryl F Zanuck. *Dir* Gregory
Ratoff. *Ass. Prod/Scr* Nunnally Johnson from
a story by John Larkin and Jerry Horwin.
Ph Carl Freund. *Ed* Louis Loeffler. *Dances*
Seymour Felix. *Mus* Louis Silvers. *Songs* Harry
Revel and Mack Gordon: I never knew
heaven could speak (*a*). *Interpolated songs* Rose
of Washington Square (*a*); California, here I
come (*c*); My man (*a*); My mammy (*c*); Ja-da
(*a*); Vamp a little (*a-b*); Pretty baby (*c*); Toot
toot tootsie (*c*); I'm just wild about Harry
(*a-d*); I'm sorry I made you cry (*a*); Rockabye
your baby with a Dixie melody (*c*); Shine on
harvest moon (*d*). *Background music* I'm always
chasing rainbows; Japanese sandman; A media
luz; I'll see you in my dreams.

Royal Wedding MGM 1951 (*93 mins*) *with*
Fred Astaire (*a*), Jane Powell (*b*), Peter
Lawford, Sarah Churchill, Keenan Wynn.
Prod Arthur Freed. *Dir* Stanley Donen. *Scr*
Alan Jay Lerner. *Ph* Robert Planck. *Art* Cedric
Gibbons, Jack Martin Smith. *Sets* Edwin B
Willis. *Ed* Albert Akst. *Sound* Douglas Shearer.
Dances Nick Castle. *Mus* Johnny Green. *Songs*
Burton Lane and Alan Jay Lerner: Too late
now (*b*); Every night at seven (*a*); Open your
eyes (*b*); I left my hat in Haiti (*a*); How could
you believe me when I said I love you (*a-b*);
Sunday jumps (*a*); The happiest day of my life
(*b*); A lovely day for a wedding.
Title in GB: *Wedding Bells*.

Second Fiddle Fox 1939 (*86 mins*) *with* Sonja
Henie (*a*), Tyrone Power (*b*), Rudy Vallee (*c*),
Mary Healy (*d*), The Brian Sisters (*e*), Edna
May Oliver, Lyle Talbot. *Prod* Gene Markey.
Dir Sidney Lanfield. *Scr* Harry Tugend from
a story by George Bradshaw. *Ph* J Peverell
Marley. *Mus* Louis Silvers. *Songs* Irving Berlin:
I poured my heart into a song (*b-c*); Back to
back (*d*); An old-fashioned tune always is new
(*c*); The song of the metronome (*e and chorus*);
When winter comes (*a-c*); I'm sorry for myself
(*c-d*).

Seven Brides For Seven Brothers MGM
1954 (*103 mins*) *with* (The Brothers) Howard
Keel (*a*), Marc Platt (*b*), Jeff Richards (*c*), Matt
Mattox (*d*), Russ Tamblyn (*e*), Tommy Rall (*f*),
Jacques D'Amboise (*g*). (The Brides) Jane
Powell (*h*), Virginia Gibson (*i*), Julie Newmeyer
(now Julie Newmar) (*j*), Nancy Kilgas (*k*).
Ruta Kilmonis (now Ruta Lee) (*l*), Betty Carr
(*m*), Norma Doggett (*n*). *Prod* Jack Cummings.
Dir Stanley Donen. *Scr* Dorothy Kingsley,
Frances Goodrich and Albert Hackett from
Stephen Vincent Benet's story 'The Sobbin'

Women'. *Ph* George J Folsey. *Art* Cedric
Gibbons, Urie McCleary. *Sets* Edwin B Willis,
Hugh Hunt. *Ed* Ralph E Winters. *Cost* Walter
Plunkett. *Sound* Douglas Shearer. *Dances*
Michael Kidd. *Mus* Adolph Deutsch, Saul
Chaplin. *Songs* Gene De Paul and Johnny
Mercer: Bless yore beautiful hide (*a*); Wonder-
ful, wonderful day (*h*); Lonesome polecat
(*b-c-d-e-f-g*); Goin' courtin' (*b-c-d-e-f-g-h*);
Sobbin' women (*a-b-c-d-e-f-g*); When you're in
love (*a-h*); June bride (*h-i-j-k-l-m-n*); It's spring,
spring, spring (*b-c-d-e-f-g-i-j-k-l-m-n*); Barn-
raising ballet (*b-c-d-e-f-g-i-j-k-l-m-n*). *Note*
Vocals for Matt Mattox dubbed by Bill Lee.

Shall We Dance? RKO 1937 *with* Fred
Astaire (*a*), Ginger Rogers (*b*), Edward Everett
Horton, Eric Blore, Harriet Hoctor (*c*), Jerome
Cowan. *Prod* Pandro S Berman. *Dir* Mark
Sandrich. *Scr* Allan Scott, Ernest Pagano,
P J Wolfson from a story by Lee Loeb and
Harold Buchman. *Ph* David Axel. *Art* Van
Nest Polglaze. *Ed* William Hamilton. *Dances*
Hermes Pan, Larry Losee. *Mus* Nathaniel
Shilkret. *Songs* George and Ira Gershwin: Shall
we dance (*a-b*); I've got beginner's luck (*a*);
Let's call the whole thing off (*a-b*); They can't
take that away from me (*a-b-c*); Promenade
('Walking the dog' sequence) (*a-b and orch.*);
Slap that bass (*a*); They all laughed (*a-b*).

Show Boat Universal 1936 (*110 mins*) *with*
Allan Jones (*a*), Irene Dunne (*b*), Paul Robeson
(*c*), Helen Morgan (*d*), Charles Winninger (*e*),
Sammy White (*f*), Queenie Smith (*g*), Hattie
McDaniel (*h*), Donald Cook. *Prod* Carl Laemmle
Jnr. *Dir* James Whale. *Scr* Oscar Hammerstein
II from his book for the Broadway show
based on Edna Ferber's novel. *Ph* John
Mescall. *Ed* Ted Hunt, Bernard W Burton.
Mus Victor Baravelle. *Songs* Jerome Kern and
Oscar Hammerstein II: You are love (*a-b*); Life
upon the wicked stage (*f-g*); Bill (lyric P G
Wodehouse) (*d*); Why do I love you (*a-b*);
Can't help lovin' that man (*d*); I have the room
above (*a-b*); I still suits me (*c-h*); Make believe
(*a-b*); Ol' man river (*c*). *Interpolated song* After
the ball (*orch.*).

Show Boat MGM 1951 (*107 mins*) *with* Howard
Keel (*a*), Kathryn Grayson (*b*), Marge and
Gower Champion (*c*), Ava Gardner (*d*),
William Warfield (*e*), Joe E Brown, Robert
Sterling, Agnes Moorhead. *Prod* Arthur Freed.
Dir George Sidney. *Scr* John Lee Mahin from
Oscar Hammerstein's book for the Broadway
show based on Edna Ferber's novel. *Ph* Charles
Rosher. *Art* Cedric Gibbons, Jack Martin
Smith. *Sets* Edwin B Willis. *Ed* John Dunning.
Sound Douglas Shearer. *Dances* Robert Alton.
Mus Adolph Deutsch. *Songs* Jerome Kern and
Oscar Hammerstein II: Make believe (*a-b*); Life
upon the wicked stage (*c*); Can't help lovin'
that man (*d*); You are love (*a-b*); Ol' man river
(*e*); I might fall back on you (*c*); Why do I love
you (*a-b*). Jerome Kern and P G Wodehouse:
Bill (*d*).
Remake of the 1936 film.
Note Vocals for Ava Gardner dubbed by
Annette Warren.

Show Business RKO 1944 (*92 mins*) *with*
George Murphy (*a*), Constance Moore (*b*),
Eddie Cantor (*c*), Joan Davis (*d*), Nancy Kelly
(*e*), Don Douglas. *Prod* Eddie Cantor. *Dir*
Edwin L Marin. *Scr* Joseph Quillan, Dorothy
Bennett from a story by Bert Granet. *Ph*
Robert De Grasse, Vernon L Walker. *Art*
Albert S D'Agostino, Jack Okey. *Sets* Darrell
Silva, Al Fields. *Ed* Theron Warth. *Sound* Jean
L Speak. *Dances* Nick Castle. *Mus* Constantin
Bakaleinikoff. *Songs* Why am I blue (*b*); It had
to be you (*a-b*); I don't want to get well (*a-c*);
I want a girl (*a-b-c-d*); They're wearing them
higher in Hawaii (*a*); Alabamy bound (*c*); The
curse of an aching heart (*c*); While strolling in
the park one day (*a-b-c-d*); You may not
remember (*e*); Making whoopee (*c*); Dinah
(*a-b-c-d*).

The Show Of Shows Warner 1929 (*129 mins*)
with Winnie Lightner (*a*), John Barrymore (*b*),
Myrna Loy (*c*), Nick Lucas (*d*), Ted Lewis and
his Orchestra (*e*), Irene Bordoni (*f*), Jack
Buchanan (*g*), Beatrice Lillie (*h*), Lupino Lane
(*i*), Douglas Fairbanks Jnr, Loretta Young,
Sid Silvers, Ben Turpin, Heinie Conklin, H B
Warner, Georges Carpentier, Louis Fazenda,
Richard Barthelmess, Marion Nixon, Dolores
Costello, Sally O'Neill, Chester Morris, Noah
Beery, Frank Fay, Harriet Lake (Ann Sothern).
Dir John Adolfi. *Ph* Bernard McGill. *Mus*
Edward Ward. *Songs* Singing in the bathtub
(*a*); Li-po-li (*c-d*); Rockabye your baby with a
Dixie melody; Your love is all I crave; Lady
luck; Jumping jack; Motion picture pirates;
If your best friends won't tell you; If I could
learn to love; Just an hour of love; Pingo
pongo; The only song I know; Your mother
and mine; My sister. *Dramatic scene* 'Richard
III' – Soliloquy (*b*).

Silk Stockings MGM 1956 (*117 mins*) *with*
Fred Astaire (*a*), Cyd Charisse (*b*), Peter Lorre
(*c*), Jules Munshin (*d*), Joseph Buloff (*e*), Janis
Paige (*f*), Tybee Afra (*g*), Betty Utti (*h*), Barrie
Chase (*i*), George Tobias, Belita. *Prod* Arthur
Freed. *Dir* Rouben Mamoulian. *Scr* Leonard
Gershe, Leonard Spiegelgass, from the book
by George S Kaufman, Leueen McGrath and
Abe Burrows for the Broadway musical based
on 'Ninotchka'. *Ph* Robert Bronner. *Art*
William A Horning, Randall Duell. *Sets* Edwin
B Willis, Hugh Hunt. *Ed* Harold F Kress.
Cost Helen Rose. *Sound* Wesley C Miller.
Dances Hermes Pan, Eugene Loring. *Mus* André
Previn, Johnny Green. *Songs* Cole Porter: Silk
stockings (*b*); Too bad (*a-c-d-e-g-h-i*); Paris
loves lovers (*a-b*); Chemical reaction (*b*); All of
you (*a-b*); Stereophonic sound (*a-f*); Without
love (*a-b*); Satin and silk (*f*); Siberia (*c-d-e*);
Fated to be mated (*a-b*); Josephine (*f*); Ritz

roll and rock (*a and chorus*); Red blues (*b-c-d-e and chorus*). *Background music* I've got you under my skin; Close; Easy to love; Love of my life; You can do no wrong.
Note Vocals for Cyd Charisse dubbed by Carole Richards.

Sing Baby Sing Fox 1936 (*87 mins*) *with* Tony Martin (*a*), Alice Faye (*b*), Dixie Dunbar (*c*), The Ritz Brothers (*d*), Adolph Menjou, Ted Healy, Gregory Ratoff. *Prod* Darryl F Zanuck, *assoc* B G De Sylva. *Dir* Sidney Lanfield. *Scr* Milton Sperling, Jack Yellen, Harry Tugend. *Ph* J Peverell Marley. *Ed* Barbara MacLean. *Mus* Louis Silvers. *Songs* Lew Pollack and Jack Yellen: Sing baby sing (*a-b*); Love will tell (*b*). Louis Alter and Sidney Mitchell: You turned the tables on me (*a-b*). Richard Whiting and Walter Bullock: When did you leave heaven (*a-b*). Ed Farley, Mike Riley and Red Hodgson: The music goes round and around (*d*).

The Singing Fool Warner 1928 *with* Al Jolson (*a*), Davy Lee, Betty Bronson, Josephine Dunn, Reed Howes, Edward Martindel. *Prod* Darryl F Zanuck. *Dir* Lloyd Bacon. *Scr* C Graham Baker. *Ph* Byron Haskin. *Ed* Ralph Dawson, Harold McCord. *Dances* Larry Ceballos. *Mus* Al Goodman. *Songs* Al Jolson, Billy Rose, B G De Sylva, Lew Brown and Ray Henderson: Sonny Boy; I'm sitting on top of the world; My mammy; It all depends on you; There's a rainbow round my shoulder (all (*a*)). Part-talkie.

Singin' In The Rain MGM 1952 (*103 mins*) *with* Gene Kelly (*a*), Debbie Reynolds (*b*), Donald O'Connor (*c*), Cyd Charisse (*d*), Jean Hagen, Millard Mitchell, Douglas Fowley, Rita Moreno. *Prod* Arthur Freed. *Dir/Dances* Stanley Donen, Gene Kelly. *Scr* Betty Comden and Adolph Green. *Ph* Harold Rosson. *Art* Cedric Gibbons, Randall Duell. *Sets* Edwin B Willis, Jacque Mapes. *Ed* Adrienne Fazan. *Cost* Walter Plunkett. *Sound* Douglas Shearer. *Mus* Lennie Hayton. *Songs* Nacio Herb Brown and Arthur Freed: Singin' in the rain (*a-b-c*); All I do is dream of you (*a-b and chorus*); Make 'em laugh (*c*); You are my lucky star (*a-b*); You were meant for me (*a-b*); Fit as a fiddle (*a-c*); Would you (*a-b*); Good morning (*a-b-c*); Broadway Ballet, interp. Broadway melody *and* Broadway rhythm (*a-d*); Beautiful girl (*unknown tenor*). Montage of: Should I, I've got a feeling you're fooling, The wedding of the painted doll (*chorus and orch.*). Roger Edens, Betty Comden and Adolph Green: Moses (*a-c*).

The Singing Kid 1st National 1936 (*85 mins*) *with* Al Jolson (*a*), Sybil Jason (*b*), Claire Dodd (*c*), Winifred Shaw (*d*), The Yacht Club Boys

(e), Cab Calloway and his Orchestra (f), Edward Everett Horton (g), Allen Jenkins (h). *Prod* Robert Lord. *Dir* William Keighley. *Scr* Robert Lord, Warren Duff, Pat C Flick. *Ph* George Barnes. *Art* Carl Jules Weyl. *Ed* Tom Richards. *Cost* Orry-Kelly. *Dances* Bobby Connolly. *Mus* Leo F Forbstein. *Songs* Harold Arlen and E Y Harburg: I live to sing-a (a-f); Here's looking at you (a); Keep that hi-de-ho in your soul (f); My, how this country has changed (e); You're the cure for what ails me (a-b-g-h); Save me sister (a-d-f).

Sing Me A Love Song 1st National 1936 (*79 mins*) *with* James Melton (a), Patricia Ellis, Zasu Pitts (b), Hugh Herbert, Ann Sheridan, Allen Jenkins, Walter Catlett. *Prod* Samuel Bischoff. *Dir* Ray Enright. *Scr* Jerry Wald, Sig Herzig from a story by Harry Sauber. *Ph* Arthur Todd. *Art* Anton Grot. *Ed* Thomas Pratt. *Cost* Milo Anderson. *Mus* Leo F Forbstein. *Songs* Harry Warren and Al Dubin: The little house that love built (a); Summer night (a and chorus); That's the least you can do for the lady (a). *Interpolated songs* Carry me back to the lone prairie (a-b); Shortnin' bread (a); Your eyes have told me so (a).
Title in GB: *Come Up Smiling*.

The Sky's The Limit RKO 1943 (*89 mins*) *with* Fred Astaire (a), Joan Leslie (b), Robert Benchley, Robert Ryan, Elizabeth Patterson, Freddie Slack and his Orchestra (c) with Ella Mae Morse (d). *Prod* David Hempstead, *assoc* Sherman Todd. *Dir* Edward H Griffith. *Scr* Frank Fenton, Lynn Root. *Ph* Russell Metty. *Art* Albert S D'Agostino, Carroll Clark. *Sets* Darrell Silvera, Claude Carpenter. *Ed* Roland Gross. *Sound* Terry Kellum, James Stewart. *Dances* Fred Astaire. *Mus* Leigh Harline. *Songs* Harold Arlen and Johnny Mercer: My shining hour (a); Hector the victory garden man (c-d); I've got a lot in common with you (a-b); One for my baby (a); Cuban sugar mill (c-d).

Some Like It Hot Ashton-Mirisch/United Artists 1959 (*121 mins*) *with* Marilyn Monroe (a), Tony Curtis, Jack Lemmon, Joe E Brown, Pat O'Brien, George Raft, Joan Shawlee, Nehemiah Persoff. *Prod/Dir/Scr* Billy Wilder. *Co-Scr* I A L Diamond. *Ph* Charles Lang Jnr. *Art* Ted Haworth. *Sets* Edward G Boyle. *Ed* Arthur Schmidt. *Cost* Orry-Kelly. *Sound* Fred Lau. *Mus* Adolph Deutsch, *assoc* Matty Malneck. *Songs* As played by 'Sweet Sue and the Society Syncopators': Runnin' wild (a); I wanna be loved by you (a); Sweet Sue; I'm thru' with love (a); By the beautiful sea (chorus); Down among the sheltering palms (chorus). Studio Orchestra: Sugar blues; La cumparsita; Stairway to the stars (Park Avenue fantasy); Sweet

Georgia Brown; Some like it hot; Randolph Street rag.

Something To Shout About Columbia 1943 (*89 mins*) *with* Don Ameche (a), Janet Blair (b), Hazel Scott (c), Jack Oakie (d), William Gaxton, Cobina Wright Jnr, Cyd Charisse. *Prod/Dir* Gregory Ratoff. *Scr* Lou Breslow, Edward Eliscu from the Cole Porter Broadway show. *Ph* Franz F Planer. *Art* Lionel Banks. *Sets* Fay Babcock. *Ed* Otto Meyer. *Cost* Jean Louis. *Sound* Lodge Cunningham. *Dances* David Lichine. *Mus* Morris Stoloff. *Songs* Cole Porter: Something to shout about (b); I always knew (a-b); It might have been (b); You'd be so nice to come home to (a-b).

A Song Is Born Goldwyn/RKO 1948 (*113 mins*) *with* Danny Kaye (a), Virginia Mayo (b), Benny Goodman (c), Lionel Hampton (d), Charlie Barnet (e), Tommy Dorsey (f), Louis Armstrong (g), Mel Powell (h), The Golden Gate Quartet (i), Buck and Bubbles (j), Phil Moore (k), Benny Carter (l), Russo and The Samba Kings (m), The Page Cavanagh Trio (n), Hugh Herbert, Steve Cochran, J Edward Bromberg, Felix Bressart, Mary Field, Ludwig Stossel. *Prod* Samuel Goldwyn. *Dir* Howard Hawks. *Scr* Harry Tugend from Billy Wilder and Charles Brackett's script for 'Ball of Fire' based on the story 'From A to Z' by Wilder and Thomas Monroe. *Ph* Gregg Toland. *Art* Perry Ferguson, George Jenkins. *Sets* Julia Heron. *Ed* Daniel Mandell. *Cost* Irene Sharaff. *Sound* Fred Lau. *Mus* Emil Newman, Hugo Friedhofer. *Songs* Gene De Paul and Don Raye: A song was born (b-c-d-e-f-g-h-i); Daddy-o (b-n). *Interpolated instrumentals* Anitra's boogie (c-j); Flying home (c-d-e-f-g-h); Stealin' apples (c); I'm getting sentimental over you (f); Muskrat ramble (h); Hawk's nest (d); Mozart clarinet concerto (c and quartet); Cherokee (e); Marie (f); Jericho (i); Anvil chorus (ensemble). *Note* Vocals for Virginia Mayo dubbed by Jeri Sullivan.

The Sound Of Music Argyle/Fox 1965 (*174 mins*) *with* Julie Andrews (a), Christopher Plummer (b), Peggy Wood (c), Anna Lee (d), Charmian Carr (e), Marni Nixon (f), Daniel Truhitte (g), Portia Nelson (h), Richard Haydn (i), Eleanor Parker (j), Evadne Baker (k). *Prod* Robert Wise, Saul Chaplin. *Dir* Robert Wise. *Scr* Ernest Lehman from the book by Howard Lindsay and Russel Crouse for the Broadway show based on the life of the Trapp Family Singers. *Ph* Ted McCord. *Art* Boris Leven. *Sets* Walter M Scott, Ruby Levitt. *Ed* William Reynolds. *Cost* Dorothy Jeakins. *Sound* Murray Spivack, Bernard Freericks. *Dances* Marc Breaux and Dee Dee Wood. *Mus* Irwin Kostal.

Songs Richard Rodgers and Oscar Hammerstein II: The sound of music (*a*); Praeludium/Morning hymn and Alleluia (*chorus*); Maria (*d-f-h-k*); Sixteen going on seventeen (*e-g*); My favourite things (*a*); Climb every mountain (*c*); The lonely goatherd (*a*); Do-re-mi (*a and children*); Edelweiss (*a-b and children*); So long, farewell (*children*). New songs with music and lyrics by Rodgers: I have confidence in me (*a*); Something good (*a-b*).

Note Vocals for Christopher Plummer dubbed by Bill Lee.

South Pacific Magna/Fox 1958 (*171 mins*) *with* Rossano Brazzi (voice of Giorgio Tozzi) (*a*), Mitzi Gaynor (*b*), John Kerr (voice of Bill Lee) (*c*), Ray Walston (*d*), Juanita Hall (voice of Muriel Smith) (*e*), France Nuyen (*f*), The Ken Darby Singers (as the sailors) (*g*). *Prod* Buddy Adler. *Dir* Joshua Logan. *Scr* Paul Osborn from the book by Oscar Hammerstein II and Joshua Logan for the Broadway show based on James Mitchener's 'Tales Of The South Pacific'. *Ph* Leon Shamroy, L B Abbot. *Art* Lyle R Wheeler, Paul S Fox. *Sets* John De Cuir, Walter M Scott. *Ed* Robert Simpson. *Sound* Fred Hynes. *Dances* Le Roy Prinz. *Mus* Alfred Newman, *assoc* Ken Darby. *Songs* Richard Rodgers and Oscar Hammerstein II: Dites-moi (*a-b and children*); A cockeyed optimist (*a-b*); Twin soliloquies (*a-b*); Bloody Mary (*g*); My girl back home (*a-c*); Some enchanted evening (*a-b*); There is nothing like a dame (*d-g*); Bali Ha'i (*e*); I'm gonna wash that man outa my hair (*b*); I'm in love with a wonderful guy (*a-b*); Younger than springtime (*c*); Happy talk (*e-f*); Honey bun (*b-d*); Carefully taught (*c*); This nearly was mine (*a-b*).

Springtime In The Rockies Fox 1942 (*91 mins*) *with* Betty Grable (*a*), John Payne (*b*), Carmen Miranda (*c*), Harry James and his Orchestra (*d*), Cesar Romero, Charlotte Greenwood (*e*), Edward Everett Horton, Jackie Gleason. *Prod* William Le Baron. *Dir* Irving Cummings. *Scr* Walter Bullock, Ken Englund from Jacques Thery's adaptation of a story by Philip Wylie. *Ph* Ernest Palmer, Henri Jaffa. *Ed* Robert Simpson. *Sound* Roger Heman. *Dances* Hermes Pan. *Mus* Alfred Newman. *Songs* Harry Warren and Mack Gordon: I had the craziest dream (*a-d*); A poem set to music (*d*); Run little raindrop run (*a*); I like to be loved by you (*c*); Pan-American jubilee. *Interpolated song* Tic tac do meu Coracao (*c*).

Stage Struck 1st National 1936 (*86 mins*) *with* Dick Powell (*a*), Joan Blondell (*b*), Jeanne Madden (*c*), The Yacht Club Boys (*d*), Warren William, Frank McHugh, Carol Hughes. *Prod* Robert Lord. *Dir/Dances* Busby Berkeley. *Scr*

Tom Buckingham, Pat C Flick, Robert Lord. *Ph* Byron Haskin. *Art* Robert Haas. *Ed* Tom Richards. *Cost* Orry-Kelly. *Mus* Leo F Forbstein. *Songs* Harold Arlen and E Y Harburg: Fancy meeting you (*a-c*); In your own quiet way (*a-c*). The Yacht Club Boys: Income tax (*d*); The body beautiful (*d*).

Star! Fox 1968 (*174 mins*) *with* Julie Andrews (*a*), Daniel Massey (*b*), Bruce Forsyth (*c*), Beryl Reid (*d*), Garrett Lewis (*e*), Richard Crenna, Michael Craig. *Prod* Saul Chaplin. *Dir* Robert Wise. *Scr* William Fairchild. *Ph* Ernest Laszlo, Emil Kosa Jnr. *Art* Boris Leven. *Sets* Walter M Scott, Howard Bristol. *Ed* William Reynolds. *Cost* Donald Brooks. *Sound* Murray Spivack, Bernard Freericks, Douglas O Williams. *Dances* Michael Kidd. *Mus* Lennie Hayton. *Songs* Star! (Jimmy Van Heusen and Sammy Cahn) (*a*); Piccadilly (*a-c-d*); The physician (*a*); Forbidden fruit (*b*); Oh it's a lovely war (*a and chorus*); Someone to watch over me (*a*); My ship (*a*); Dear little boy (*a-b*); Do do do (*a*); Has anybody seen our ship (*a-b*); In my garden of joy (Saul Chaplin) (*a and chorus*); Limehouse blues (*a-e*); Parisian pierrot (*a*); Someday I'll find you (*a-b*); Burlington Bertie from Bow (*a*); 'n' everything (*a-e*); The Saga of Jenny (*a and chorus*).

Biography of Gertrude Lawrence.

A Star Is Born Transcona/Warner 1954 (*181 mins*) *with* Judy Garland, James Mason, Jack Carson, Charles Bickford, Tommy Noonan, James Brown, Lucy Marlowe. *Prod* Sid Luft, *assoc* Vern Alves. *Dir* George Cukor. *Scr* Moss Hart based on the 1937 film by Dorothy Parker, Alan Campbell and Robert Carson. *Ph* Sam Leavitt. *Art* Malcolm Bert. *Sets* George James Hopkins. *Ed* Folmer Blangsted. *Cost* Jean Louis, Mary Ann Nyberg,

Irene Sharaff. *Sound* Charles B Lang, David Forrest. *Dances* Richard Barstow. *Mus* Ray Heindorf. *Songs* Harold Arlen and Ira Gershwin: The man that got away; Gotta have me go with you; Here's what I'm here for*; It's a new world; Someone at last; Lose that long face*. Leonard Gershe: Born in a trunk (interp. I'll get by; You took advantage of me; Black bottom; Swanee; The peanut vendor; My melancholy baby). All sung by Judy Garland.
Note * Songs deleted before release.

Stars Over Broadway Warner 1935 (*89 mins*) *with* James Melton (*a*), Jane Froman (*b*), Jean Muir (*c*), Pat O'Brien, Frank McHugh, Marie Wilson, Frank Fay. *Prod* Samuel Bischoff. *Dir* William Keighley. *Scr* Jerry Wald, Julius J Epstein, Pat C Flick. *Ph* George Barnes. *Art* Carl Jules Weyl. *Ed* Bert L'Orle. *Cost* Orry-Kelly. *Dances* Busby Berkeley, Bobby Connolly. *Mus* Leo F Forbstein. *Songs* Harry Warren and Al Dubin: You let me down (*b*); Where am I (*a-b*); At your service madame (*a*). *Interpolated songs* Carry me back to the lone prairie (*a*); Ave Maria (*c*); 'Martha' – M'appari (Flotow) (*a*); 'Aida' – Celeste Aida (Verdi) (*a*).

Star Spangled Rhythm Paramount 1942 (*99 mins*) *with* Betty Hutton (*a*), Johnny Johnston (*b*), Dick Powell (*c*), Vera Zorina (*d*), Eddie 'Rochester' Anderson (*e*), Mary Martin (*f*), The Golden Gate Quartet (*g*), Eddie Bracken (*h*), Paulette Goddard (*i*), Veronica Lake (*j*), Dorothy Lamour (*k*), Bing Crosby (*l*), Cass Daley (*m*), Betty Jane Rhodes (*n*), Victor Moore, Bob Hope, William Bendix, Jerry Colonna, Gil Lamb, Alan Ladd, Fred Mac-Murray, Marjorie Reynolds, Diana Lynn, Sterling Holloway. *Prod* Joseph Sistrom. *Dir* George Marshall. *Scr* Melvin Frank, Norman Panama, Harry Tugend. *Ph* Leo Tover. *Art* Hans Dreier. *Sets* Ernst Fegte. *Ed* Arthur Schmidt. *Cost* Edith Head. *Dances* Danny Dare, George Balanchine. *Mus* Robert Emmett Dolan. *Songs* Harold Arlen and Johnny Mercer: That old black magic (*b-d*); Hit the road to dreamland (*c-f-g*); He loved me till the all-clear came (*m*); Old glory (*l*); A sweater, a sarong, and a peek-a-boo bang (*i-j-k*); I'm doing it for defence (*a*); Sharp as a tack; On the swing shift.
Note Vocal for Veronica Lake dubbed by Martha Mears.

State Fair Fox 1945 (*100 mins*) *with* Dick Haymes (*a*), Vivian Blaine (*b*), Jeanne Crain (*c*), Charles Winninger (*d*), Fay Bainter (*e*), Dana Andrews, Donald Meek. *Prod* William Perlberg. *Dir* Walter Lang. *Scr* Sonia Levien, Paul Green, Oscar Hammerstein II from the story by Philip Strong. *Ph* Leon Shamroy, Fred Sersen. *Art* Lyle R Wheeler, Lewis

Creber. *Sets* Thomas Little, Al Orenbach. *Ed* J Watson Webb. *Sound* Bernard Freericks, Roger Heman. *Mus* Alfred Newman, Charles Henderson. *Songs* Richard Rodgers and Oscar Hammerstein II: That's for me (*a-b*); It's a grand night for singing (*a-b-c*); It might as well be spring (*c*); Our State Fair (*a-c-d-e*); Isn't it kinda fun (*a-b*); All I owe I-o-way (*a-b-c*). Musical remake of the 1933 film.
Note Vocals for Jeanne Crain dubbed by Louanne Hogan.

State Fair Fox 1962 (*118 mins*) *with* Pat Boone (*a*), Pamela Tiffin (*b*), Ann-Margret (*c*), David Street (*d*), Bobby Darin (*e*), Alice Faye (*f*), Tom Ewell (*g*), Wally Cox. *Prod* Charles Brackett. *Dir* Jose Ferrer. *Scr* Richard Breen from the screen play of the 1945 film (as above). *Ph* William C Mellor, Emil Kosa Jnr, L B Abbott. *Art* Walter M Simonds, Jack Martin Smith. *Sets* Walter M Scott. *Ed* David Bretherton. *Dances* Nick Castle. *Mus* Alfred Newman. *Songs* Richard Rodgers and Oscar Hammerstein II: Our State Fair (*a-b-f-g*); It might as well be spring (*b*); That's for me (*a*); Isn't it kinda fun (*c-d*); It's a grand night for singing (*a-b-c-e-f*). New songs with music *and* lyrics by Richard Rodgers: Willing and eager (*a-c*); Never say no to a man (*f*); This isn't heaven (*e*); It's the little things in Texas (*f-g*); More than just a friend (*g*).
Note Vocals for Pamela Tiffin dubbed by Anita Gordon.

Step Lively RKO 1944 (*88 mins*) *with* Frank Sinatra (*a*), George Murphy (*b*), Gloria De Haven (*c*), Adolph Menjou, Anne Jeffreys, Walter Slezak, Eugene Pallette. *Prod* Robert Fellows. *Dir* Tim Whelan. *Scr* Warren Duff, Peter Milne. *Ph* Robert De Grasse. *Art* Albert S D'Agostino, Carroll Clark. *Sets* Darrell Silvera, Claude Carpenter. *Ed* Gene Milford. *Sound* Jean L Speak. *Dances* Ernst Matray. *Mus* Constantin Bakaleinikoff. *Songs* Jule Styne and Sammy Cahn: As long as there's music (*a*); Come out, come out wherever you are (*a-c*); And then you kissed me (*a*); Some other time (*c*); Where does love begin (*a-c*); Ask the Madame; Why must there be an opening song. Musical remake of the Marx Brothers' film 'Room Service'.

Stormy Weather Fox 1943 (*77 mins*) *with* Bill 'Bojangles' Robinson (*a*), Lena Horne (*b*), Fats Waller (*c*), Ada Brown (*d*), Cab Calloway and his Orchestra (*e*), Eddie 'Rochester' Anderson (*f*), Katherine Dunham and her Dancers (*g*), Flournoy Miller (*h*), The Nicholas Brothers (*i*), Dooley Wilson (*j*). *Prod* William Le Baron, *assoc* Irving Mills. *Dir* Andrew Stone. *Scr* Frederick Jackson, Ted Koehler. *Ph* Leon

Shamroy, Fred Sersen. *Art* James Basevi, Joseph C Wright. *Sets* Thomas Little, Fred J Rode. *Ed* James B Clark. *Sound* Alfred Bruzlin, Roger Heman. *Dances* Clarence Robinson. Nick Castle. *Mus* Benny Carter. *Songs* Ted Koehler, Harold Arlen, Fats Waller, Nat King Cole, Jimmy McHugh, etc.: Stormy weather (*b*); Ain't misbehavin' (*c*); There's no two ways about love (*b*); I can't give you anything but love (*c*); My my, ain't that something (*c*); Geechee Joe (*e*); That ain't right (*c-d*); Moppin' and boppin' (*c*); Rang tang tang; Linda Brown; Dat, dat, dah; I left my sugar in Salt Lake City.

The Story Of Vernon And Irene Castle RKO 1939 (*93 mins*) *with* Fred Astaire (*a*), Ginger Rogers (*b*), Edna May Oliver, Lew Fields, Walter Brennan, Janet Beecher. *Prod* Pandro S Berman, George Haight. *Dir* Henry C Potter. *Scr* Richard Sherman, Oscar Hammerstein II, Dorothy Yost from a story by Irene Castle. *Ph* Robert De Grasse. *Art* Van Nest Polglaze. *Ed* William Hamilton. *Mus* Constantin Bakaleinikoff. *Songs* The Castle walk (*a-b*); Come Josephine in my flying machine (*a*); Yama yama man (*a-b*); By the beautiful sea; Cuddle up a little closer; Only when you're in my arms; The Darktown strutters' ball; Waiting for the Robert E Lee (*a-b*); While they were dancing around; Oh you beautiful doll; Who's your lady friend; Chicago; Too much mustard. Biography of the pre-World War I dance team.

Strike Up The Band MGM 1940 (*120 mins*) *with* Judy Garland (*a*), Mickey Rooney (*b*), Paul Whiteman and his Orchestra (*c*), June Preisser (*d*), William Tracy, Larry Nunn. *Prod* Arthur Freed. *Dir/Dances* Busby Berkeley. *Scr* Fred Finklehoffe, John Monks Jnr. *Ph* Ray June. *Art* Cedric Gibbons. *Sets* Edwin B Willis. *Ed* Ben Lewis. *Sound* Douglas Shearer. *Mus* Georgie Stoll, *assoc* Roger Edens. *Songs* Roger Edens: Drummer boy (*a-b-c*); Do the conga (*a-b and ensemble*); I ain't got nobody (*a*); Nell of New Rochelle (*ensemble*). Roger Edens and Arthur Freed: Our love affair (*a-b*). George and Ira Gershwin: Strike up the band (*a-b-c*). *Interpolated song* Wonderful one (*a*).

Summer Holiday MGM 1946 (*92 mins*) *with* Mickey Rooney (*a*), Gloria De Haven (*b*), Agnes Moorhead (*c*), Marilyn Maxwell (*d*), Anne Francis (*e*), Walter Huston (*f*), Frank Morgan (*g*). *Prod* Arthur Freed. *Dir* Rouben Mamoulian. *Scr* Frances Goodrich, Albert Hackett, Ralph Blane from Eugene O'Neill's 'Ah Wilderness!'. *Ph* Charles Schoenbaum. *Art* Cedric Gibbons, Jack Martin Smith. *Sets* Edwin B Willis, Richard Pefferle. *Ed* Albert Akst. *Cost* Walter Plunkett, Irene. *Sound* Douglas Shearer, Frank B Mackenzie. *Dances*

Charles Walters. *Mus* Lennie Hayton. *Songs* Harry Warren and Ralph Blane: Afraid to fall in love (*a-b*); The Stanley Steamer (*a-b-c-f*); Independence day (*f-g and chorus*); Weary blues (*d*); All hail Danville High (*chorus*); It's our home town (*a-b-f*); The sweetest kid I've ever known (*d*).

Summer Stock MGM 1950 (*109 mins*) *with* Gene Kelly (*a*), Judy Garland (*b*), Gloria De Haven (*c*), Carleton Carpenter (*d*), Eddie Bracken (*e*), Phil Silvers (*f*), Hans Conreid (*g*). *Prod* Joe Pasternak. *Dir* Charles Walters. *Scr* George Wells, Sy Gomberg. *Ph* Robert Planck. *Art* Cedric Gibbons, Jack Martin Smith. *Sets* Edwin B Willis. *Ed* Albert Akst. *Sound* Douglas Shearer. *Dances* Nick Castle. *Mus* Johnny Green, *assoc* Saul Chaplin. *Songs* Harry Warren and Mack Gordon: If you feel like singing (*b*); Friendly star (*b*); Memory Island (*g*); Dig dig dig for your dinner (*c*); Happy harvest (*b*); Blue jean polka. Harry Warren, lyrics Jack Brooks and Saul Chaplin: You, wonderful you (*a-b*); Heavenly music (*a-f*). *Interpolated songs* Get happy (*b*); All for you (*a-b*); Portland fancy (*a-b*).
Title in GB: If You Feel Like Singing.

Sunny Side Up Fox 1929 *with* Charles Farrell (*a*), Janet Gaynor (*b*), Joe E Brown, El Brendel, Sharon Lynn, Frank Richardson. *Prod* B G De Sylva. *Dir* David Butler. *Scr* B G De Sylva, Lew Brown, Ray Henderson. *Ph* Ernest Palmer, John Schmitz. *Dances* Seymour Felix. *Mus* Howard Jackson. *Songs* De Sylva, Brown and Henderson: Sunny side up; I'm a dreamer (*b*); If I had a talking picture of you (*a-b*); Turn on the heat.

Sun Valley Serenade Fox 1941 (*86 mins*) *with* Glenn Miller and his Orchestra (*a*), Sonja Henie (*b*), John Payne (*c*), Dorothy Dandridge (*d*), The Nicholas Brothers (*e*), Lynn Bari (*f*), Ray Eberle (*g*), Tex Beneke and The Modernaires (*h*), Joan Davis, Milton Berle. *Prod* Milton Sperling. *Dir* Bruce Humberstone. *Scr* Robert Ellis, Helen Logan from a story by Art Arthur and Robert Harari. *Ph* Edward Cronjager. *Art* Richard Day, Lewis Creber. *Ed* James B Clark. *Dances* Hermes Pan. *Mus* Emil Newman. *Songs* Harry Warren and Mack Gordon: It happened in Sun Valley (*a-g-h*); Chattanooga choo-choo (*a-d-e-h*); The farmer in the dell (*a*); The kiss polka (*a-b-c*); I know why (*a-b-c-f*). *Interpolated instrumentals* Moonlight serenade (*a*); In the mood (*a*).
Note Vocals for Lynn Bari dubbed by Lorraine Elliot.

Sweet Adeline Warner-Vitaphone 1394 (*90 mins*) *with* Irene Dunne (*a*), Phil Regan (*b*), Winifred Shaw (*c*), Donald Woods, Hugh

Herbert, Ned Sparks, Louis Calhern. *Prod*
Edward Chodorov. *Dir* Mervyn Le Roy. *Scr*
Erwin S Gelsey from the Broadway show by
Jerome Kern and Oscar Hammerstein II. *Ph*
Sol Polito. *Art* Robert Haas. *Ed* Harold Mc-
Lernon. *Cost* Orry-Kelly. *Dances* Bobby
Connolly. *Mus* Leo F Forbstein. *Songs* Jerome
Kern and Oscar Hammerstein II: Don't ever
leave me (*a*); Why was I born (*a*); We were so
young (*a-b*); Spring is here; My husband's first
wife; Out of the blue; Here am I; Naughty boy;
'T was not so long ago.

Sweet Charity Universal 1968 *with* Shirley
Maclaine (*a*), Sammy Davis Jnr (*b*), Ricardo
Montalban (*c*), John McMartin (*d*), Stubby
Kaye (*e*), Chita Rivera (*f*), Paula Kelly (*g*),
Barbara Bouchet (*h*), Suzanne Charney (*i*).
Prod Robert Arthur. *Dir/Dances* Bob Fosse.
Scr Peter Stone from the book by Neil Simon
for the Broadway show based on Federico
Fellini's 'Nights Of Cabiria'. *Ph* Robert Surtees.
Art Alexander Golitzen, George C Webb. *Sets*
Jack D Moore. *Ed* Stuart Gilmore. *Cost* Edith
Head. *Sound* Len Peterson, Waldon O Watson,
William Russell, Ronald Pierce. *Mus* Ralph
Burns. *Songs* Cy Coleman and Dorothy Fields:
Sweet Charity; Big spender (*g-f and chorus*);
Where am I going (*a*); My personal property
(*a*); Rich man's frug (*i and chorus*); If my friends
could see me now (*a*); There's gotta be
something better than this (*a-f-g*); The rhythm
of life (*b*); I'm a brass band today (*a and chorus*);
It's a nice face; I love to cry at weddings.

Swing Time RKO 1936 (*105 mins*) *with* Fred
Astaire (*a*), Ginger Rogers (*b*), George Metaxa,
Victor Moore, Helen Broderick, Eric Blore,
Betty Furness. *Prod* Pandro S Berman. *Dir*
George Stevens. *Scr* Howard Lindsay, Allan
Scott from a story by Erwin S Gelsey. *Ph*
David Abel. *Art* Van Nest Polglaze. *Ed* Henry
Berman. *Dances* Hermes Pan. *Mus* Nathaniel
Shilkret. *Songs* Jerome Kern and Dorothy
Fields: A fine romance (*a-b*); The way you look
tonight (*a*); Pick yourself up (*a-b*); Bojangles
of Harlem (*a*); The waltz in swingtime (*a-b*);
Never gonna dance (*a-b*).

Take A Chance Paramount 1933 *with* Lillian
Roth (*a*), Buddy Rogers (*b*), James Dunn (*c*),
June Knight (*d*), Cliff Edwards (*e*), Dorothy
Lee (*f*). *Prod* B G De Sylva. *Dir* Lawrence
Schwab, Monte Brice. *Scr* De Sylva, Schwab
and Brice from the Broadway show. *Ph* William
Steiner. *Art* Hans Dreier. *Dances* Bobby
Connolly. *Songs* Nacio Herb Brown, Richard
Whiting and B G De Sylva: Eadie was a lady
(*a*); You're an old smoothie (*a-c*); Turn out the
light (*b-c-d-e*). Vincent Youmans and B G De
Sylva: Rise and shine (*a*); Should I be sweet

(*d*). Harold Arlen, E Y Harburg and Billy
Rose: It's only a paper moon (*b-d*). Louis Alter
and Arthur Swanstrom: Come up and see me
some time (*a*). Herman Hupfeld: Night owl (*e*).
Roger Edens and E Y Harburg: New deal rhythm.

Take Me Out To The Ball Game MGM
1948 (*93 mins*) *with* Gene Kelly (*a*), Esther
Williams (*b*), Frank Sinatra (*c*), Betty Garrett
(*d*), Jules Munshin (*e*), Edward Arnold. *Prod*
Arthur Freed. *Dir* Busby Berkeley. *Scr* George
Wells, Harry Tugend. *Ph* George J Folsey.
Art Cedric Gibbons, Daniel B Cathcart. *Sets*
Edwin B Willis. *Ed* Blanche Sewell. *Sound*
Douglas Shearer. *Dances* Stanley Donen, Gene
Kelly. *Mus* Adolph Deutsch, *assoc* Roger Edens.
Songs Roger Edens, Betty Comden and Adolph
Green: Strictly U.S.A. (*a-b-c-d*); The right girl
for me (*c*); The hat my father wore on St
Patrick's day (*a*); It's fate, baby, it's fate (*c-d*);
Yes indeedy (*a-c*); O'Brian to Ryan to Goldberg
(*a-c-e*). Harold Arlen and E Y Harburg: Last
night when we were young (*c*) (deleted before
release). *Interpolated song* Take me out to the
ball game (*a-b-c*).
Title in GB: *Everybody's Cheering*.

Tea for Two Warner 1950 (*98 mins*) *with*
Doris Day (*a*), Gordon Macrae (*b*), Gene Nelson
(*c*), Eve Arden (*d*), Virginia Gibson (*e*), Patrice
Wymore (*f*), Billy De Wolfe (*g*). *Prod* William
Jacobs. *Dir* David Butler. *Scr* Harry Clork
based on the book by Otto Harbach and Frank
Mandel for the Broadway show 'No, No,
Nanette'. *Ph* Wilfrid M Cline. *Art* Douglas
Bacon. *Sets* Lyle B Reifsnider. *Ed* Irene Morra.
Cost Leah Rhodes. *Sound* Dolph Thomas, David
Forrest. *Dances* Le Roy Prinz, Eddie Prinz.
Mus Ray Heindorf. *Songs* Vincent Youmans,
lyrics by (1) Irving Caesar: Tea for two (*a-b-c*);
I want to be happy (*a-b*). (2) Otto Harbach:
No, no, Nanette (*a-c*). (3) Ira Gershwin: Oh
me, oh my (*a-c-f*). (4) Anne Caldwell: I know
that you know (*a-b-c*). *Interpolated songs* The call
of the sea (*b*); Do, do, do (*a-b*); Crazy rhythm
(*c-f*); Charleston (*e-g and chorus*); I only have
eyes for you (*b-e*).

Thanks A Million Fox 1935 (*87 mins*) *with*
Fred Allen, Dick Powell (*a*), Ann Dvorak (*b*),
The Yacht Club Boys (*c*), Patsy Kelly (*d*), Paul
Whiteman and his Orchestra (*e*) with Ramona
(*f*) and Rubinoff (*g*). *Prod* Darryl F Zanuck.
Dir Roy Del Ruth. *Scr* Nunnally Johnson. *Ph*
J Peverell Marley. *Ed* William Lambert. *Mus*
Arthur Lange. *Songs* Arthur Johnston and Gus
Kahn: Thanks a million (*a*); I'm sitting high
on a hilltop (*a*); Sugar plum (*b-d*); New
Orleans (*e-f*); I've got a pocketful of sunshine
(*a*); Sing brother. Bert Kalmar and Harry
Ruby: What a beautiful night.

Thank Your Lucky Stars Warner 1943 (*127
mins*) *with* Dinah Shore (*a*), Bette Davis (*b*),
Dennis Morgan (*c*), Joan Leslie (*d*), Eddie
Cantor (*e*), Ann Sheridan (*f*), John Garfield
(*g*), Alexis Smith (*h*), Spike Jones and his City
Slickers (*i*), Olivia De Haviland, Errol Flynn,
Edward Everett Horton, Ida Lupino, Alan
Hale, George Tobias. *Prod* Arthur Schwartz,
Mark Hellinger. *Dir* David Butler. *Scr* Melvin
Frank, Norman Panama, James V Kern from
a story by Arthur Schwartz and Everett
Freeman. *Ph* Arthur Edeson. *Art* Anton Grot,
Leo F Kuter. *Sets* Walter T Tilford. *Ed* Irene
Morra. *Sound* Francis J Scheid, David Forrest.
Dances Le Roy Prinz. *Mus* Leo F Forbstein.
Songs Arthur Schwartz and Frank Loesser:
Thank your lucky stars (*a*); They're either too
young or too old (*b*); How sweet you are (*a*);
I'm riding for a fall (*c-d*); Love isn't born, it's
made (*f*); The dreamer (*a*); Goodnight good
neighbour (*c*); Ice cold Katy; I'm going North;
That's what you jolly well get; We're staying
home tonight. Harold Arlen and Johnny
Mercer: Blues in the night (*g*).

That Midnight Kiss MGM 1949 (*96 mins*)
with Mario Lanza (*a*), Kathryn Grayson (*b*),
Jose Iturbi (*c*), Amparo Iturbi (*d*), Keenan
Wynn (*e*), J Carroll Naish (*f*), Marjorie Reynolds.
Prod Joe Pasternak. *Dir* Norman Taurog. *Scr*
Bruce Manning, Tamara Hovey. *Ph* Robert
Surtees. *Art* Cedric Gibbons, Preston Ames.
Sets Edwin B Willis. *Ed* Gene Ruggiero. *Sound*
Douglas Shearer. *Mus* Charles Previn. *Songs*
Bronislau Kaper and Bob Russell: I know, I
know, I know (*a*). *Interpolated music* They didn't
believe me (*a-b*); Down among the sheltering
palms (*e and quartet*); Santa Lucia (*f*); Una
furtiva lagrima (*a-c*); Revolutionary etude
(Chopin) (*c-d*); Caro nome (Verdi) (*b-c*); Celeste
Aida (Verdi) (*a*); Piano Concerto – excerpt
(Liszt) (*c*); Lucia (Donizetti) (*a*); Finale based
on themes from Tchaikovsky's Fifth Symphony
(*a-b*).

That Night In Rio Fox 1941 (*90 mins*) *with*
Alice Faye (*a*), Don Ameche (*b*), Carmen
Miranda (*c*), The Flores Brothers (*d*), S Z
Sakall, J Carroll Naish, Leonid Kinskey. *Prod*
Fred Kohlmar. *Dir* Irving Cummings. *Scr*
George Seaton, Hal Long, Bess Meredyth from
a story by Rudolph Lothar and Hans Adler.
Ph Leon Shamroy, Ray Rennahan. *Art* Richard
Day. *Ed* Walter Thompson. *Dances* Hermes Pan.
Mus Alfred Newman. *Songs* Harry Warren and
Mack Gordon: Boa noite (*a-b-c*); Chica chica
boom chic (*a-b-c*); I-yi-yi-yi-yi-yi- like you very
much (*c*); They met in Rio (*a-b*); The Baron is in
conference (*chorus*). *Interpolated song* Cae cae (*c*).

There's No Business Like Show Business

Fox 1954 (*117 mins*) *with* Ethel Merman (*a*),
Dan Dailey (*b*), Mitzi Gaynor (*c*), Donald
O'Connor (*d*), Johnnie Ray (*e*), Marilyn
Monroe (*f*), Hugh O'Brian. *Prod* Sol C Siegel.
Dir Walter Lang. *Scr* Henry and Phoebe
Ephron from a story by Lamar Trotti. *Ph*
Leon Shamroy. *Art* Lyle R Wheeler, John De
Cuir. *Ed* Robert Simpson. *Cost* Miles White,
Billy Travilla. *Dances* Robert Alton, Jack Cole.
Mus Lionel Newman, Alfred Newman, *assoc*
Ken Darby. *Songs* Irving Berlin: There's no
business like show business (*a-b-c-d-e-f*);
When the midnight choo-choo leaves for
Alabam (*a-b-c-d*); Let's have another cup of
coffee (*a*); Play a simple melody (*a-b*); After
you get what you want (*f*); You'd be surprised
(*b*); A sailor's not a sailor (*a-c*); Alexander's
ragtime band (*a-b-c-d-e*); A pretty girl is like a
melody (*a-b*); Remember (*chorus*); Heat wave
(*f*); If you believe (*e*); I can make you laugh
(*d*) (deleted before release); Lazy (*c-d-f*); A man
chases a girl (*c-d*); Marie (*trio*); But I ain't got a
man (*f*) (deleted before release). *Background music*
Puttin' on the Ritz; Let's face the music;
Cheek to cheek.

This is The Army Warner/1st National 1943
(*121 mins*) *with* Kate Smith (*a*), Joan Leslie (*b*),
George Murphy (*c*), Irving Berlin (*d*), Frances
Langford (*e*), Gertrude Niesen (*f*), 'This Is The
Army' Revue Company (*g*), Ronald Reagan,
Rosemary De Camp, George Tobias, Alan
Hale, Joe Louis. *Prod* Jack L Warner, Hal B
Wallis. *Dir* Michael Curtiz. *Scr* Claude Binyon,
Casey Robinson. *Ph* Bert Glennon, Sol Polito.
Art John Koenig, John Hughes. *Sets* George
James Hopkins. *Ed* George Amy. *Cost*
Orry-Kelly. *Sound* C A Riggs. *Dances* Le Roy
Prinz, Robert Sidney. *Mus* Leo F Forbstein,
assoc Ray Heindorf. *Songs* Irving Berlin: This is

*Thoroughly Modern Millie/Mary Tyler Moore
and Julie Andrews*

the army, Mr Jones; I left my heart at the Stage Door Canteen; American eagles; With my head in the clouds; Mandy; The army's made a man of me; I'm getting tired so I can sleep; Poor little me, I'm on KP; How about a cheer for the navy; What the well-dressed man in Harlem will wear; Ladies of the chorus; This time (all (*g*)); God bless America (*a*); Oh how I hate to get up in the morning (*d*). Based on the American Army stage revue.

Thoroughly Modern Millie Universal 1967 *with* Julie Andrews (*a*), Mary Tyler Moore (*b*), James Fox (*c*), Beatrice Lillie (*d*), Carol Channing (*e*), John Gavin (*f*), Jack Soo (*g*), Pat Morita (*h*), Ann Dee (*i*). *Prod* Ross Hunter. *Dir* George Roy Hill. *Scr* Richard Morris. *Ph* Russell Metty. *Art* Alexander Golitzen, George C Webb. *Sets* Howard Bristol. *Ed* Stuart Gilmore. *Cost* Jean Louis. *Sound* Waldon O Watson, William Russell, Ronald Pierce. *Dances* Joe Layton. *Mus* Elmer Bernstein, André Previn. *Songs* Jimmy Van Heusen and Sammy Cahn: Thoroughly modern Millie (*a*); The Tapioca (*a-c*). Van Heusen and Jay Thompson: Jimmy (*a*). *Interpolated songs* Baby face (*a*); Do it again (*e*); Trinkt le chaim – Jewish wedding song (*a-b*); Jazz baby (*e*); Poor butterfly (*a*); Japanese sandman (*g-h*); Rose of Washington Square (*i*); Stumblin' (*a-b*); I'm sitting on top of the world (*unidentified girl*). *Background music* Charmaine; I can't believe that you're in love with me.

Thousands Cheer MGM 1943 (*126 mins*) *with* Lena Horne (*a*), Gene Kelly (*b*), June Allyson (*c*), Gloria De Haven (*d*), Judy Garland (*e*), Eleanor Powell (*f*), Kathryn Grayson (*g*), Jose Iturbi (*h*), Mickey Rooney (*i*), Red Skelton (*j*), Ann Sothern (*k*), Kay Kyser and his Orchestra (*l*), Hazel Scott (*m*), Marilyn Maxwell (*n*),

Virginia O'Brien (*o*), Lucille Ball (*p*), Bob Crosby (*q*), Ben Blue, John Boles, Mary Astor, Frank Morgan, Don Loper. *Prod* Joe Pasternak. *Dir* George Sidney. *Scr* Paul Jarrico, Richard Collins. *Ph* George J Folsey. *Art* Cedric Gibbons, Daniel B Cathcart. *Sets* Edwin B Willis, Jacques Mesereau. *Ed* George Boemler. *Cost* Irene. *Sound* Douglas Shearer. A N Fenton. *Mus* Herbert Stothart, *assoc* Roger Edens. *Songs* Ralph Freed, Burton Lane, E Y Harburg, Ralph Blane, Walter Ruick, etc.: Daybreak (*g-h*); Three letters in the mailbox (*g*); I dug a ditch in Wichita (*g-l*); The joint is really jumpin' in Carnegie Hall (*e-h*); In a little Spanish town (*c-d-o*); Honeysuckle Rose (*a*); The United Nations on the march (*g-h*); Let there be music; Just as long as I know Katy's waiting; I'm lost, you're lost; Why don't we try; Sempre libera (Verdi) (*g*).

Three For The Show Columbia 1955 *with* Betty Grable (*a*), Jack Lemmon (*b*), Marge Champion (*c*), Gower Champion (*d*), Myron McCormick, Paul Harvey. *Prod* Jonie Taps. *Dir* Henry C Potter. *Scr* Edward Hope, Leonard Stern from Somerset Maugham's 'Too Many Husbands'. *Ph* Arthur Arling. *Art* Walter Holtscher. *Sets* William Kiernan. *Ed* Viola Lawrence. *Cost* Jean Louis. *Sound* John Livadry. *Dances* Jack Cole. *Mus* Morris Stoloff. *Songs* George and Ira Gershwin: Someone to watch over me (*c*); I've got a crush on you (*a-b*); Others: Down boy (*a*); How come you do me like you do (*a*); Which one; Swan Lake ballet (*c-d*).

Three Little Words MGM 1950 (*102 mins*) *with* Fred Astaire (*a*), Red Skelton (*b*), Arlene Dahl (*c*), Gloria De Haven (*d*), Vera-Ellen (*e*), Debbie Reynolds (*f*), Carleton Carpenter (*g*), Phil Regan (*h*), Gale Robbins, Keenan Wynn. *Prod* Jack Cummings. *Dir* Richard Thorpe. *Scr* George Wells. *Ph* Harry Jackson. *Art* Cedric Gibbons, Urie McCleary. *Sets* Edwin B Willis. *Ed* Ben Lewis. *Sound* Douglas Shearer. *Dances* Hermes Pan. *Mus* André Previn. *Songs* Bert Kalmar and Harry Ruby: Three little words (*a*); Nevertheless (*a-b-e*); I love you so much (*c*); Where did you get that girl (*a-e*); Who's sorry now (*d*); I wanna be loved by you (*f-g*); My sunny Tennessee (*a-b*); So long oo-long (*a-b*); Thinking of you (*a-e*); Come on Papa; Mr and Mrs Hoofer at home (*a-e*); All alone Monday; Hooray for Captain Spaulding; She's mine, all mine.

Biography of Bert Kalmar and Harry Ruby. *Note* Vocals for Vera-Ellen dubbed by Anita Ellis, and for Debbie Reynolds by Helen Kane.

Till The Clouds Roll By MGM 1946 (*120 mins*) *with* Robert Walker (*a*), Judy Garland

Three for the Show/Betty Grable

(*b*), Lena Horne (*c*), June Allyson (*d*), Kathryn Grayson (*e*), Gower Champion (*f*), Virginia O'Brien (*g*), Frank Sinatra (*h*), Lucille Bremer (*i*), Cyd Charisse (*j*), Ray McDonald (*k*), Van Heflin (*l*), Angela Lansbury (*m*), Van Johnson (*n*), Tony Martin (*o*), Johnny Johnston (*p*), Dinah Shore (*q*), Esther Williams (*r*). *Prod* Arthur Freed. *Dir* Richard Whorf, Vincente Minnelli. *Scr* Jean Holloway, George Wells, Myles Connolly. *Ph* Harry Stradling, George J Folsey. *Art* Cedric Gibbons. *Sets* Edwin B Willis. *Ed* Albert Akst. *Cost* Irene. *Sound* Douglas Shearer. *Dances* Robert Alton. *Mus* Lennie Hayton. *Songs* Jerome Kern: Till the clouds roll by (*d-e-g-h-i-k-o-p*); Look for the silver lining (*b*); Ol' man river (*b*); Who (*b*); Who cares if my boat goes upstream (*e-o*); Can't help lovin' that man (*c*); Cleopatterer (*d*); Life upon the wicked stage (*g*); I won't dance (*i-n*); Leave it to Jane (*d*); Make believe (*e-o*); All the things you are (*o*); Long ago; She didn't say yes; The land where good songs go; How'd you like to spoon with me (*m*); They didn't believe me (*q*); Why was I born (*c*); The last time I saw Paris (*q*); Smoke gets in your eyes; Go little boat; Sunny (*b*); Yesterdays; One more dance; A fine romance; 'Mark Twain Suite'-polka. Biography of composer Jerome Kern.

The Time, The Place And The Girl
Warner 1946 *with* Dennis Morgan (*a*), Jack Carson (*b*), Martha Vickers (*c*), Carmen Cavallaro (*d*), Janis Paige (*e*), The Condos Brothers, S Z Sakall, Alan Hale. *Prod* Alex Gottlieb. *Dir* David Butler. *Scr* Francis Swann, Agnes C Johnson, Lynn Starling. *Ph* William V Skall. *Art* Hugh Reticker. *Ed* Irene Morra. *Dances* Le Roy Prinz. *Mus* Leo F Forbstein, *assoc* Ray Heindorf, Frederick Hollander. *Songs* Arthur Schwartz and Leo Robin: Through a

thousand dreams (*a-c-d*); A gal in calico (*a-b-c*); Oh but I do (*a*); A rainy night in Rio (*a-b-c-e*); I happened to walk down First Street. A solid citizen of the solid South. Remake of the 1929 film.

Tin Pan Alley Fox 1940 (*94 mins*) *with* Alice Faye (*a*), Betty Grable (*b*), John Payne (*c*), Jack Oakie (*d*), Billy Gilbert (*e*), The Nicholas Brothers (*f*), The Brian Sisters (*g*), The Roberts Brothers (*h*). *Prod* Kenneth McGowan. *Dir* Walter Lang. *Scr* Robert Ellis, Helen Logan. *Ph* Leon Shamroy. *Art* Richard Day. *Ed* Walter Thompson. *Cost* Travis Banton. *Sound* Eugene Grossman, Roger Heman. *Dances* Seymour Felix. *Mus* Alfred Newman. *Songs* Harry Warren and Mack Gordon: You say the sweetest things (*a-c-d*). *Interpolated songs* Goodbye Broadway (*d and chorus*); America I love you (*a-c-g and ensemble*); Honeysuckle Rose (*b and chorus*); K-K-K-Katy (and parodies thereof) (*d and chorus*); The sheik of Araby (*a-b-e-f*); On Moonlight Bay (*a*); Moonlight and roses (*b and chorus*). *Background and source music* Smiles; Pack up your troubles in your old kitbag; Oh you beautiful doll; Memories; Pretty baby; In old Chicago; Over there.

The Toast Of New Orleans MGM 1950 (*97 mins*) *with* Mario Lanza (*a*), Kathryn Grayson (*b*), David Niven, James Mitchell (*c*), Rita Moreno (*d*), J Carroll Naish, Clinton Sundberg. *Prod* Joe Pasternak. *Dir* Norman Taurog. *Scr* George Wells, Sy Gomberg. *Ph* William Snyder. *Art* Cedric Gibbons, Daniel B Cathcart. *Sets* Edwin B Willis. *Ed* Gene Ruggiero. *Sound* Douglas Shearer. *Mus* Georgie Stoll, Johnny Green. *Songs* Nicholas Brodszky and Sammy Cahn: The toast of New Orleans (*a*); Be my love (*a-b*); Bayou lullaby (*a-b*); The Tina-Lina (*a-c-d*); I'll never love you (*a*);

Boom-biddy-boom-boom (*a*). *Interpolated arias* 'Carmen' – Flower song (Bizet) (*a*); 'Mignon' – Je suis Titania (Thomas) (*b*); 'Madame Butterfly' – Love duet (Puccini) (*a-b*).

Top Hat RKO 1935 (*105 mins*) *with* Fred Astaire (*a*), Ginger Rogers (*b*), Helen Broderick, Edward Everett Horton, Eric Blore, Erik Rhodes. *Prod* Pandro S Berman. *Dir* Mark Sandrich. *Scr* Dwight Taylor, Allan Scott. *Ph* David Abel, Vernon Walker. *Art* Van Nest Polglaze. *Sets* Carroll Clark. *Ed* William Hamilton. *Cost* Bernard Newman. *Sound* Hugh McDowell Jnr. *Dances* Hermes Pan, Fred Astaire. *Mus* Max Steiner. *Songs* Irving Berlin: Top hat, white tie and tails (*a*); Isn't this a lovely day (*a-b*); The Piccolino (*a-b and chorus*); No strings (*a*); Cheek to cheek (*a-b*); Get thee behind me Satan (deleted before release, and later revived for *Follow the Fleet*).

Twenty Million Sweethearts Warner 1934 (*89 mins*) *with* Dick Powell (*a*), Ginger Rogers (*b*), The Mills Brothers (*c*), Ted Fio Rito and his Orchestra (*d*), Pat O'Brien, Allen Jenkins. *Prod* Samuel Bischoff. *Dir* Ray Enright. *Scr* Warren Duff, Harry Sauber. *Ph* Sid Hickox. *Ed* Clarence Kolster. *Cost* Orry-Kelly. *Mus* Leo F Forbstein. *Songs* Harry Warren and Al Dubin: I'll string along with you (*a-b*); Fair and warmer (*a-d*); How'm I doin' (hey hey) (Fowler-Redman) (*c*); What are your intentions; Out for no good.

Two Girls And A Sailor MGM 1944 (*124 mins*) *with* June Allyson (*a*), Gloria De Haven (*b*), Van Johnson, Tom Drake, Jimmy Durante (*c*), Lena Horne (*d*), Carlos Ramirez (*e*), Harry James and his Orchestra (*f*) with Helen Forrest (*g*), Xavier Cugat and his Orchestra (*h*) with Lina Romay (*i*), Jose Iturbi (*j*), Amparo Iturbi (*k*), Virginia O'Brien (*l*), Gracie Allen (*m*), Albert Coates (*n*). *Prod* Joe Pasternak. *Dir* Richard Thorpe. *Scr* Richard Connell, Gladys Lehman. *Ph* Robert Surtees. *Art* Cedric Gibbons, Paul Groesse. *Sets* Edwin B Willis, John Bonar. *Ed* George Boemler. *Cost* Irene. *Sound* Douglas Shearer. *Dances* Sammy Lee. *Mus* Georgie Stoll. *Songs* Ralph Freed, Jimmy McHugh, Sammy Fain: My mother told me (*b*); A love like ours (*a-b*); In a moment of madness (*f-g*); The young man with the horn (*a-f*); You, dear (*f*); Castles in the air (*f*). *Interpolated songs* Paper doll (*d*); Sweet and lovely (*a-b-f-g*); The thrill of a new romance (*b*); Granada (*e-h*); Inka dinka doo (*c-f*); Take it easy (*h-i-l*); Estrellita (*f*); A-tisket a-tasket (*a-b*); Did you ever have the feeling that you wanted to stay (*c*); Flash (*f*); Concerto for index finger (*m-n*); Charmaine (*f*); Babalu (*h*); Ritual fire dance (*j-k*); Who will be with you when I'm far

away (*c-f*); Dardanella (*orch.*); Wonderful one let's dance (*orch.*).

The Unsinkable Molly Brown Marten/MGM 1964 (*128 mins*) *with* Debbie Reynolds (*a*), Harve Presnell (*b*), Ed Begley, Jack Kruschen, Hermione Baddeley, Harvey Lembeck, Grover Dale. *Prod* Lawrence Weingarten, Roger Edens. *Dir* Charles Walters. *Scr* Helen Deutsch from Meredith Willson's Broadway show based on a story by Richard Morris. *Ph* Daniel L Fapp. *Art* George W Davis, Preston Ames. *Ed* Frederick Steinkamp. *Sound* Wesley C Miller. *Dances* Peter Gennaro. *Mus* Robert Armbruster. *Songs* Meredith Willson: Belly up to the bar, boys (*a*); I'll never say no (*a-b*); Colorado my home (*b*); I ain't down yet (*a-b*); Up where the people are (*a*); He's my friend (*b*); The beautiful people of Denver (*a*); I've already started (*b*); I may never fall in love with you (*a*); Soliloquy (*b*).

Up In Arms Avalon/RKO 1944 (*106 mins*) *with* Danny Kaye (*a*), Dinah Shore (*b*), Constance Dowling, Dana Andrews, Louis Calhern, Lyle Talbot, The Goldwyn Girls (incl. Virginia Mayo). *Prod* Samuel Goldwyn, *assoc* Don Hartman. *Dir* Elliott Nugent. *Scr* Don Hartman, Robert Pirosh, Allan Boretz based on 'The Nervous Wreck' by Owen Davis. *Ph* Ray Rennahan. *Art* Perry Ferguson, McClure Capps, Stewart Chaney. *Sets* Howard Bristol. *Ed* Daniel Mandell, James Newcom. *Sound* Fred Lau. *Dances* Danny Dare. *Mus* Ray Heindorf, Louis Forbes. *Songs* Harold Arlen and Ted Koehler: Fall out for freedom (*chorus*); Now I know (*b*); Tess's torch song (*a-b*). Sylvia Fine and Max Liebman: Manic-Depressive Presents (The lobby scene) (*a*); Malady in 4F (*a*).

The Vagabond King Paramount 1930 (*104 mins*) *with* Dennis King (*a*), Jeannette MacDonald (*b*), Lillian Roth (*c*), Theresa Allen, Warner Oland, O P Heggie. *Prod* Jesse L Lasky, *assoc* Bachman. *Dir* Ludwig Berger. *Scr* Herman J Mankiewicz from the Friml operetta based on Justin McCarthy's 'If I Were King'. *Ph* Henry Gerrard, Ray Rennahan. *Songs* Rudolf Friml, W H Post and Brian Hooker: The song of the vagabonds (*a and chorus*); Only a rose (*a-b*); Some day (*b*); Valse Huguette (*c*); Love me tonight (*a-b*). Sam Coslow, Newell Chase and Leo Robin: If I were King (*a*); King Louis (*c*); Mary, Queen of Heaven.

The Vagabond King Paramount 1956 (*88 mins*) *with* Oreste (*a*), Kathryn Grayson (*b*), Rita Moreno (*c*), Leslie Nielsen, Cedric Hardwicke, William Prince. *Prod* Pat Duggan. *Dir* Michael Curtiz. *Scr* Ken Englund, Noel

Langley, origin as above. *Ph* Robert Burks.
Art Hal Pereira, Henry Bumstead. *Ed* Arthur
Schmidt. *Cost* Edith Head. *Mus* Victor Young.
Songs Rudolf Friml, W H Post and Brian
Hooker: The song of the vagabonds (*a and
chorus*); Only a rose (*a-b*); Some day (*b*); Valse
Huguette (*c*); Love me tonight (*a-b*). Rudolf
Friml and Johnny Burke: This same heart;
Bon jour.

The Varsity Show Warner 1937 *with* Dick
Powell (*a*), Fred Waring and his Pennsylvanians
(*b*), Priscilla Lane (*c*), Rosemary Lane (*d*), Buck
and Bubbles (*e*), Johnny 'Scat' Davis (*f*), Ted
Healy, Carole Landis. *Prod* Louis F Edelman.
Dir William Keighley. *Scr* Warren Duff,
Richard Macaulay, Jerry Wald, Sig Herzig.
Ph Sol Polito, George Barnes. *Ed* George Amy.
Cost Orry-Kelly. *Dances* Busby Berkeley. *Mus*
Leo F Forbstein. *Songs* Richard Whiting and
Johnny Mercer: Love is on the air tonight
(*a-b*); Have you got any castles, baby (*c*); Old
King Cole (*f*); We're working our way through
college (*a and chorus*); On with the dance (*d*);
You've got something there (*a-d*); Moonlight
on the campus (*a*); When your college days
are gone (*b and ensemble*); Little fraternity pin;
Let that be a lesson to you.

Wake Up And Live Fox 1937 *with* Walter
Winchell, Ben Bernie and his Orchestra (*a*),
Alice Faye (*b*), Jack Haley (*c*), Joan Davis (*d*),
Leah Ray (*e*), The Condos Brothers (*f*), Grace
Bradley (*g*), The Brewster Twins (*h*), Ned
Sparks, Patsy Kelly. *Prod* Darryl F Zanuck,
assoc Kenneth McGowan. *Dir* Sidney Lanfield.
Scr Harry Tugend, Jack Yellen. *Ph* Edward
Cronjager. *Art* Mark-Lee Kirk. *Ed* Robert
Simpson. *Dances* Jack Haskell. *Mus* Louis
Silvers. *Songs* Harry Revel and Mack Gordon:
Wake up and live (*b-c*); There's a lull in my life
(*b*); Never in a million years (*b-c*); I'm bubbling
over (*g-h*); It's swell of you (*b-c*); Oh but I'm
happy; I love you too much, muchacha.
Note Vocals for Jack Haley dubbed by Buddy
Clark.

A Weekend In Havana Fox 1941 (*80 mins*)
with Alice Faye (*a*), John Payne (*b*), Carmen
Miranda (*c*), Cesar Romero (*d*), Cobina Wright
Jnr, Leonid Kinskey, George Barbier. *Prod*
William Le Baron. *Dir* Walter Lang. *Scr* Karl
Tunberg, Darrel Ware. *Ph* Ernest Palmer. *Art*
Richard Day. *Ed* Allen McNeil. *Dances* Hermes
Pan. *Mus* Alfred Newman. *Songs* Harry Warren
and Mack Gordon: A weekend in Havana (*c*);
Tropical magic (*a*); When I love I love (*c*); The
Nango (*a and chorus*); The man with the lollipop
song; Rebola a bola (*c*): James V Monaco and
Mack Gordon: Romance and rumba.

We're Not Dressing Paramount 1934 (*63 mins*)

with Bing Crosby (*a*), Carole Lombard, Ethel
Merman (*b*), Leon Errol (*c*), Burns and Allen,
Ray Milland, Jay Henry. *Prod* Benjamin
Glazer. *Dir* Norman Taurog. *Scr* Horace
Jackson, Francis Martin, George Marion Jnr
from a story series by Benjamin Glazer,
suggested by J M Barrie's 'The Admirable
Crichton'. *Ph* Charles Lang. *Art* Hans Dreier.
Ed Stuart Heisler. *Mus* Nathaniel Finston.
Songs Harry Revel and Mack Gordon: She
reminds me of you (*a*); Love thy neighbour
(*a-b-c*); Goodnight lovely little lady (*c*); We've
never seen a mermaid (*chorus*); May I (*a*); It's
just an old Spanish custom (*b-c*); The right girl
for me (*a*); It's the animal in me (*b*); Once in
a blue moon (*a*).

The West Point Story Warner 1950 (*107
mins*) *with* James Cagney (*a*), Virginia Mayo
(*b*), Doris Day (*c*), Gordon Macrae (*d*), Gene
Nelson (*e*), Alan Hale Jnr. *Prod* Louis F
Edelman. *Dir* Roy Del Ruth. *Scr* John Monks
Jnr, Irving Wallace, Charles Hoffman. *Ph* Sid
Hickox. *Art* Charles H Clarke. *Sets* Armor E
Marlowe. *Ed* Owen Marks. *Cost* Milo Anderson,
Marjorie Best. *Sound* Francis J Scheid. *Dances*
Le Roy Prinz, Eddie Prinz, Al White, Johnny
Boyle Jnr. *Mus* Ray Heindorf, *assoc* Hugh
Martin. *Songs* Jule Styne and Sammy Cahn:
You love me (*c-d*); 10,432 sheep (*c*); It could
only happen in Brooklyn (*a-b*); Long before I
knew you; By the kissing rock (*c-d*); Military
polka. *Interpolated song* Fine and dandy (*c*).
Title in GB: *Fine And Dandy*.

West Side Story Mirisch-Seven Arts/United
Artists 1961 (*155 mins*) *with* Natalie Wood (*a*),
Richard Beymer (*b*), Russ Tamblyn (*c*), George
Chakiris (*d*), Rita Moreno (*e*), Tucker Smith
(*f*), Sue Oakes, Simon Oakland. *Prod* Robert
Wise, Saul Chaplin. *Dir* Robert Wise, Jerome
Robbins. *Scr* Ernest Lehman from Arthur
Laurents' book for the Broadway show. *Ph*
Daniel Fapp. *Art/Sets* Boris Leven, Victor A
Gangelin. *Ed* Thomas Stanford. *Cost* Irene
Sharaff. *Sound* Fred Hynes, Gordon E Sawyer.
Dances Jerome Robbins. *Mus* Johnny Green,
assoc Saul Chaplin, Irwin Kostal, Sid Ramin.
Songs Leonard Bernstein and Stephen Sondheim:
Prologue (*Jets and Sharks*); The Jet song (*c and
Jets*); Something's coming (*b*); Dance at the
gym (*ensemble*); Maria (*b*); America (*d-e and
Sharks*); Tonight (*a-b*); Gee Officer Krupke
(*c and Jets*); I feel pretty (*a*); One hand, one
heart (*a-b*); Quintet (*a-b-e and Jets*); The rumble
(*ensemble*); Cool (*f and Jets*); A boy like that/I
have a love (*a-e*); Somewhere (*a-b*).
Note Vocals for Natalie Wood dubbed by
Marni Nixon, for Richard Beymer by Jim
Bryant and for Rita Moreno by Betty Wand.

Where Do We Go From Here? Fox 1945
with Fred MacMurray (*a*), Joan Leslie (*b*), June
Haver (*c*), Carlos Ramirez (*d*), Phil Silvers
(*e*), Fortunio Bonanova (*f*), Alan Mowbray.
Prod William Perlberg. *Dir* Gregory Ratoff.
Scr Morrie Ryskind. *Ph* Leon Shamroy. *Art*
Lyle R Wheeler, Leland Fuller. *Ed* J Watson
Webb. *Sound* Roger Heman. *Dances* Fanchon.
Mus Emil Newman, Charles Henderson. *Songs*
Kurt Weill and Ira Gershwin: All at once (*a*);
If love remains (*a-b*); The Nina, the Pinta and
the Santa Maria (*d-f*); The song of the
Rhineland; Christopher Columbus.

Where's Charley? Warner 1952 (*97 mins*) *with*
Ray Bolger (*a*), Allyn McLerie (*b*), Mary
Germaine (*c*), Robert Shackleton (*d*), Horace
Cooper (*e*). *Prod* Gerry Blattner. *Dir* David
Butler. *Scr* John Monks Jnr from the Broadway
show by George Abbott based on 'Charley's
Aunt' by Brandon Thomas. *Ph* Erwin Hillier.
Art David Folkes. *Ed* Reginald Mills. *Sound*
Harold King. *Dances* Michael Kidd. *Mus* Robert
Farnon. *Songs* Frank Loesser: Once in love with
Amy (*a*); My darling, my darling (*c-d*); Make a
miracle (*a-b*); Lucia (*a-e*); Where's Charley?
(*a-d*); Better get out of here (*a-b-c-d*); The New
Ashmolean Marching Society and Students
Conservatory Band (*a and chorus*). *Ballets* Red
rose cotillon; South America.

White Christmas Paramount 1954 (*120 mins*)
with Bing Crosby (*a*), Danny Kaye (*b*), Rose-
mary Clooney (*c*), Vera-Ellen (*d*), Dean Jagger,
Mary Wickes. *Prod* Robert Emmett Dolan.
Dir Michael Curtiz. *Scr* Norman Panama,
Melvin Frank, Norman Krasna. *Ph* Loyal
Griggs. *Art* Hal Pereira, Roland Anderson.
Sets Sam Comer, Grace Gregory. *Ed* Frank
Bracht. *Cost* Edith Head. *Sound* Hugo Grenz-
bach, John Cope. *Dances* Robert Alton. *Mus*

Joseph J Lilley. *Songs* Irving Berlin: White
Christmas (*a-b-c-d*); The old man (*a-b and
chorus*); I wish I was back in the army (*a-b-c-d*);
The best things happen while you're dancing
(*a-b*); Sisters (*a-b-c-d*); Love, you didn't do
right by me (*c*); What can you do with a
general (*a*); Abraham (*d*); Snow (*a-b-c-d*);
Choreography (*b*); Mandy (*d and ensemble*).
Montage Heat wave, Let me sing and I'm
happy, Blue skies (*a-b*); I want to see a minstrel
show (*a-b*); Count your blessings instead of
sheep (*a-c*); A crooner, a comic (*a-b*) (deleted
before release).
Note Vocals for Vera-Ellen dubbed by Trudy
Stevens.

Whoopee Goldwyn/United Artists 1930 (*94
mins*) *with* Eddie Cantor (*a*), Claire Dodd (*b*),
Betty Grable (*c*), Eleanor Hunt, Paul Gregory,
Ethel Shutta, The Goldwyn Girls. *Prod* Samuel
Goldwyn, Florenz Ziegfeld. *Dir* Thornton
Freeland. *Scr* William Conselman from William
Anthony McGuire's story for the Ziegfeld
Broadway show. *Ph* Gregg Toland, Lee
Garmes, Ray Rennahan. *Ed* Stuart Heisler.
Sound Oscar Lagerstrom. *Dances* Busby Berkeley.
Mus Alfred Newman. *Songs* Walter Donaldson
and Gus Kahn: Makin' whoopee (*a*); A girl
friend of a boy friend of mine (*b*); My baby
just cares for me (*a*); Stetson. Nacio Herb
Brown and Edward Eliscu: I'll still belong
to you (*a*).

With A Song In My Heart Fox 1952 (*117
mins*) *with* Susan Hayward (*a*), David Wayne
(*b*), Thelma Ritter, Rory Calhoun, Una Merkel,
Richard Allan (*c*), Max Showalter (*d*), Leif
Ericson, Robert Wagner. *Prod/Scr* Lamar
Trotti. *Dir* Walter Lang. *Ph* Leon Shamroy.
Art Lyle R Wheeler. *Sets* Thomas Little,
Walter M Scott. *Ed* J Watson Webb. *Sound*
A L Van Kirbach, Roger Heman. *Dances* Billy
Daniels. *Mus* Alfred Newman, *assoc* Ken Darby.
Songs With a song in my heart (*a-c*); Mont-
parnasse (*b and chorus*); Hoe that corn (*b-d*). All
other songs by (*a*): Tea for two; Blue moon;
It's a good day; Embraceable you; Get happy;
I'll walk alone; They're either too young or too
old; That old feeling; The right kind of lovin';
I'm through with love; Indiana; On the great
white way; American Medley (*with full chorus*)
interpolating – The eyes of Texas; The stein
song; Carry me back to old Virginny; Dixie;
Alabamy bound; America the beautiful;
Chicago; Give my regards to Broadway;
California here I come.
Biography of Jane Froman, who dubbed the
singing for Susan Hayward.

The Wizard Of Oz MGM 1939 (*101 mins*)
with Judy Garland (*a*), Ray Bolger (*b*), Jack

Haley (*c*), Bert Lahr (*d*), Billie Burke (*e*), The Singer Midgets (*f*), Frank Morgan, Margaret Hamilton. *Prod* Mervyn Le Roy, Arthur Freed. *Dir* Victor Fleming. *Scr* Florence Ryerson. Noel Langley, Edgar Allan Woolf from Frank L Baum's story. *Ph* Harold Rosson. *Art* Cedric Gibbons, William Horning. *Sets* Edwin B Willis. *Ed* Blanche Sewell. *Sound* Douglas Shearer. *Dances* Bobby Connolly. *Mus* Georgie Stoll, Herbert Stothart, *assoc* Ken Darby. *Songs* Harold Arlen and E Y Harburg: Over the rainbow (*a*); If I only had a brain (*b-c-d*); Come out wherever you are (*e*); Ding dong the witch is dead (*a-e-f*); If I were king of the forest (*d*); In the merry old land of Oz (*f and ensemble*); Munchkinland (*f*); We're off to see the wizard (*a-b-c-d*); The jitterbug (*a-b-c-d*) (deleted before release).

Wonder Bar 1st National 1934 (*84 mins*) *with* Dick Powell (*a*), Al Jolson (*b*), Dolores Del Rio (*c*), Ricardo Cortez (*d*), Kay Francis, Hugh Herbert, Fifi D'Orsay. *Prod* Darryl F Zanuck. *Dir* Lloyd Bacon. *Scr* Earl Baldwin from a story by Karl Farkas and Geza Berezeg. *Ph* Sol Polito. *Ed* George Amy. *Cost* Orry-Kelly. *Dances* Busby Berkeley. *Mus* Leo F Forbstein. *Songs* Harry Warren and Al Dubin: Wonder bar (*a-b*); Goin' to Heaven on a mule (*b*); Why do I dream those dreams (*a*); Don't say goodnight (*a-c-d and chorus*); Vive la France; Tango del Rio (*c-d*).

Words And Music MGM 1948 (*119 mins*) *with* Tom Drake (*a*), Mickey Rooney (*b*), Perry Como (*c*), Mel Tormé (*d*), Judy Garland (*e*), Betty Garrett (*f*), Lena Horne (*g*), Ann Sothern (*h*), Allyn McLerie (*i*), June Allyson (*j*), Gene Kelly (*k*), Vera-Ellen (*l*), Cyd Charisse (*m*), Janet Leigh (*n*), Marshall Thompson (*o*), The Blackburn Twins (*p*). *Prod* Arthur Freed. *Dir* Norman Taurog. *Scr* Fred Finklehoffe from a story by Guy Bolton and Jean Holloway. *Ph* Charles Rosher, Harry Stradling. *Art* Cedric Gibbons, Jack Martin Smith. *Sets* Edwin B Willis, Richard Pefferle. *Ed* Albert Akst, Ferris Webster. *Cost* Helen Rose, Valles. *Sound* Douglas Shearer. *Dances* Robert Alton, Gene Kelly. *Mus* Lennie Hayton. *Songs* Richard Rodgers and Lorenz Hart: Blue room (*c-m*); Manhattan (*a-b-o*); Johnny one note (*e*); Mountain greenery (*c-i*); There's a small hotel (*f*); Slaughter on 10th Avenue (*k-l*); I wish I were in love again (*b-e*); Spring is here (*b*); Blue moon (*d*); With a song in my heart (*a-c*); Way out west (*f*); On your toes, This can't be love, The girl friend (*m*); Someone should tell them, Thou swell (*j-p*); A tree in the park, A little birdie told me so, Where's that Rainbow (*b*); The lady is a tramp (*g*); Where or when (*g*). *Background music* My heart stood still; Here in

my arms; Dancing on the ceiling; I married an angel.
Biography of Richard Rodgers and Lorenz Hart.

Yankee Doodle Dandy Warner/1st National 1942 (*126 mins*) *with* James Cagney (*a*), Joan Leslie (*b*), Walter Huston (*c*), Rosemary De Camp (*d*), Jeanne Cagney (*e*), Irene Manning (*f*), Frances Langford (*g*), Eddie Foy Jnr, Richard Whorf, S Z Sakall. *Prod* Hal B Wallis, *assoc* William Cagney. *Dir* Michael Curtiz. *Scr* Robert Buckner, Edmund Joseph. *Ph* James Wong Howe. *Art* Carl Jules Weyl. *Ed* George Amy. *Cost* Milo Anderson. *Sound* Nathan Levinson, Everett A Brown, *Dances* Le Roy Prinz, John Boyle, Seymour Felix. *Mus* Leo F Forbstein. *Songs* George M Cohan: Yankee doodle boy (*a and chorus*); 45 minutes from Broadway (*a*); Harrigan (*a-b*); Your home town (*a-c-d-e*); You're a grand old flag (*a and chorus*); Mary's a grand old name (*a-b-f*); Little Nellie Kelly (*g*); Give my regards to Broadway (*a and chorus*); So long Mary (*f and chorus*); You're a wonderful girl (*a-c-d-e*); Over there (*a-g*); The barbers' ball (*a-c-d-e*). *Interpolated songs* The love nest (*g*); Off the record (*a*).
Biography of George M Cohan.

Yolanda And The Thief MGM 1945 (*107 mins*) *with* Fred Astaire (*a*), Lucille Bremer (*b*). Leon Ames, Frank Morgan, Mildred Natwick. *Prod* Arthur Freed. *Dir* Vincente Minnelli. *Scr* Irving Brecher, from a story by Ludwig Bemelmans and Jacques Thery. *Ph* Charles Rosher. *Art* Cedric Gibbons, Jack Martin Smith. *Sets* Edwin B Willis, Richard Pefferle. *Ed* Albert Akst. *Cost* Irene Sharaff. *Sound* Douglas Shearer. *Dances* Eugene Loring. *Mus* Lennie Hayton. *Songs* Harry Warren and Arthur Freed: Angel (*b*); Coffee time (Carnival ballet) (*a-b and chorus*); Yolande (*a*); This is a day for love (*b*); Will you marry me (Dream ballet) (*a-b*); Candlelight.

You Can't Have Everything Fox 1937 *with* Tony Martin (*a*), Alice Faye (*b*), The Ritz Brothers (*c*), Don Ameche (*d*), Louis Prima and his Band (*e*), Gypsy Rose Lee (*f*), Rubinoff (*g*). *Prod* Darryl F Zanuck, *assoc* Lawrence Schwab. *Dir* Norman Taurog. *Scr* Harry Tugend, Jack Yellen, Karl Tunberg from a story by Gregory Ratoff. *Ph* Lucien Andriot. *Art* Duncan Cramer. *Ed* Hansen Fritsch. *Dances* Larry Losee. *Mus* David Buttolph. *Songs* Harry Revel and Mack Gordon: You can't have everything (*b*); Afraid to dream (*a-b-d*); The loveliness of you (*a-b*); Danger, love at work (*b-e*); Please pardon us, we're in love (*b*). Louis Prima: Rhythm on the radio (*e*); It's a southern holiday (*e*).

You'll Never Get Rich Columbia 1941
(*88 mins*) *with* Fred Astaire (*a*), Rita Hayworth
(*b*), Robert Benchley, Frieda Inescourt, John
Hubbard, Osa Massen, Cliff Nazarro. *Prod*
Gene Markey, Samuel Bischoff. *Dir* Sidney
Lanfield. *Scr* Michael Fessier, Ernest Pagano.
Ph Philip Tannura. *Art* Lionel Banks. *Ed* Otto
Meyer. *Cost* Jean Louis. *Dances* Robert Alton.
Mus Morris Stoloff. *Songs* Cole Porter: So near
and yet so far (*a-b*); Since I kissed my baby
goodbye (*a*); Dream dancing (*a-b*); Astairable
rag (*a*); Wedding cakewalk (*a*); Shooting the
works for Uncle Sam; Boogie woogie barcarolle.
Note Vocals for Rita Hayworth dubbed by
Nan Wynn.

Young At Heart Arwin/Warner 1954 (*117
mins*) *with* Frank Sinatra (*a*), Doris Day (*b*),
Dorothy Malone, Gig Young, Ethel Barrymore,
Elizabeth Fraser, Alan Hale Jnr, Robert Keith.
Prod Henry Blanke. *Dir* Gordon Douglas.
Scr Julius J Epstein, Lenore Coffee from the
1938 screenplay by Liam O'Brien for Fannie
Hurst's 'Four Daughters'. *Ph* Ted McCord.
Art John Beckman. *Sets* William Wallace. *Ed*
William Ziegler. *Cost* Howard Shoup. *Sound*
Leslie Hewitt, David Forrest. *Mus* Ray Hein-
dorf. *Songs* Young at heart (*a*); Someone to
watch over me (*a*); There's a rising moon (*b*);
Ready, willing and able (*b*); You my love
(*a-b*); Just one of those things (*a*); Hold me in
your arms (*b*); On wings of song (*string quartet*).

You Were Never Lovelier Columbia 1942
(*97 mins*) *with* Fred Astaire (*a*), Rita Hayworth
(*b*), Xavier Cugat and his Orchestra, with Lina
Romay (*c*), Adolphe Menjou, Larry Parks,
Leslie Brooks, Isobel Elsom. *Prod* Louis F
Edelman. *Dir* William A Seiter. *Scr* Michael
Fessier, Delmer Daves, Ernest Pagano from an
Argentinian film story by Carlos Olivari, Sixto
Pondal Rios. *Ph* Ted Tetzlaff. *Art* Lionel
Banks. *Ed* William Lyons. *Cost* Jean Louis.
Dances Val Raset, Fred Astaire. *Mus* Leigh
Harline. *Songs* Jerome Kern and Johnny
Mercer: You were never lovelier (*a-b*); I'm old
fashioned (*a-b*); The Shorty George (*a-b*);
Wedding in the spring (*a*); On the beam (*a*);
Dearly beloved (*a-b*). *Interpolated song* Chiu
chiu (*c*).
Note Vocals for Rita Hayworth dubbed by
Nan Wynn.

The Ziegfeld Follies MGM 1944 (*100 mins*)
with Judy Garland (*a*), Lena Horne (*b*), Fred
Astaire (*c*), Gene Kelly (*d*), Red Skelton (*e*),
Esther Williams (*f*), Keenan Wynn (*g*), Kathryn
Grayson (*h*), Cyd Charisse (*i*), James Melton
(*j*), Lucille Bremer (*k*), Lucille Ball (*l*), Victor
Moore (*m*), Virginia O'Brien (*n*), William
Powell (*o*), Edward Arnold (*p*), Marion Bell

(*q*), Pamela Britton (*r*), Fanny Brice (*s*), Hume
Cronyn (*t*). *Prod* Arthur Freed. *Dir* George
Sidney, Robert Lewis, Lemuel Ayres, Vincente
Minnelli. *Ph* George J Folsey, Charles Rosher,
Art Cedric Gibbons, Merrill Pye, Jack Martin
Smith. *Sets* Edwin B Willis, Mac Alper. *Ed*
Albert Akst. *Cost* Helen Rose, Irene Sharaff.
Sound Douglas Shearer. *Dances* Robert Alton,
Eugene Loring, Charles Walters. *Mus* Lennie
Hayton, *assoc* Roger Edens, Kay Thompson.
Songs Harry Warren and Arthur Freed: This
heart of mine (*c-k*). Earl Brent and Freed:
There's beauty everywhere (*b*): Roger Edens
and Brent: Bring on the beautiful girls (*c-l*);
Bring on the wonderful men (*n*). Edens and
Kay Thompson: The star gives an interview
(*a and chorus*). Ralph Blane and Hugh Martin:
Love (*b*). George and Ira Gershwin: The
Babbitt and the Bromide (*c-d*). Philip Braham
and Douglas Furber: Limehouse blues (*c-k-r*).
Verdi: 'La Traviata' – excerpt (*j-q*). Unknown
music for Water ballet (*f*). *Sketches* Guzzler's
Gin (*e*); Number please (*g*); Sweepstake ticket
(*s-t*); Pay the two dollars (*m-p*).

Ziegfeld Girl MGM 1941 (*131 mins*) *with*
Tony Martin (*a*), Judy Garland (*b*), Charles
Winninger (*c*), Hedy Lamarr, Lana Turner,
Edward Everett Horton, Jackie Cooper (*d*) Al
Shean (*e*), Rosario and Antonio (*f*), Dan Dailey.
Prod Pandro S Berman. *Ass. Prod/Dir*
Robert Z Leonard. *Scr* Sonya Levien, Marguerite
Roberts from a story by William Anthony
McGuire. *Ph* Ray June. *Art* Cedric Gibbons,
Daniel B Cathcart. *Sets* Edwin B Willis. *Ed*
Blanche Sewell. *Cost* Adrian. *Sound* Douglas
Shearer. *Dances* Busby Berkeley. *Mus* Herbert
Stothart, *assoc* Roger Edens. *Songs* Roger Edens:
Ziegfeld girl (*title music*); Minnie from Trinidad
(*b and chorus*); I thought I'd split my sides (*b-c*);
Trinidad (*a and chorus*); Caribbean love song
(*a and chorus*); We must have music (*b*) (deleted
before release, later used as short feature).
Nacio Herb Brown and Gus Kahn: You
stepped out of a dream (*a-d and chorus*).
Interpolated songs I'm always chasing rainbows
(*b-c*); Mr Gallagher and Mr Shean (*c-e*); You
never looked so beautiful before (*b and chorus*);
Whispering (*vocal trio*); You (*chorus*); The kids
from Seville (*f*); You gotta pull strings (*chorus*)
Rose room (*orch.*).

Index of Names

I fully appreciate that many of the directors and actors listed here may have made non-musical films of importance. However, for the purposes of this book I have discussed their work only in relation to musicals. The listing of films for each individual is complete as far as musical films are concerned. Titles in *bold type* are included in the Select Filmographies, pages 101–54. Dates of films are abbreviated to the last two digits.

Dick Powell and Ruby Keeler/The Golddiggers of 1933

Abbott, George. Bn 1889. Broadway writer-director-producer who visited Hollywood occasionally to film his stage hits. Musicals *Why Bring That Up?* 29/*Too Many Girls* 40/**The Pajama Game** 57/**Damn Yankees** 58.

Adamson, Harold. Bn 10 December 1906, Greenville, N.J. Lyricist mainly in collaboration with Burton Lane (33–35) and Jimmy McHugh (37–38/43–48) (*q.v.*). Other musicals **The Great Ziegfeld** 36/*Hold That Ghost* 41/*Hit Parade Of 1943* 43/*Bathing Beauty, Hollywood Canteen* 44/**Gentlemen Prefer Blondes** 53.

Adler, Larry. Bn 1914, Baltimore, Md. Harmonica-player who made guest appearances in several musicals *Many Happy Returns, Operator 13* 34/**The Big Broadcast Of 1937** 36/*The Singing Marine* 37/*Music For Millions* 44/*Three Daring Daughters* 47.

Adler, Richard. Bn 3 August 1921, New York. Composer-lyricist. Together with Jerry Ross (bn 3 September 1926, New York) wrote several Broadway shows, promising well for the future, but after Ross's death on 11 November 1955 Adler left the stage for TV advertising. Now Presidential consultant on entertainment. Musicals **The Pajama Game** 57/**Damn Yankees** 58.

Ager, Milton. Bn 6 October 1893. Composer. Ex-vaudeville pianist who wrote many songs for early talkies. Musicals *Honky Tonk* 29/*They Learned About Women, Chasing Rainbows,* **The King Of Jazz** 30/*Listen Darling* 38/*Andy Hardy Meets Debutante* 40.

Akst, Harry. Bn 15 August 1894, New York. Composer. Former pianist-bandleader who wrote many popular songs of the thirties. Later accompanist to Al Jolson, Eddie Fisher *et al.* Wrote for musicals *On With The Show, Broadway Babies, Is Everybody Happy?* 29/*Leathernecking, No, No, Nanette, So Long Letty, Song Of The Flame, Song Of The West, Dancing Sweeties* 30/**Palmy Days, Bright Lights, June Moon** (appeared) 31/**The Kid From Spain** 32/*Broadway Bad* 33/*Marie Galante, Stand Up And Cheer* 34/*Paddy O'Day, After The Dance, Bright Lights* 35/*Can This Be Dixie?, The Music Goes Round* 36/*Sing And Be Happy, The Holy Terror* 37/*Rascals, Walking Down Broadway* 38/*Harvest Melody, Chatterbox, Lady Of Burlesque* 43/*This Time For Keeps* 46. Died 31 March 1963.

Alberghetti, Anna Maria. Bn 15 May 1936, Pesaro, N. Italy. Teenage soprano who arrived in Hollywood just as teenage sopranos were going out. Musicals *Here Comes The Groom* 51/*The Stars Are Singing* 53/*Ten Thousand Bedrooms* 57/*Cinderfella* 60.

Alexander, Jeff. Bn 2 July 1910, Seattle, Wash. Composer-conductor. Scored dramatic films and TV series, musical director for musicals *The Affairs Of Dobie Gillis* 53/**Kismet** 55/*Jailhouse Rock* 57/*Kid Galahad* 62/*Double Trouble* 67.

Alexander, Ross. Bn 1907, New York. Juvenile lead with Warners in the Busby Berkeley era. Musicals **Flirtation Walk** 34/*Going Highbrow, Shipmates Forever, We're In The Money* 35/**Ready, Willing And Able** 37. Killed in a shooting accident 1937.

Allen, Barbara Jo. Bn 1904. Actress. Appeared in films as the fluttery Vera Vague, the character she portrayed on the Bob Hope radio show. Musicals *Sing, Dance, Plenty Hot* (début), *Melody And Moonlight, Melody Ranch* 40/*Ice-Capades, Kiss The Boys Goodbye* 41/*Priorities On Parade, Ice-Capades Revue* 42/*Get Going* 43/*Cowboy Canteen, The Girl Rush, Rosie The Riveter, Moon Over Las Vegas* 44/*Lake Placid Serenade* 45/*Earl Carroll's Sketchbook* 46/*The Opposite Sex* 56.

Allyson, June. Bn 7 October 1920, Lucerne, N.Y., as Ella Geisman. Actress. Perennial cute teenager of MGM wartime musicals, who was (much) later allowed to be a dramatic actress. Musicals *Best Foot Forward* (début), **Girl Crazy, Thousands Cheer,** *Meet The People* 43/**Two Girls And A Sailor,** *Music For Millions* 44/*Two Sisters From Boston* 45/**Till The Clouds Roll By** 46/*Good News* 47/**Words And Music** 48/**The Glenn Miller Story** 54/*You Can't Run Away From It, The Opposite Sex* 56.

Alter, Louis. Bn 18 June 1902, Haverhill, Mass. Composer. Concert pianist, accompanied vaudeville stars, wrote light music and many songs for Broadway and Hollywood. Musicals **Hollywood Revue,** *Lord Byron Of Broadway* 29/*Chasing Rainbows* 30/**Take A Chance** 33/*Dizzy Dames, Going Highbrow, Rainmakers* 35/**Sing Baby Sing,** *Rainbow On The River* 36/*Make A Wish, Vogues Of 1938* 37/*Las Vegas Nights* 41/*Living In A Big Way, Breakfast In Hollywood* 46/**New Orleans** 47.

Alton, Robert. Bn 28 January 1906, Bennington, Vermont, as Robert Alton Hart. Dance director long with MGM, who occasionally produced and directed. Musicals *Strike Me Pink* 35/*You'll Never Get Rich* 41/**Broadway Rhythm** 43/**Ziegfeld Follies,** *Bathing Beauty* 44/**The Harvey Girls** 45/**Till The Clouds Roll By** 46/**The Pirate** 47/**The Barkleys Of Broadway, Easter Parade, Words And Music, Kissing Bandit** 48/**Annie Get Your Gun** 49/*Pagan Love Song* (dir) 50/**Show Boat** 51/**The Belle Of New York** 52/**Call Me Madam,** *I Love Melvin* 53/**There's No Business Like Show Business, White Christmas, The Country Girl** 54/**The Girl Rush** (prod) 55. Died 1957.

Ameche, Don. Bn 31 May 1910, Kenosha, Wisconsin, as Dominic Amici. Actor-singer who starred in many Fox musicals. Latterly on Broadway and TV. Musicals *In Old Chicago, One*

In A Million, **You Can't Have Everything**
37 | *Happy Landing, Josette,* **Alexander's Rag-
time Band** 38 | *Swanee River, The Three
Musketeers* 39 | **Down Argentine Way,** *Lillian
Russell* 40 | **Moon Over Miami, That Night
In Rio,** *Kiss The Boys Goodbye* 41 | **Something To
Shout About** 43 | *Greenwich Village* 44 | *Slightly
French* 49.

Anderson, Eddie 'Rochester'. Bn 18
September 1905, Oakland, Calif. Negro
comedian first noticed as Jack Benny's 'valet',
but whose screen appearances were limited in
scope by Hollywood's treatment of coloured
performers. Musicals *Green Pastures, The Music
Goes Round* 36 | *Melody For Two* 37 | **The Gold-
diggers In Paris,** *Thanks For The Memory* 38 |
Man About Town, Honolulu 39 | *Buck Benny Rides
Again, Love Thy Neighbour* 40 | **The Birth Of
The Blues,** *Kiss The Boys Goodbye* 41 | **Cabin In
The Sky, Star Spangled Rhythm,** *Cairo* 42 |
What's Buzzin' Cousin?, **Stormy Weather,
Broadway Rhythm** 43 | *I Love A Bandleader* 45 |
The Show-Off 46.

Andrews, Julie. Bn 1 October 1935, Walton-
on-Thames, Surrey, as Julia Elizabeth Wells.
Actress-singer. Former juvenile prodigy who
emerged as the biggest box-office star of the
sixties after not being asked to repeat her stage
success in *My Fair Lady*. Musicals **Mary
Poppins** 64 | **The Sound Of Music** 65 |
Thoroughly Modern Millie 67 | **Star!** 68 |
Darling Lili 69.

Andrews Sisters. Bn in Minneapolis,
Laverne 6 July 1915, Maxine 3 February 1918,
Patti 16 February 1920. A hit on radio and
records in the late thirties, the harmony group
practically lived on the Universal lot during the
war years. Musicals *Argentine Nights* 40 | *Buck
Privates, Hold That Ghost, In The Navy, Ride 'em
Cowboy* 41 | *Give Out Sisters, Private Buckaroo,
What's Cookin'?* 42 | *Always A Bridesmaid, How's
About It? Swingtime Johnny* 43 | *Follow The Boys,
Hollywood Canteen, Moonlight And Cactus* 44 |
Make Mine Music (soundtrack) 46 | *The Road
To Rio* 47 | *Melody Time* (soundtrack) 48. A
recent reunion was broken up by the death of
Laverne in 1967.

Ann-Margret. Bn 28 April 1941, Sweden, as
Ann-Margret Olson. Actress-singer-dancer of
recent musicals **State Fair** 62 | **Bye Bye Birdie**
63 | *The Pleasure Seekers, Viva Las Vegas* 64.

Anthony, Ray. Bn 22 January 1922,
Cleveland. Ohio. Trumpeter-bandleader. Guest
artist, with and without his band, in several
musicals **Sun Valley Serenade** (as member of
Glenn Miller band) 41 | **Daddy Long Legs** 55 |
The Girl Can't Help It, This Could Be The Night
57 | **The Five Pennies,** *Girls' Town* 59.

Archainbaud, George. Bn 7 May 1890, Paris.
Director. Actor and stage manager in Europe,
film director since 1915. Early talkies under
contract to Radio, and directed several
Paramount musicals in late thirties. Musicals
Broadway Scandals 29 | *Hideaway Girl, The Thrill
Of A Lifetime* 37 | *Thanks For The Memory, Her
Jungle Love* 38 | *Some Like It Hot* 39. Died 1959.

Arden, Eve. Bn 30 April 1912, Mill Valley,
Calif., as Eunice Quedens. Actress. Ex-Ziegfeld
Follies show girl whose cool cynicism made her
a Hollywood scene-stealer. Musicals *The
Cocoanut Grove, Having Wonderful Time* 38 | *At
The Circus* 39 | *No, No, Nanette* 40 | **Ziegfeld
Girl,** *San Antonio Rose* 41 | *Hit Parade Of 1943,
Let's Face It* 43 | **Cover Girl,** *Pan-Americana,
Patrick The Great* 44 | *Earl Carroll's Vanities* 45 |
The Kid From Brooklyn 46 | *Song Of Scheherazade*
47 | **One Touch Of Venus** 48 | *My Dream Is
Yours* 49 | **Tea For Two** 50. Latterly on TV.

Arlen, Harold. Bn 15 February 1905,
Buffalo, N.Y., as Hyman Arluck. Composer.
Former singer-pianist who made his name with
the Cotton Club revues and wrote many
musicianly songs for films. Musicals **The Big
Broadcast,** *Manhattan Parade* 32 | **Take A
Chance** 33 | *Let's Fall In Love* 34 | *Strike Me Pink*
35 | **The Singing Kid, Stage Struck** 36 | *Artists
And Models,* **The Golddiggers Of 1937** 37 |
**Babes In Arms, The Wizard Of Oz, At The
Circus,** *Love Affair* 39 | *Andy Hardy Meets
Debutante* 40 | **Blues In The Night** 41 | **Cabin In
The Sky, Star Spangled Rhythm, Rio Rita**
42 | **The Sky's The Limit** 43 | **Up In Arms** 44 |
Here Come The Waves, *Out Of This World* 45 |
Casbah 48 | *The Petty Girl, My Blue Heaven* 50 |
Mr Imperium 51 | **The Farmer Takes A Wife**
52 | **A Star Is Born, The Country Girl** 54 |
Gay Purr-ee, *I Could Go On Singing* 62.

Armstrong, Louis. Bn 4 July 1900, New Orleans. Trumpeter-bandleader. Guest appearances in many films have made Satchmo one of the few great jazz musicians known to the public at large. Musicals *Pennies From Heaven* 36 / *Every Day's A Holiday, Artists And Models* 37 / *Doctor Rhythm, Going Places* 38 / **Cabin In The Sky** 42 / *Atlantic City, Hollywood Canteen, Jam Session* 44 / **New Orleans** 47 / *A Song Is Born* 48 / *The Strip, Here Comes The Groom* 51 / *Glory Alley* 52 / **The Glenn Miller Story** 54 / **High Society** 56 / **The Five Pennies** 59 / *Jazz On A Summer's Day* 60 / *Paris Blues* 61 / *Girl Crazy* 66 / **Hello Dolly!** 69. Died July 1971.

Arnaz, Desi. Bn 2 March 1917, Santiago, Cuba, as Desiderio Alberto Arnaz y de Acha III. Bandleader, actor, on radio, records and screen, latterly TV producer. Musicals *Too Many Girls* 40 / *Four Jacks And A Jill* 41 / *Cuban Pete* 46 / *Holiday In Havana* 49.

Arnheim, Gus. Bn 11 September 1897, Philadelphia. Society bandleader of the twenties and thirties who did feature spots in films and wrote many popular songs. Musicals *Broadway, Street Girl* 29 / *Puttin' On The Ritz* 30 / **Cuban Love Song,** *Flying High* 31 / *Trocadero* 44. Died 19 January 1955.

Arzner, Dorothy. Bn 1900, San Francisco. Director, one of the few women to make it from studio stenographer to the top. Musicals **Paramount On Parade** 30 / *Dance Girl Dance* 40.

Astaire, Fred. Bn 10 May 1899, Omaha, as Frederick Austerlitz. Dancer-singer-actor-composer-author-choreographer and all-round genius of the Hollywood musical. Started on the stage with his sister Adele as a child, and despite threats of retirement for the past twenty-five years is still active at seventy at the time of writing. Musicals *Dancing Lady* (début), *Flying Down To Rio* 33 / *The Gay Divorcee, Roberta* 34 / *Top Hat, Follow The Fleet* 35 / *Swing Time* 36 / *Shall We Dance?, A Damsel In Distress* 37 / *Carefree* 38 / *The Story Of Vernon And Irene Castle* 39 / *Second Chorus, The Broadway Melody Of 1940* 40 / *You'll Never Get Rich* 41 / *You Were Never Lovelier, Holiday Inn* 42 / *The Sky's The Limit* 43 / *The Ziegfeld Follies* 44 / *Yolanda And The Thief* 45 / *Blue Skies* 46 / *Easter Parade, The Barkleys Of Broadway* 48 / *Let's Dance* 49 / *Three Little Words* 50 / *Royal Wedding* 51 / *The Belle Of New York* 52 / *The Band Wagon* 53 / *Daddy Long Legs* 55 / *Funny Face, Silk Stockings* 56 / *Finian's Rainbow* 68.

Auer, John H. Bn 3 August 1906. Budapest, Hungary. Director-producer of Republic and RKO 'B' musicals *Frankie And Johnny* (dir) 35 / *Rhythm In The Clouds* (dir) 37 / *Outside Of Paradise* (dir) 38 / *Hit Parade Of 1941* (dir) 40 / *Johnny Doughboy* (prod-dir), *Moonlight Masquerade* (prod-dir) 42 / *Tahiti Honey* (prod-dir) 43 / *The Girl Rush* (prod), *Music In Manhattan* (prod-dir), *Pan-Americana* (prod-dir), *Seven Days Ashore* (prod-dir) 44 / *Beat The Band* (dir) 46 / *Hit Parade Of 1951* (prod-dir) 50.

Avalon, Frankie. Bn 18 September 1940, Philadelphia, as Frank Avallone. Singer-actor featured in the pop musicals of the sixties *Jamboree* 57 / *Beach Party, Bikini Beach* 64 / *Ski Party, I'll Take Sweden* 65 / *Fireball 500* 66.

Bacon, Lloyd. Bn 1890, San Jose, Calif.

entertainer, who only made one musical (and that with a dubbed voice) *Carmen Jones* 54.

Belita. Bn 21 October 1924, Nether Wallop, as Gladys Jepson-Turner. Actress-skater-dancer-swimmer who made a few musicals and retired in the mid-fifties. Musicals *Ice-Capades* 41/*Silver Skates* 43/*Lady Let's Dance* 44/ *Suspense* 46/*Invitation To The Dance* 54/**Silk Stockings** 56.

Benny, Jack. Bn 14 February 1894, Chicago, as Benny Kubelsky. Comedian. A double hit on radio and films in the thirties, and still active on TV. A master of comedy timing, he willingly shared the limelight with colleagues such as Rochester, Dennis Day, Phil Harris, whom he helped to stardom. Musicals **Hollywood Revue** (début) 29/*Chasing Rainbows* 30/*Mr Broadway* 33/*Transatlantic Merry-Go-Round* 34/ **The Broadway Melody Of 1936** 35/ *College Holiday,* **The Big Broadcast Of 1937** 36/ *Artists And Models* 37/*Artists And Models Abroad* 38/*Man About Town* 39/*Buck Benny Rides Again, Love Thy Neighbour* 40/*Hollywood Canteen* 44/*Somebody Loves Me* 52/*Beau James* 57.

Bergen, Edgar. Bn 1903. Ventriloquist. Long-time radio favourite with his dummies Charlie McCarthy and Mortimer Snerd, around whom a number of films were created. Latterly in dramatic roles on screen and TV and gained reflected glory as father of actress Candice Bergen. Musicals **The Goldwyn Follies** 38/ *Charlie McCarthy Detective* 39/*Here We Go Again* 42/*Around The World, Stage Door Canteen* 43/*Song Of The Open Road* 44/*Fun And Fancy Free* 46.

Berkeley, Busby. Bn 29 November 1895, Los Angeles, as William Berkeley Enos. Choreographer-director. Former stage actor and director who came to Hollywood to direct

dance sequences in **Whoopee** and stayed to create a new form of musical cinema. His expert handling of chorus numbers in the early Warner musicals and his unique conception of symmetrical patterns shot from overhead camera angles opened up the sound stages for other directors. Criticized as vulgar, over-stylized and garish, 'the Busby Berkeley era' was still one of the most important milestones, bringing musicals back after the 1931–2 film slump and giving them a dimension previously lacking. Musicals (dance director credits; * indicates director credit) **Whoopee** 30/ **Palmy Days** 31/**The Kid From Spain** 32/ **Roman Scandals, The Golddiggers Of 1933, Footlight Parade, Forty-Second Street** 33/ **Dames, Wonder Bar,** *Fashions Of 1934* 34/ **The Golddiggers Of 1935*, Stars Over Broadway, I Live For Love*, In Caliente,** *Bright Lights** 35/*Stage Struck** 36/*The Singing Marine,* **Hollywood Hotel*, The Gold-diggers Of 1937, The Varsity Show** 37/*The Garden Of The Moon*,* **The Golddiggers In Paris** 38/*Babes In Arms*, Broadway Serenade* 39/ **Strike Up The Band*** 40/*Babes On Broad-way*,* **Lady Be Good, Ziegfeld Girl** 41/*For Me And My Gal*, Born To Sing* 42/**Girl Crazy, The Gang's All Here*** 43/*Cinderella Jones** 46/**Take Me Out To The Ball Game*,** *Romance On The High Seas* 48/*Two Weeks With Love* 50/*Call Me Mister, Two Tickets To Broadway* 51/*Small Town Girl, Easy To Love* 53/ **Rose Marie*** 54/**Billy Rose's Jumbo** (2nd unit dir) 62.

Berle, Milton. Bn 12 July 1908, New York, as Milton Berlinger. Comedian. Starred on TV in experimental days of 1929 and still at it, regarding films as a sideline. Musicals *Radio City Revels, New Faces Of 1937* 37/**Sun Valley Serenade,** *Tall, Dark And Handsome* 41/ *Always Leave Them Laughing* 49/*Let's Make Love* 60.

Berlin, Irving. Bn 11 May 1888, Temun, Russia, as Israel Baline. Composer-lyricist. Probably the greatest popular song-writer of all, who provided words and music for some of the most outstanding songs, shows and films *Hallelujah* 29/**Mammy,** *Puttin' On The Ritz* 30/ *Reaching For The Moon* 31/*Kid Millions* 34/**Top Hat, Follow The Fleet** 35/*On The Avenue* 37/**Carefree, Alexander's Ragtime Band** 38/ **Second Fiddle** 39/*Louisiana Purchase* 41/ **Holiday Inn** 42/**This Is The Army** (also appeared) 43/**Blue Skies** 46/**Easter Parade** 48/ **Annie Get Your Gun** 49/**Call Me Madam** 53/ **There's No Business Like Show Business, White Christmas** 54.

Berman, Pandro S. Bn 28 March 1905,

Pittsburgh. Producer, initially responsible for the Astaire-Rogers musicals, who virtually deserted the *genre* when he left RKO for MGM. Musicals *Melody Cruise* 33 | *Hips Hips Hooray,* **The Gay Divorcee, Roberta** 34 | **Top Hat, Follow The Fleet,** *In Person, I Dream Too Much* 35 | *That Girl From Paris,* **Swing Time** 36 | **Shall We Dance?,** *A Damsel In Distress* 37 | *Having Wonderful Time,* **Carefree** 38 | **The Story Of Vernon And Irene Castle** 39 | **Ziegfeld Girl** 41 | **Rio Rita** 42 | *Living In A Big Way* 46 | *Jailhouse Rock* 57.

Bernhardt, Curtis. Bn 1899, Germany. Director. Former stage actor who worked briefly in England before going to Hollywood in 1940. Musicals *Juke Girl, Happy Go Lucky* 42 | **The Merry Widow** 52 | *Interrupted Melody* 55.

Bernie, Ben. Bn 30 May 1891, Bayonne, N.J., as Benjamin Woodruff Ancel. Bandleader, long known on radio as 'The Old Maestro'. Conducted a 'feud' with Walter Winchell which they took into films. Musicals *Shoot The Works* 34 | *Stolen Harmony* 35 | **Love And Hisses, Wake Up And Live** 37. Died 20 October 1943.

Bernstein, Leonard. Bn 25 August 1918, Lawrence, Mass. Composer-pianist-conductor. A classically trained musician and youthful prodigy who expressed his understanding of popular music by writing two outstanding musicals **On The Town** 49 | **West Side Story** 61. Now conductor of the N.Y. Philharmonic.

Binyon, Claude. Bn 1905. Director-scenarist. Directed musicals **And The Angels Sing** 44 | *Aaron Slick From Punkin Crick* 51 | *Here Come The Girls* 52. Wrote others including *Sing You Sinners* 38 | **This Is The Army** 43 | *You Can't Run Away From It* 56 | **Pepe** 60.

Blaine, Vivian. Bn 21 November 1921, Newark, N.J., as Vivian Stapleton. Singer-

actress. Ex-dance band singer who starred in Fox musicals of the mid-forties, then returned to the stage. Musicals *Jitterbugs* 43 | *Greenwich Village, Something For The Boys* 44 | *Nob Hill, State Fair,* Doll Face 45 | *If I'm Lucky, Three Little Girls In Blue* 46 | *Skirts Ahoy* 52 | **Guys And Dolls** 55.

Blair, Janet. Bn 23 April 1921, Altoona, Pa. Actress. Discovered singing with Hal Kemp band and signed by Columbia. Left Hollywood for the stage and returned in the sixties. Musicals *Broadway* 42 | **Something To Shout About** 43 | *Tonight And Every Night* 44 | *Tars And Spars* 46 | *The Fabulous Dorseys* 47 | *The One And Only Genuine Original Family Band* 68.

Blane, Ralph. Bn 26 July 1914, Broken Arrow, Oklahoma. Composer-lyricist. Former singer who met Hugh Martin (*q.v.*) in a Broadway show. They formed a singing group and collaborated on many stage and screen musicals (listed under Martin). Blane worked with other composers on songs for *No Leave No Love, Easy To Wed, The Thrill Of A Romance* 45 | **Summer Holiday** 46 | *One Sunday Afternoon* 48 | *My Dream Is Yours* 49 | *My Blue Heaven* 50 | *Skirts Ahoy* 52 | *The French Line* 53.

Blondell, Joan. Bn 30 August 1909, New York. Actress, who was almost a fixture in Warner musicals of the Berkeley era as the wise-cracking but warm-hearted chorus girl. Still active as the wise-cracking but warm-hearted matron. Musicals *Blonde Crazy* 31 | *Big City Blues* 32 | **Footlight Parade, The Gold-diggers Of 1933,** *Broadway Bad* 33 | **Dames** 34 | *We're In The Money, Broadway Gondolier* 35 | **Stage Struck,** *Colleen* 36 | *The King And The Chorus Girl,* **The Golddiggers Of 1937** 37 | *The East Side Of Heaven* 38 | *Two Girls On Broadway* 40 | *The Opposite Sex* 56 | *This*

Could Be The Night 57 / *Stay Away Joe* 68.

Blue, Ben. Bn 1901. Comedian, lean and swarthy, often found in musicals by Paramount in the thirties and MGM in the forties *College Rhythm* 34 / *College Holiday* 36 / *Artists And Models,* **High, Wide And Handsome, The Big Broadcast Of 1938,** *The Thrill Of A Lifetime, Turn Off The Moon* 37 / *College Swing, The Cocoanut Grove* 38 / *Paris Honeymoon* 39 / *For Me And My Gal, Panama Hattie* 42 / **Thousands Cheer, Broadway Rhythm** 43 / **Two Girls And A Sailor** 44 / *Easy To Wed, Two Sisters From Boston* 45 / *My Wild Irish Rose* 47 / *One Sunday Afternoon* 48.

Blyth, Ann. Bn 16 August 1928, Mt. Kisco, N.Y. Actress. Teenage soprano in Universal musicals, later in MGM operettas. Now strictly dramatic. Musicals *A Chip Off The Old Block* (début), *Babes On Swing Street, Bowery To Broadway, The Merry Monahans* 44 / *Top O' The Morning* 48 / *The Great Caruso* 51 / *The Student Prince,* **Rose Marie** 54 / **Kismet** 55 / **The Helen Morgan Story** 57.

Boles, John. Bn 28 October 1900. Actor. Romantic singing lead of the earliest musicals. Later went dramatic and retired in 1943. Musicals *The Desert Song,* **Rio Rita** 29 / *One Heavenly Night, Song Of The West,* **The King Of Jazz** 30 / *Careless Lady* 32 / *My Lips Betray* 33 / *Bottoms Up, Music In The Air, Stand Up And Cheer* 34 / *Rose Of The Rancho, Redheads On Parade, Curly Top, The Littlest Rebel* 35 / *Romance In The Dark* 38 / **Thousands Cheer** 43. Died February 1969.

Bolger, Ray. Bn 10 January 1903, Dorchester, Mass. Dancer. Reached stardom in 'George White's Scandals' as one of the greatest stage dancers, whose eccentric routines often obscured his superb technique. Only occasionally in films, but usually with memorable results. Musicals **The Great Ziegfeld** 36 / **Rosalie** 37 / *Sweethearts* 38 / **The Wizard Of Oz** 39 / *Sunny* 40 / *Four Jacks And A Jill* 41 / *Stage Door Canteen* 43 / *Holiday In Mexico,* **The Harvey**

Girls 45 / *Make Mine Laughs,* **Look For The Silver Lining** 49 / **April In Paris, Where's Charley?** 52 / *Babes In Toyland* 61.

Boone, Pat. Bn 1 June 1934, Jacksonville, Florida. Well-scrubbed ex-rock and roll singer who turned to ballads and 'nice' films as an example to modern youth, but never quite regained his earlier success. Musicals *April Love* 57 / *Mardi Gras* 58 / *All Hands On Deck* 61 / **State Fair** 62 / *The Main Attraction* 63.

Borzage, Frank. Bn 1894, Salt Lake City. Director. To Hollywood 1913 as extra, then featured player and star of Westerns. Started directing in silent films and later did a few musicals **Flirtation Walk** 34 / *Shipmates Forever* 35 / *Hearts Divided* 36 / *Seven Sweethearts* 42 / *His Butler's Sister, Stage Door Canteen* 43. Died 1962.

Boswell, Connie. Bn 3 December, New Orleans. Singer. With her sisters Martha and Vet formed the top harmony group of the early thirties; later a star in her own right, although crippled by polio. Guest appearances in musicals (Boswell Sisters) **The Big Broadcast** 32 / *Moulin Rouge, Transatlantic Merry-Go-Round* 34 / (Connie solo) *Artists And Models* 37 / *Kiss The Boys Goodbye* 41 / *Syncopation* 42 / *Swing Parade Of 1946* 46.

Boyle, Jack. Dance director-dancer, mostly in 'B' musicals **Yankee Doodle Dandy** 42 / *Melody Parade, Spotlight Scandals* (also appeared), *My Best Gal* (appeared only) 43 / *Bowery To Broadway* 44 / *Sunbonnet Sue* 45 / *High School Hero* 46 / *Ladies Of The Chorus, Mary Lou* 48 / **The West Point Story** 50.

Bracken, Eddie. Bn 7 February 1920, Astoria, Long Island. Actor. Comedy stand-by of many a Paramount musical, and diffident lead in slight comedies. Musicals *Too Many*

Joan Blondell / Dames; Ray Bolger / Babes in Toyland

Girls 40 / *Sweater Girl*, **The Fleet's In, Star Spangled Rhythm,** *Happy Go Lucky* 42 / *Rainbow Island* 44 / *Bring On The Girls, Out Of This World, Duffy's Tavern* 45 / *Ladies' Man* 47 / **Summer Stock** 50 / *Two Tickets To Broadway* 51 / *About Face* 52.

Brackett, Charles. Bn 1892. Producer-writer, long-time collaborator with Billy Wilder as director and co-writer. Only musical from this team was *The Emperor Waltz* 48. Later executive with 20th Century-Fox, for whom Brackett produced *inter alia* musicals **The King And I** 56 / *High Time* 60 / **State Fair** 62. Died 9 March 1969.

Bradley, Grace. Singer-actress, well featured in Paramount musicals of the mid-thirties *The Way To Love* 33 / *Old Man Rhythm, Rose Of The Rancho, Stolen Harmony* 35 / **Anything Goes, Three Cheers For Love** 36 / **The Big Broadcast Of 1938,** *Sitting On The Moon,* **Wake Up And Live** 37 / *There's Magic In Music* 41.

Breaux, Marc, and **Wood,** Dee Dee. Choreographers. Husband and wife team who were Broadway dancers when invited by director Michael Kidd to join him in choreography for 'Subways Are For Sleeping'. Dee Dee Wood did choreography for screen version of *Li'l Abner*, 59, both did so for **Mary Poppins** 64 / **The Sound Of Music** 65 / **The Happiest Millionaire** 67 / **Chitty Chitty Bang Bang** 68.

Breen, Bobby. Bn *c.* 1923. Boy singer around whom RKO built a series of schmaltzy musicals in the mid-thirties. Later went into cabaret. Musicals *Let's Sing Again, Rainbow On The River* 36 / *Make A Wish, Hawaii Calls* 37 / *Breaking The Ice* 38 / *Fisherman's Wharf, Way Down South* 39 / *Johnny Doughboy* 42.

Bremer, Lucille. Bn *c.* 1923. Dancer who had a brief run with MGM in top musicals, but never progressed. Musicals **Meet Me In St Louis, The Ziegfeld Follies** 44 / **Yolanda And The Thief** 45 / **Till The Clouds Roll By** 46.

Brice, Fanny. Bn 29 October 1891, New York, as Fanny Borach. Actress. One of the greatest Broadway stars of the Ziegfeld era who never quite made the same impact on the screen. Her biography *Funny Girl* was filmed in 1968 with Barbra Streisand as Fanny Brice, but had been suggested thirty years earlier in *Rose Of Washington Square*. Musicals *My Man* 28 / *Be Yourself* 30 / **The Great Ziegfeld** 36 / *Everybody Sing* 38 / **The Ziegfeld Follies** 44 /. Died in semi-retirement 29 May 1951.

Bricusse, Leslie. Bn 1933. Composer-lyricist-scenarist. Was actor-writer with Beatrice Lillie before teaming up with Anthony Newley for stage work. Now writes own music and lyrics for film musicals **Doctor Dolittle** (also screenplay) 67 / *Goodbye Mr Chips* 69.

Brisson, Carl. Bn 24 December 1895, Copenhagen, Denmark, as Carl Brisson Petersen. Former boxer who starred in operetta and films in London before going to Hollywood. Musicals *Murder At The Vanities* 34 / *All The King's Horses, Ship Cafe* 35. Returned to cabaret and starred until his death on 26 September 1958.

Brodszky, Nicholas. Bn 1905, Russia. Composer of theme music for British films of the forties and songs for MGM musicals of the fifties **The Toast Of New Orleans** 50 / *Rich, Young And Pretty* 51 / *Because You're Mine, Latin Lovers, Small Town Girl* 53 / *The Student Prince, The Flame And The Flesh* 54 / **Love Me Or Leave Me, Serenade** 55 / **Meet Me In Las Vegas,** *Let's Be Happy, The Opposite Sex* 56 / *Ten Thousand Bedrooms* 57. Died 24 December 1958.

Brooks, Jack. Bn 14 February 1912, Liverpool. Composer-lyricist. Contract writer of many songs for Universal musicals, later with MGM and Paramount *Melody Lane, San Antonio Rose, Don't Get Personal* 41 / *Allergic To Love* 44 / *Song Of The Sarong, That Night With You, That's The Spirit, The Naughty Nineties, Frisco Sal, Here Come The Co-eds* 45 / *Song Of Scheherazade, I'll Be Yours* 47 / *The Countess Of Monte Cristo, Mexican Hayride* 48 / *Yes Sir, That's My Baby* 49 / **Summer Stock** 50 / *Son Of Paleface* 52 / *Artists And Models* 56 / *Cinderfella* 60 / *Ladies' Man* 61.

Brown, Nacio Herb. Bn 22 February 1896, Deming, New Mexico. Composer (usually in association with Arthur Freed) of the greatest song hits from MGM musicals over a twenty-year span **The Broadway Melody, Hollywood Revue,** *Lord Byron Of Broadway, Marianne* 29 / **Whoopee,** *Good News, Montana Moon, One Heavenly Night* 30 / *Going Hollywood,* **Take A Chance** 33 / *Student Tour, Hollywood Party* 34 / **The Broadway Melody Of 1936,** *A Night At The Opera* 35 / *San Francisco* 36 / *Thoroughbreds Don't Cry,* **The Broadway Melody Of 1938** 37 / *Ice Follies Of 1939,* **Babes In Arms** 39 / *Andy Hardy Meets Debutante, Little Nellie Kelly, Two Girls On Broadway* 40 / *Ziegfeld Girl* 41 / *Born To Sing* 42 / *Wintertime, Swing Fever* 43 / *Greenwich Village, Hot Rhythm, Holiday In Mexico* 45 / *On An Island With You* 47 / *The Kissing Bandit* 48 / *Pagan Love Song* 50 / **Singin' In The Rain** 52. Died 28 September 1964.

Bruce, Virginia. Bn 29 September 1910, Minneapolis. Actress. Former extra in silents who sang and danced in several early musicals before 'going dramatic'. Musicals *Why Bring That Up* (featured début), **The Love Parade**

29 | *Safety In Numbers*, **Paramount On Parade**
30 | *Here Comes The Band* 35 | **Born To Dance,**
The Great Ziegfeld 36 | *Let Freedom Ring* 39 |
Pardon My Sarong 42 | *Brazil* 44.

Bullock, Walter. Bn 6 May 1907, Shelburn,
Indiana. Lyricist. Wrote (mostly with Harold
Spina and Alfred Newman) many songs for Fox
musicals of the thirties *Blue Skies* 29 | *Coronado*
35 | **Sing Baby Sing** 36 | *Sally Irene And Mary,*
Nobody's Baby, New Faces Of 1937, 52nd Street
37 | *Happy Landing, Just Around The Corner,*
Little Miss Broadway 38 | *The Bluebird, The Three*
Musketeers 39 | *Hit Parade Of 1941* 40 | *Moon Over*
Her Shoulder 41. Died 19 August 1953.

Burke, Joe. Bn 16 March 1884, Phila-
delphia. Composer, who collaborated with Al
Dubin on songs for early Warner-Vitaphone-
1st National musicals *Sally,* **The Golddiggers**
Of Broadway 29 | *Big Boy, Cuckoos, Dancing*
Sweeties, Hold Everything, Oh Sailor Behave, She
Couldn't Say No, Top Speed 30 | *Blessed Event,*
Crooner 32 | *Palooka* (music by Ann Ronell) 34.
Died 9 June 1950, but some of his songs were
revived the following year in *Painting The*
Clouds With Sunshine 51.

Burke, Johnny. Bn 3 October 1908, Antioch,
Calif. Lyricist who provided most of the words
Bing Crosby ever sang in films, in faithful
partnerships with Arthur Johnston (36–7),
James V. Monaco (38–41), Jimmy Van Heusen
(40–53) (*q.v.* for credits). Only other musicals
involved additional lyrics to Strauss in *Emperor*
Waltz, 48, and Friml in **The Vagabond King**
56. Died 25 February 1964.

Burns, Bob. Bn 2 August 1896, Van Buren,
Arkansas. Burly home-spun comic known as
'The Arkansas Philosopher' in vaudeville and
for his home-made 'bazooka' in Paramount
musicals *Rhythm On The Range,* **The Big**

Broadcast Of 1937 36 | *Mountain Music, Radio*
City Revels, Waikiki Wedding 37 | *Tropic Holiday*
38 | *Belle Of The Yukon* 44. Retired into real
estate. Date of death unknown.

Burns and Allen. Comedians. George Burns
(bn 20 January 1896) met, teamed with, and
married Gracie Allen (bn 26 July 1902) in 1922.
Great radio favourites, they made many
Paramount musicals **The Big Broadcast** 32 |
International House, College Humour 33 | **We're Not**
Dressing, *Many Happy Returns* 34 | **The Big**
Broadcast Of 1936, *Here Comes Cookie, Love In*
Bloom 35 | **The Big Broadcast Of 1937,** *College*
Holiday 36 | **A Damsel In Distress** 37 | *College*
Swing 38 | *Honolulu* 39. Gracie did a solo spot in
Two Girls And A Sailor 44. On TV regularly
till Gracie's death in 1964.

Butler, David. Bn 1895, San Francisco.
Director-producer. Long associated with
musicals, best known for his Warner (Doris
Day) films of the forties and fifties. Also
produced films marked *. Musicals **Fox Follies**
Of 1929, Sunny Side Up 29 | *Just Imagine,*
High Society Blues 30 | *Delicious* 31 | *My Weakness*
33 | *Bottoms Up* 34 | *Bright Eyes, Captain January,*
The Little Colonel, The Littlest Rebel 35 | *Pigskin*
Parade 36 | *Ali Baba Goes To Town, You're A*
Sweetheart 37 | *Straight Place And Show, Kentucky*
Moonshine, East Side Of Heaven 38 | *That's Right*
You're Wrong 39 | *You'll Find Out*, *If I Had My*
*Way** 40 | *Playmates** 41 | *The Road To Morocco* 42 |
Thank Your Lucky Stars 43 | *Shine On Harvest*
Moon 44 | **The Time, The Place And The**
Girl, *Two Guys From Milwaukee* 46 | *My Wild*
Irish Rose 47 | *Two Guys From Texas* 48 | **It's A**
Great Feeling, Look For The Silver Lining
49 | **Tea For Two,** *The Daughter Of Rosie*
O'Grady 50 | *Lullaby Of Broadway, Painting The*
Clouds With Sunshine 51 | **April In Paris,**
Where's Charley?, *By The Light Of The Silvery*
Moon 52 | **Calamity Jane** 53 | *The Right Approach*
61.

Buzzell, Edward (Eddie). Bn 13 November
1896, Brooklyn, N.Y. Director. Also producer,
author and actor in Broadway musicals. On
screen starred in *Little Johnny Jones,* 29, then
turned director for *The Girl Friend* 35 | **At The**
Circus, *Honolulu* 39 | *Go West* 40 | *Ship Ahoy* 42 |
Best Foot Forward 43 | *Easy To Wed* 45 | *Neptune's*
Daughter 49.

Caesar, Irving. Bn 4 July 1895. Lyricist for
many Broadway shows and occasional films
No, No, Nanette 30 | **George White's Scandals**
34 | *Curly Top* 35 | *Stowaway* 36 | *No, No, Nanette*
40 | **Tea For Two** 50.

Cagney, James. Bn 17 July 1901, New York.
Actor. The 'little tough guy' of the screen, but
also a capable song-and-dance man who had

been in vaudeville at seventeen. Did a number of musicals for Warners, gaining an Oscar for *Yankee Doodle Dandy*, but only guest spots thereafter. Musicals *Blonde Crazy* 31 | *Footlight Parade* 33 | *Something To Sing About* 37 | *The Strawberry Blonde* 41 | *Yankee Doodle Dandy* 42 | *The West Point Story* 50 | *Starlift* 51 | *Love Me Or Leave Me*, *The Seven Little Foys* 55.

Cahn, Sammy. Bn 18 June 1913, New York. Lyricist. Ex-violinist who ran a dance band with Saul Chaplin, with whom he started writing in 1935. Wrote hundreds of film songs in regular association with Chaplin (40–3), Jule Styne (43–50), Nicholas Brodszky (50–6) and Jimmy Van Heusen (56 to date) (*q.v.* for credits). With other composers wrote for *Lady Of Burlesque* 43 | *Always Leave Them Laughing* 49 | *She's Working Her Way Through College* 51 | *April In Paris* 52 | *Peter Pan, Three Sailors And A Girl* (also prod) 53 | *Pete Kelly's Blues* 55 | *The Court Jester* 56.

Calloway, Cab. Bn 24 December 1907, Rochester, N.Y. Negro bandleader whose extrovert showmanship earned him guest spots in musicals *The Big Broadcast* 32 | *International House* 33 | *The Singing Kid* 36 | *Manhattan Merry-Go-Round* 37 | *Stormy Weather* 43 | *Sensations Of 1945* 44 | *St Louis Blues* 58.

Canova, Judy. Bn 20 November 1916, Jacksonville, Florida. Comedienne, wide-mouthed, pigtailed and stentorian, whose country-based 'B' musicals rated higher as entertainment than art *Going Highbrow*, *In Caliente* 35 | *The Thrill Of A Lifetime, Artists And Models* 37 | *Scatterbrain* 40 | *Sis Hopkins, Puddin' Head* 41 | *Joan Of Ozark, Sleepytime Gal, True To The Army* 42 | *Sleepy Lagoon, Chatterbox* 43 | *Louisiana Hayride* 44 | *Hit The Hay, Singin' In The Corn* 46 | *Variety Girl* 47 | *Honey Chile* 51.

Cantor, Eddie. Bn 31 January 1892, New York as Edward Iskowitz. Comedian. Blackface entertainer of the Ziegfeld Follies who went to Hollywood to film *Kid Boots* 26, and met the onset of sound with a series of historic Goldwyn spectaculars. Concentrated more on radio than films in later years. Musicals *Glorifying The American Girl* 29 | *Whoopee* 30 | *Palmy Days* 31 | *The Kid From Spain* 32 | *Roman Scandals* 33 | *Kid Millions* 34 | *Strike Me Pink* 35 | *Ali Baba Goes To Town* 37 | *Thank Your Lucky Stars* 43 | *Hollywood Canteen, Show Business* (also prod) 44 | *If You Knew Susie* (also prod) 47. Also dubbed the soundtrack songs for Keefe Brasselle in his biography, *The Eddie Cantor Story* 53. On TV from 1950 but had to retire after a heart attack two years later. Died 10 October 1964.

Carle, Frankie. Bn 25 March 1903, Provi-

dence, Rhode Island, as Francisco Carlone. Pianist-bandleader-composer who featured in minor musicals of the forties *Riverboat Rhythm*, *The Sweetheart Of Sigma Chi*, *Variety Time* 46 | *Mary Lou* 48 | *My Dream Is Yours, Make Mine Laughs, Make Believe Ballroom* 49 | *Footlight Varieties* 51.

Carlisle, Kitty. Bn 2 September 1914, New Orleans. Broadway operetta star who partnered Bing Crosby and Allan Jones in a few musicals, but preferred stage work. Musicals *Murder At The Vanities, Here Is My Heart, She Loves Me Not* 34 | *A Night At The Opera* 35 | *Larceny With Music* 43 | *Hollywood Canteen* 45.

Carlisle, Mary. Actress who was Paramount's blonde love interest in Crosby and college musicals of the thirties *College Humour, The Sweetheart Of Sigma Chi* 33 | *College Rhythm, Million Dollar Ransom, Palooka* 34 | *The Old Homestead* 35 | *Double Or Nothing* 37 | *Doctor Rhythm* 38 | *Hawaiian Nights* 39 | *Dance Girl Dance* 40.

Carmichael, Hoagy. Bn 22 November 1899, Bloomington, Indiana, as Hoagland Howard Carmichael. Pianist-composer-singer-actor who made individual contributions to many films, but few musicals (only *Young Man With A Horn* 50, and *The Las Vegas Story* 52). Wrote songs for musicals *Anything Goes* 36 | *Every Day's A Holiday* 37 | *College Swing, Sing You Sinners, Thanks For The Memory* 38 | *St Louis Blues* 39 | *Road Show* 40 | *True To Life* 43 | *The Stork Club* 46 | *Here Comes The Groom* 51 | *The Las Vegas Story* 52 | *Gentlemen Prefer Blondes* 53, *Three For the Show* 55.

Carminati, Tullio. Bn 1894, Italy. Actor-singer who played romantic leads in operetta films of the mid-thirties *Moulin Rouge*, *One Night Of Love* 34 | *Let's Live Tonight, Paris In The Spring* 35. Died 26 February 1971.

Caron, Leslie. Bn 1 July 1931, Paris. Actress-dancer. Star of the Ballets des Champs Élysées in her teens, taken to Hollywood by Gene Kelly for *An American In Paris* 51. Brilliant dancing and *gamine* charm promised a bright future, but she deserted musicals for dramatic acting. Other musicals *Glory Alley* 52 | *Lili* 53 | *Daddy Long Legs* 55 | *Gigi* 58.

Carroll, John. Bn 1907, Mandeville, Louisiana, as Julian La Faye. Actor. Tall, dark, Latin type, of robust voice and physique, who got sidetracked into Westerns and action 'B' movies after a good start in musicals *Hi Gaucho* (début) 36 | *Go West* 40 | *Lady Be Good, Sunny* 41 | *Rio Rita* 42 | *Hit Parade Of 1943* 43 | *Fiesta* 46 | *Hit Parade Of 1951* 50 | *The Farmer Takes A Wife* 52.

Carroll, Nancy. Bn 19 November 1906, New

York, as Ann La Hiff. Baby-faced actress of silents who co-starred with Buddy Rogers in early musicals *Dance Of Life, Illusion, Shopworn Angel, Sweetie* 29 / *Follow Through, Close Harmony,* **Paramount On Parade,** *Honey* 30 / *Transatlantic Merry-Go-Round* 34 / *After The Dance* 35 / *That Certain Age* 38. Died 1965.

Carson, Jack. Bn 27 October 1910, Carmen, Manitoba. Actor. Ex-vaudeville comic and dancer who became a regular in Warner films, portraying hero or heel with equal facility. Musicals *Having Wonderful Time, She's Got Everything,* **Carefree** 38 / *Love Thy Neighbour* 40 / *The Strawberry Blonde,* **Blues In The Night,** *Navy Blues* 41 / *Shine On Harvest Moon, Hollywood Canteen, Two Guys From Milwaukee,* **The Time, The Place And The Girl** 46 / *Love And Learn* 47 / *April Showers, Two Guys From Texas,* **Romance On The High Seas** 48 / *My Dream Is Yours,* **It's A Great Feeling** 49 / **Dangerous When Wet** 53 / **Red Garters, A Star Is Born** 54. Died from cancer June 1963.

Carter, Everett. Bn 28 April 1919, New York. Lyricist. Contract writer with Universal in the early forties, providing lyrics to music of Milton Rosen (*q.v.* for credits). With other composers wrote for *Allergic To Love, Bowery To Broadway, Ghost Catchers* 44. Left films to become music professor.

Castle, Nick. Bn 21 March 1910, Brooklyn. Choreographer. Stage dancer at seventeen, to films ten years later and worked as dance director *You're A Sweetheart, One In A Million,* **Love And Hisses** 37 / *Hold That Co-ed, Little Miss Broadway, Rascals, Rebecca Of Sunnybrook Farm, Straight Place And Show* 38 / *Swanee River* 39 / *Young People,* **Down Argentine Way** 40 / *Buck Privates, Hold That Ghost, Ride 'em Cowboy* 41 / *The Mayor Of 44th Street,*

Moonlight Masquerade, **Orchestra Wives,** *Hellzapoppin', Joan Of Ozark, Johnny Doughboy* 42 / **Stormy Weather,** *What's Buzzin' Cousin?* 43 / *Something For The Boys,* **Show Business** 44 / *Mexicana, Nob Hill* 45 / *The Thrill Of Brazil, Earl Carroll's Sketchbook* 46 / *Lulu Belle* 48 / *Nancy Goes To Rio, You're My Everything* 49 / **Summer Stock** 50 / *Rich Young and Pretty,* **Royal Wedding** 51 / *Everything I Have Is Yours, Here Come The Girls, Skirts Ahoy, Stars And Stripes Forever* 52 / **Red Garters** 54 / *The Seven Little Foys* 55 / **Anything Goes,** *Bundle Of Joy* 56 / **State Fair** 62. Died 28 August 1968.

Cavallaro, Carmen. Bn 6 May 1913, New York. Pianist and bandleader who guested briefly in films of the forties. Known as 'The Poet Of The Piano', he followed Eddie Duchin professionally and was a natural to dub Tyrone Power's 'playing' in the title role of *The Eddie Duchin Story* 56. Other musicals *Hollywood Canteen* 44 / **Billy Rose's Diamond Horseshoe,** *Out Of This World* 45 / **The Time, The Place And The Girl** 46.

Ceballos, Larry. Dance director who began with Warners during the early days of sound. Musicals **The Singing Fool** 28 / **The Golddiggers Of Broadway,** *On With The Show, Smiling Irish Eyes* 29 / *Sitting Pretty* 33 / *The Music Goes Round* 36 / **One Night In The Tropics,** *Sing Dance Plenty Hot* 40 / *Follies Girl* 43 / *A Song For Miss Julie* 45 / *Queen Of Burlesque* 46 / *Copacabana* 47.

Champion, Gower. Bn 22 June 1921, Geneva, Illinois. Dancer-choreographer-director. Best known for dancing features in MGM musicals with his wife Marge (bn 2 September 1921, Los Angeles as Marge Belcher) **Mr Music** 50 / **Show Boat** 51 / *Everything I Have Is Yours,* **Lovely To Look At** 52 / **Give A Girl A Break** 53 / *Jupiter's Darling,* **Three For The Show** 55. Gower made his début solo in **Till The Clouds Roll By** 46, and later choreographed *The Girl Most Likely* 57. Now a successful Broadway director.

Chaplin, Saul. Bn 19 February 1912, Brooklyn. Composer-arranger-musical director-producer. Ran a dance band with Sammy Cahn (*q.v.*) with whom he began writing songs. They worked together for Columbia pictures from 1940–3, and when Cahn left Chaplin remained as musical associate, moving to MGM 1950 as musical director and associate producer. Now one of Hollywood's top producers. Wrote songs for *Argentine Nights* 40 / *Go West Young Lady, Rookies On Parade, Sing For Your Supper, Time Out For Rhythm, Two Latins From Manhattan* 41 / *Crazy House, Ever Since Venus,*

Marge and Gower Champion / Jupiter's Darling

Redhead From Manhattan 43 / Cowboy Canteen,
Kansas City Kitty, Louisiana Hayride 44 / Meet Me
On Broadway 45 / Two Blondes And A Redhead 47 /
The Countess Of Monte Cristo 48. Associate M.D. /
arranger for **Cover Girl** 44 / **The Jolson Story**
46 / **Jolson Sings Again** 49 / **Summer Stock** 50 /
An American In Paris 51 / **Lovely To Look
At,** Everything I Have Is Yours 52 / **Kiss Me Kate,
Give A Girl A Break** 53 / **Seven Brides For
Seven Brothers** 54 / **High Society** 56 / I Could
Go On Singing 63. Co-produced **Les Girls** 57 /
Merry Andrew 58 / **Can Can** 60 / **West Side
Story** (also M.D.) 61 / **The Sound Of Music** 65 /
Star! 68.

Charisse, Cyd. Bn 8 March 1923, Amarillo,
Texas, as Tula Ellice Finklea. Actress-dancer.
With Ballet Russe as a teenager, later brought
her brilliant balletic style to MGM. Seldom
starred but often outshone the titular stars of
her films **Something To Shout About**
(début), **Ziegfeld Follies** 44 / **The Harvey
Girls** 45 / Fiesta, **Till The Clouds Roll By** 46 /
On An Island With You, The Unfinished Dance 47 /
Words And Music, The Kissing Bandit 48 /
Singin' In The Rain 52 / **The Band Wagon,**
Easy To Love 53 / **Deep In My Heart** 54 /
Brigadoon, It's Always Fair Weather 55 /
Meet Me In Las Vegas, Silk Stockings 56 /
Black Tights 60. Later did a song-and-dance act
in clubs with husband Tony Martin, then made
a comeback in straight roles.

Cherkose, Eddie. Bn 25 May 1912, Detroit.
Lyricist who wrote for a number of minor
musicals (Republic, Monogram, Universal) in
the early forties Charlie McCarthy Detective 39 /
It's A Date, Melody Ranch 40 / Almost Married,
Angels With Broken Wings, Puddin' Head, Nice
Girl, Sweetheart Of The Campus, Rookies On
Parade 41 / Rhythm Parade 42 / Larceny With
Music, Melody Parade, Never A Dull Moment,
Crazy House 43 / A Wave A Wac And A
Marine 44. US Army, then into radio.

Chevalier, Maurice. Bn 12 September 1888,
Menilmontant, France. Actor. Dancing partner
of Mistinguett at twenty-one, star of stage
musicals in Paris and London, and became one
of Hollywood's earliest heart-throbs in his
forties, via a series of Lubitsch romances for
Paramount. In France during war; returning
to USA at seventy became as big a star as
ever, dispensing white-haired benevolence and
nostalgic charm. Musicals Innocents Of Paris
(début), **The Love Parade** 29 / Playboy Of Paris,
The Big Pond, **Paramount On Parade** 30 / The
Smiling Lieutenant 31 / **Love Me Tonight, One
Hour With You** 32 / Bedtime Story, The Way To
Love 33 / **The Merry Widow** 34 / Folies Bergère
35 / **Gigi** 58 / Black Tights, **Can Can, Pepe** 60 / In

Search Of The Castaways 63 / I'd Rather Be Rich 64 /
The Aristocats (soundtrack) 70.

Claire, Bernice. Bn Oakland, Calif., as
Bernice Jahnigan. Actress, formerly on stage,
under contract to 1st National in early musicals
No, No, Nanette (début), Song Of The Flame,
Spring Is Here, Top Speed 30 / Moonlight And
Pretzels 33.

Clare, Sidney. Bn 15 August 1892, New
York. Lyricist. Ex-vaudeville comedian and
dancer who turned out many songs for Fox
musicals with Harry Akst, Richard Whiting,
Oscar Levant and others Hit The Deck, Jazz
Heaven, Street Girl, Tanned Legs 29 / Jimmy And
Sally 33 / Transatlantic Merry-Go-Round, 365
Nights In Hollywood 34 / Bright Eyes, The Littlest
Rebel, Music Is Magic, Paddy O'Day, She Learned
About Sailors 35 / Can This Be Dixie?, Song And
Dance Man 36 / The Holy Terror, Sing And Be
Happy 37 / Walking Down Broadway, Hold That
Co-ed, Kentucky Moonshine, Rascals 38 / Pack Up
Your Troubles 39 / Singin' In The Corn 46 / **Hit The
Deck** 55.

Clark, Petula. Bn 15 November 1933, West
Ewell, Surrey. Actress in British films as a
child. Retired on marriage, returned as pop
singer. Musicals **Finian's Rainbow** 68 /
Goodbye Mr Chips 69.

Cline, Edward F. Bn 1892, Kenosha,
Wisconsin. Director, ex-Sennett, who
specialized in comedy, but directed some
musicals, many for Universal Leathernecking 30 /
It's A Great Life 36 / Hawaii Calls 37 / Breaking
The Ice 38 / Moonlight And Cactus 40 / Give Out
Sisters, Private Buckaroo, What's Cookin'? 42 /
Crazy House, Swingtime Johnny 43 / Ghost Catchers,
Hat Check Honey, Night Club Girl, Slightly
Terrific 44 / See My Lawyer, Penthouse Rhythm 45.
Died 1961.

Clooney, Rosemary. Bn 23 May 1928, Maysville, Kentucky. Singer, ex-dance bands and radio. Paramount's white hope of the fifties, but returned to TV. Musicals *Here Come The Girls* 52 / *The Stars Are Singing* 53 / **Red Garters, White Christmas, Deep In My Heart** 54.

Cohan, George M. Bn 3 July 1878. Composer-lyricist-author-producer-director-actor-singer, the one-man band of Broadway before and during World War I. His musical play *Little Johnny Jones* was filmed 29, and he wrote and starred in *The Phantom President* 32, but didn't really like Hollywood, until he helped Warners make his biopic, **Yankee Doodle Dandy** 42, in which he was portrayed by James Cagney. Died shortly afterwards, on 5 October 1942.

Cole, Jack. Bn 1914. Dance director who staged routines for many Columbia and Fox musicals. Appeared in **Moon Over Miami** 41. Choreographed *Eadie Was A Lady,* **Cover Girl,** *Tonight And Every Night* 44 / **The Jolson Story,** *Tars and Spars, The Thrill of Brazil* 46 / *Down To Earth* 47 / *On The Riviera, Meet Me After The Show* 51 / **The Farmer Takes A Wife, The Merry Widow** 52 / **Gentlemen Prefer Blondes,** *The 'I Don't Care' Girl* 53 / **There's No Business Like Show Business** 54 / **Kismet, Three For The Show, Gentlemen Marry Brunettes** 55 / **Les Girls** 57 / *Let's Make Love* 60.

Cole, Nat 'King'. Bn 17 March 1919, Montgomery, Alabama. Negro singer-pianist who started doing guest spots with his Trio in wartime musicals *Here Comes Elmer* 43 / *Pin-Up Girl, Stars On Parade, Swing In The Saddle* 44 / *See My Lawyer* 45 / *Breakfast In Hollywood* 46 / *Make Believe Ballroom* 49 / *Small Town Girl* 53 / *St Louis Blues* 58. Died from cancer 15 February 1965.

Coleman, Cy. Bn 14 June 1929, New York. Pianist-composer of several recent Broadway shows. Only one filmed to date is **Sweet Charity** 68.

Colonna, Jerry. Bn 17 September 1904, Boston. Moustached, pop-eyed comic; former jazz trombonist and studio musician who started doing comedy while playing on the Fred Allen radio show. Strictly a cameo player in musicals **Rosalie** 37 / *Little Miss Broadway, The Garden Of The Moon, College Swing* 38 / **Naughty But Nice** 39 / *Melody And Moonlight,* **The Road To Singapore** 40 / *Ice-Capades, Sis Hopkins, You're The One* 41 / *True To The Army, Ice-Capades Revue, Priorities On Parade,* **Star Spangled Rhythm** 42 / *Atlantic City* 44 / **Make Mine Music** (soundtrack) 46 / *The Road To Rio* 47 / *Alice In Wonderland* (soundtrack) 51 / **Meet Me In Las Vegas** 56 / **The Road To Hong Kong** 62.

Columbia Pictures. Not one of the leaders in the field, Columbia has yet produced some worthwhile musicals, of which **One Night Of Love** 34, was probably more than any other film responsible for elevating the company from 'Poverty Row' to the upper echelon of Hollywood production companies. During the thirty-four years he ran the studio, founder Harry Cohn was a far from benevolent despot, but he allied a keen business sense with a respect for men like Frank Capra, and along with the pot-boilers produced many prestige pictures. Such films are now the norm for Columbia, which also has interests in TV and records through subsidiaries. Musically, Columbia's place in posterity is assured by such films as **Cover Girl** 44 / **The Jolson Story** 46 / **Pal Joey** 57 / **Funny Girl** 68.

Columbo, Russ. Bn 1908 as Ruggerio de Rudolpho Columbo. Bandleader-composer and later singing actor who threatened to rival Crosby in musicals like *Street Girl* 29 / *Broadway Thru' A Keyhole* 33 / *Moulin Rouge, Wake Up And Dream* 34. Died in a shooting accident 2 September 1934.

Comden and Green. Lyricists-scenarists-singers. Betty Comden (bn 3 May 1915, New York) and Adolph Green (bn 2 December 1915, New York) worked together in the early forties as 'The Revuers' (Judy Holliday was a third member). Wrote 'On The Town' 44, adapted it for the screen and stayed at MGM to write lyrics and screenplays for *Good News* 47 / **The Barkleys Of Broadway, Take Me Out To The Ball Game** 48 / **On The Town** 49 / **Singin' In The Rain** 52 / **The Band Wagon** 53 / **It's Always Fair Weather** 55 / **Bells Are Ringing** 60.

Como, Perry. Bn 18 May 1912, Canonsburg, Pa. Singer (ex-Ted Weems band) who made a hit in early forties. Did several films for Fox but got out of his contract and headed back to radio and TV, his real *métier*. Musicals *Something For The Boys* 44 / *Doll Face* 45 / *If I'm Lucky* 46 / **Words And Music** 48.

Connolly, Bobby. Bn 1890 Dance director with Warner (34–8) and MGM (39–43). Worked on **Take A Chance,** *Moonlight And Pretzels* 33 / **Flirtation Walk, Sweet Adeline** 34 / *Sweet Music, Broadway Hostess,* **Stars Over Broadway,** *Go Into Your Dance* 35 / *Colleen, Cain and Mabel,* **The Singing Kid** 36 / *The King And The Chorus Girl,* **Ready, Willing And Able** 37 / *Swing Your Lady, Fools For Scandal* 38 / **The Wizard Of Oz,** *Honolulu,* **At The Circus** 39 / *Two Girls on*

Broadway, **The Broadway Melody Of 1940** 40/
For Me And My Gal, Ship Ahoy 42/**I Dood
It** 43.

Conrad, Con. Bn 18 June 1891, New York,
as Conrad K. Dober. Composer on Broadway
before Fox called him West in 1929. Wrote the
first Academy Award winning song 'The
Continental' and teamed successfully with Herb
Magidson from 1934 until his death on 28
September 1938. Musicals *Fox Follies Of 1929,
Broadway* 29/ *New Fox Follies Of 1931, Happy
Days, Let's Go Places* 30/**Palmy Days** 31/*I Like
It That Way, Gift Of Gab,* **The Gay Divorcee**
34/*Reckless, King Solomon Of Broadway, Here's To
Romance* 35/**The Great Ziegfeld** 36.

Coslow, Sam. Bn 27 December 1902, New
York. Composer-lyricist-producer. Wrote many
songs for Paramount in the thirties, regularly in
collaboration with Arthur Johnston (*q.v.* for
credits) and with other co-writers for *Why
Bring That Up?, The Dance Of Life* 29/**Para-
mount On Parade, Honey, The Vagabond
King** 30/*College Humour, College Coach, The
Way to Love, Hello Everybody, Too Much Harmony*
33/*Murder at the Vanities, The Belle of the
Nineties, Many Happy Returns, You Belong
To Me* 34/*Goin' To Town, Coronado, All The
King's Horses* 35/*Poppy, Klondike Annie,* **100
Men and a Girl,** *Rhythm On The Range*
36/*Mountain Music, Swing High Swing Low,
This Way Please, Hideaway Girl, Every
Day's A Holiday, The Thrill Of A Lifetime,
Turn Off The Moon, Champagne Waltz* 37/*You
And Me, Love On Toast, Romance In The Dark*
38/*St Louis Blues* 39. Wrote for and produced
Dreaming Out Loud 40/*Out Of This World* 45/
Copacabana 47.

Crosby, Bing. Bn 2 March 1901, Tacoma,
Wash., as Harry Lillis Crosby. Probably
the best-loved entertainer in the history of
popular music. Bing's screen career has ranged
from *The King of Jazz* 30, as one of Paul
Whiteman's Rhythm Boys, and a series of Mack
Sennett shorts, through his many Paramount
starring vehicles of the thirties and the
memorable 'Road' series with Hope and
Lamour, to his acceptance as a serious actor of
Academy Award-winning stature. One of
Hollywood's richest actors, he has (justifiably)
coasted in recent years, though his TV and
film production company is highly active.
Musicals **The King of Jazz,** *Check And Double
Check* 30/*Reaching For The Moon* 31/**The Big
Broadcast** 32/*Too Much Harmony, Going Holly-
wood, College Humour* 33/*Here Is My Heart, She
Loves Me Not,* **We're Not Dressing** 34/**The
Big Broadcast Of 1936, Mississippi,** *Two For
Tonight* 35/**Anything Goes,** *Pennies From*

Heaven, Rhythm On The Range 36/*Double Or
Nothing, Waikiki Wedding* 37/*Sing You Sinners,
The East Side Of Heaven, Doctor Rhythm* 38/*Paris
Honeymoon, The Star Maker* 39/*Rhythm On The
River, If I Had My Way,* **The Road To Singa-
pore** 40/**The Birth Of The Blues,** *The Road To
Zanzibar* 41/*The Road To Morocco,* **Holiday Inn,
Star Spangled Rhythm** 42/*Dixie* 43/**Going
My Way** 44/*The Bells Of St Mary's, Duffy's
Tavern,* **Here Come The Waves,** *Out Of This
World* (soundtrack), *The Road To Utopia* 45/
Blue Skies 46/*Welcome Stranger, Variety Girl,
The Road To Rio* 47/*A Connecticut Yankee At
King Arthur's Court, The Emperor Waltz* 48/
Top O' The Morning, Ichabod And Mr Toad
(soundtrack) 49/**Mr Music,** *Riding High* 50/
Here Comes The Groom 51/*Just For You, The
Road To Bali* 52/*Little Boy Lost* 53/**White
Christmas, The Country Girl** 54/**Anything
Goes, High Society** 56/*Say One For Me* 59/
Pepe, *High Time, Let's Make Love* 60/**The Road
To Hong Kong** 62/**Robin And The Seven
Hoods** 64.

Crosby, Bob. Bn 23 August 1913, Spokane,
Wash. Bandleader-actor-singer. Lived in the
shadow of brother Bing as a singer, but made
his name leading one of the best jazz big bands
of the thirties. Appeared on screen with the
band and as an actor in musicals *Let's Make
Music* 40/*Rookies On Parade, Sis Hopkins* 41/
Presenting Lily Mars 42/*Reveille With Beverley,*
Thousands Cheer 43/*Kansas City Kitty, Meet
Miss Bobby-Sox, My Gal Loves Music, Pardon My
Rhythm, The Singing Sheriff* 44/*Two Tickets To
Broadway* 51/*The Road To Bali* 52/*When You're
Smiling* 53/**The Five Pennies** 59.

Crosland, Alan. Bn 10 August 1894, New
York. Director. Former actor-stage manager
who joined Edison Co. in 1912, and became a

talkie pioneer in early musicals **The Jazz Singer** 27/*On With The Show* 29/*Big Boy, The Song Of The Flame, Viennese Nights* 30/*Children Of Dreams* 31/*King Solomon Of Broadway* 35. Died 1936.

Cugat, Xavier. Bn 1 January 1900, Gerona, Spain, as Francisco De Asis Javier Cugat De Bru Y Deulofeo. Bandleader whose colourful Latin American orchestra and singers were a standard feature of MGM musicals of the forties *Go West Young Man* 36/**You Were Never Lovelier** 42/*The Heat's On, Stage Door Canteen* 43/*Bathing Beauty*, **Two Girls And A Sailor** 44/*Holiday In Mexico, No Leave No Love, The Thrill Of A Romance, Weekend At The Waldorf* 45/*This Time For Keeps* 46/*Luxury Liner, On An Island With You* 47/*A Date With Judy* 48/*Neptune's Daughter* 49.

Cukor, George. Bn 7 July 1899, New York. Director. Former stage director-actor-producer who came to films 1929 as dialogue director (*All Quiet On The Western Front*, etc.). Specialized in big-scale films, often of great feminine appeal, but also did a few stylish musicals **One Hour With You** (part-dir) 32/**A Star Is Born** 54/*Les Girls* 57/*Let's Make Love, Song Without End* 60/**My Fair Lady** 64.

Cummings, Irving. Bn 9 October 1888, New York. Director who came to silent films as an actor. As a director of musicals with Fox he cut his teeth on Shirley Temple vehicles *Curly Top* 35/*Poor Little Rich Girl* 36/*Vogues Of 1938, Merry-Go-Round of 1938* 37/*Little Miss Broadway, Just Around The Corner* 38/**Down Argentine Way**, *Lillian Russell* 40/*Louisiana Purchase*, **That Night In Rio** 41/*My Gal Sal*, **Springtime In The Rockies** 42/*Sweet Rosie O'Grady* 43/*The Dolly Sisters* 46; produced *Double Dynamite* 51, directed by Irving Cummings Jnr. Died 1959.

Cummings, Jack. Bn 1900, New Brunswick, Canada. Producer on MGM staff since 1934, responsible for many routine, and some top, musicals **Born To Dance** 36/**Broadway Melody Of 1938** 37/*Listen Darling* 38/*Honolulu* 39/*Go West*, **The Broadway Melody Of 1940**, *Two Girls On Broadway* 40/*Ship Ahoy* 42/**I Dood It, Broadway Rhythm** 43/*Bathing Beauty* 44/*Easy To Wed* 45/**It Happened In Brooklyn**, *Fiesta* 46/*Neptune's Daughter* 49/*Two Weeks With Love*, **Three Little Words** 50/*Excuse My Dust, Texas Carnival* 51/**Kiss Me Kate, Give A Girl A Break** 53/**Seven Brides For Seven Brothers** 54/*Interrupted Melody* 55/**Can Can** 60/*Viva Las Vegas* 64.

Curtis, Tony. Bn 3 June 1925, Brooklyn, N.Y., as Bernard Schwarz. Expert light comedy actor who made the grade from second feature swashbuckling via dramas, epics, crazy comedies and occasional musicals *So This Is Paris* 54/**Some Like It Hot** 59/*Pepe* 60.

Curtiz, Michael. Bn 24 December 1888, Budapest, Hungary, as Mihaly Kertesz. Director who gave some of the best years of his life to Warner Brothers. A widely varied output included many top musicals of the forties and fifties *Mammy* 30/*Bright Lights* 31/*Bright Lights* (prod) 35/**Yankee Doodle Dandy** 42/**This Is The Army** 43/**Night And Day** 45/**Romance On The High Seas** 48/*My Dream Is Yours* (prod-dir) 49/*Young Man With A Horn* 50/**I'll See You In My Dreams** 51/**The Jazz Singer** 52/**White Christmas** 54/**The Vagabond King, The Best Things In Life Are Free** 56/**The Helen Morgan Story** 57/*King Creole* 58. Died 1962.

Dailey, Dan. Bn 14 December 1915, New York. Actor. Ex-vaudeville song-and-dance man who starred in many first-rate musicals, later turning to dramatic roles in films and TV. Musicals *Hullabaloo* 40/**Lady Be Good, Ziegfeld Girl**, *Moon Over Her Shoulder* 41/*Give Out Sisters, Panama Hattie* 42/**Mother Wore Tights** 47/*Give My Regards To Broadway, You Were Meant For Me, You're My Everything, When My Baby Smiles At Me* 48/*My Blue Heaven, I'll Get By* 50/*Call Me Mister* 51/*Meet Me At The Fair* 52/**There's No Business Like Show Business** 54/**It's Always Fair Weather** 55/**The Best Things In Life Are Free, Meet Me In Las Vegas** 56/*Pepe* 60.

Dale, Virginia. Bn *c.* 1918. Actress-singer-dancer who co-starred briefly in Paramount musicals *Love Thy Neighbour, Dancing On A Dime, Buck Benny Rides Again* 40/*Kiss The Boys Goodbye, Las Vegas Nights* 41/**Holiday Inn** 42.

Daley, Cass. Bn 17 July 1915, Philadelphia, as Katherine Daley. Comedienne. Former dance band singer whose toothy grin and raucous comedy songs helped (or not, according to taste) several Paramount musicals *The Fleet's In, Star Spangled Rhythm* 42 | *Crazy House, Riding High* 43 | *Out Of This World, Duffy's Tavern* 45 | *Ladies' Man, Variety Girl* 47 | *Here Comes The Groom* 51 | *Red Garters* 54.

Damone, Vic. Bn 12 June 1928, Brooklyn, as Vito Rocco Farinola. Singer. Sinatra-influenced performer whose record success took him into MGM musicals of the early fifties *Rich, Young And Pretty* (début), *The Strip* 51 | *Deep In My Heart, Athena* 54 | *Kismet, Hit The Deck* 55 | *Meet Me in Las Vegas* (guest) 56. Flirted with dramatic acting on screen and TV, but has concentrated mainly on night-club work.

Dandridge, Dorothy. Bn 9 November 1920, Cleveland. Ohio. Negro singer-actress, in films as a child. Became a competent dramatic performer, achieving her greatest success in her last two musicals with someone else's singing voice. Musicals *A Day At The Races* 37 | *Sun Valley Serenade* 41 | *Hit Parade Of 1943* 43 | *Atlantic City* 44 | *Carmen Jones* 54 | *Porgy And Bess* 59. Later a success as a sophisticated cabaret star, she died in 1965.

Daniels, Bebe. Bn 14 January 1901, Dallas, Texas, as Virginia Daniels. Actress. In films at five, starring in Harold Lloyd comedies at thirteen, Mack Sennett bathing beauty, star of silents and early musicals, then married Ben Lyon and settled in England from the mid-thirties. The Lyons became a national institution via radio and TV, ending after a near-fatal illness following which she concentrated on writing. Musicals *Rio Rita* 29 | *Dixiana* 30 | *Reaching For The Moon* 31 | *Forty-Second Street, The Cocktail Hour* 33 | *Music Is Magic* 35. Died 16 March 1971.

Daniels, Billy. Dance director, not to be confused with photographer and singer of the same name. Worked on many musicals, mainly for Paramount *Duffy's Tavern, Masquerade In Mexico* 45 | *Monsieur Beaucaire, The Stork Club* 46 | *Welcome Stranger, The Road To Rio, The Perils Of Pauline, Ladies' Man* 47 | *One Touch Of Venus, The Paleface* 48 | *With A Song In My Heart* 52 | *Scared Stiff, The French Line* 53.

Darby, Ken. Bn 13 May 1909, Hebron, Nebraska. Composer-conductor-arranger. Founded The King's Men vocal group 1929, later joined Disney as musical associate. Has been associated with many important musicals as choral director and assistant musical director *The Wizard Of Oz* 39 | *Make Mine Music* 46 | *Melody Time* 48 | *Golden Girl* 51 | *The King And*

I, Carousel 56 | *South Pacific* 58 | *Porgy And Bess* 59 | *The Flower Drum Song* 61 | *Camelot* 67 | *Finian's Rainbow* 68 are only a few.

Davis, Joan. Bn 23 May 1907, St Paul, Minn. Comedienne. Life-long vaudevillian who came to films via Sennett shorts in 1934, and brought her specialized brand of gawky comedy to many top musicals and 'B' features. Musicals *Thin Ice, Sing And Be Happy, Sally Irene And Mary, Life Begins In College, Love And Hisses, On The Avenue, The Holy Terror, Wake Up and Live, You Can't Have Everything* 37 | *Hold That Co-ed, Josette, Just Around The Corner, My Lucky Star* 38 | *Tail Spin* 39 | *Sun Valley Serenade, Hold That Ghost, Two Latins From Manhattan* 41 | *Yokel Boy* 42 | *Around The World* 43 | *Show Business, Beautiful But Broke, Kansas City Kitty* 44 | *George White's Scandals Of 1945* 45 | *If You Knew Susie* 47 | *Make Mine Laughs* 49. Left films for TV and 'I Married Joan' of fond memory. Died 23 May 1961.

Davis, Johnny 'Scat'. Trumpeter-scat singer-comedian featured with Fred Waring in mid-thirties. Later made guest appearances, and played in minor comedies. Musicals *The Varsity Show, Hollywood Hotel* 37 | *A Cowboy From Brooklyn, The Garden Of The Moon, Mr Chump* 38 | *Knickerbocker Holiday, You Can't Ration Love* 44.

Davis, Sammy, Jnr. Bn 8 December 1925, New York. Negro entertainer who sang and

danced in *Rufus Jones For President* 31, at the age of six before spending his life in vaudeville with his uncle Will Mastin's Trio. A Broadway hit in 1956, he has rarely realized his potential in films. Musicals **Meet Me In Las Vegas** (soundtrack only) 56/**Porgy And Bess** 59/ **Pepe** 60/**Robin And The Seven Hoods**, The Threepenny Opera 64/**Sweet Charity** 68.

Day, Dennis. Bn 21 May 1917, New York, as Eugene Dennis McNulty. Tenor singer who gained fame on the Jack Benny radio show and appeared occasionally in musicals *Buck Benny Rides Again* 40/*The Powers Girl* 42/*Sleepy Lagoon* 43/*Music In Manhattan, Knickerbocker Holiday* 44/*Melody Time* (soundtrack) 48/*Make Mine Laughs* 49/*I'll Get By* 50/*Golden Girl* 51.

Day, Doris. Bn 3 April 1924, Cincinnati, as Doris Kappelhoff. Actress-singer. Former Bob Crosby-Les Brown vocalist who went straight to the top in her first film with a bouncy vocal style, fresh young personality and natural ability; all of which saw her through an endless series of Warner musicals. Recently a box-office winner in Ross Hunter comedies, but most fondly remembered for musicals **Romance On The High Seas** (début) 48/*My Dream Is Yours,* **It's A Great Feeling** 49/**Young Man With A Horn, The West Point Story, Tea For Two** 50/*I'll See You In My Dreams, Starlift, Lullaby Of Broadway* 51/**April In Paris,** *By The Light Of The Silvery Moon, On Moonlight Bay* 52/ **Calamity Jane** 53/*Lucky Me,* **Young At Heart** 54/**Love Me Or Leave Me** 55/**The Pajama Game** 57/**Billy Rose's Jumbo** 62.

De Haven, Gloria. Bn 23 July 1923, Los Angeles. Actress. Ex-dance band singer who brightened up several of MGM's wartime

musicals but never developed as a legitimate screen actress. Latterly on stage, TV and in night-clubs. Musicals *Best Foot Forward,* **Broadway Rhythm, Thousands Cheer** 43/ **Step Lively, Two Girls And A Sailor** 44/ **Summer Holiday** 46/*Yes Sir That's My Baby* 49/*I'll Get By,* **Summer Stock, Three Little Words** 50/*Two Tickets to Broadway* 51/*Down Among The Sheltering Palms* 52/*So This Is Paris* 54/*The Girl Rush* 55.

Del Rio, Dolores. Bn 3 August 1905, Durango, Mexico, as Lolita Dolores Asunsolo. Ageless star of silents and the Berkeley era, still active after several comebacks. Musicals **Flying Down To Rio** 33/**Wonder Bar** 34/ **I Live For Love, In Caliente** 35/*Flaming Star* 60.

Del Ruth, Roy. Bn 1895, Philadelphia. Director. Former journalist who turned scenarist for Mack Sennett in 1915 and started directing two years later with Ben Turpin two-reelers. Specialized in comedies and musicals for Warner, Fox and MGM **The Golddiggers Of Broadway,** The Desert Song 29/*Hold Everything, The Life Of The Party* 30/*Blessed Event* 32/*Kid Millions* 34/**Thanks A Million,** Folies Bergère, **The Broadway Melody Of 1936** 35/**Born To Dance** 36/**The Broadway Melody Of 1938, On The Avenue** 37/*Happy Landing, My Lucky Star* 38/*Tail Spin, The Star Maker* 39/*The Chocolate Soldier* 41/**Broadway Rhythm, Dubarry Was A Lady** 43/*Always Leave Them Laughing* 49/**The West Point Story** 50/ *Starlift, On Moonlight Bay* 51/*About Face* 52/ *Three Sailors And A Girl* 53. Died 1961.

Delta Rhythm Boys. Negro close harmony quintet (very under-rated musically) who guested in 'B' musicals of the war years *Crazy House, Hi' Ya Sailor, So's Your Uncle* 43/ *Follow The Boys, Hi Good Lookin', Night Club Girl, The Reckless Age, Weekend Pass* 44/*Easy To Look At* 45.

De Paul, Gene. Bn 17 June 1919, New York. Composer-lyricist. Ex-pianist and singer who joined Universal in 1941 and formed a writing partnership with Don Raye (q.v.) that lasted ten years. In 1954 joined forces with Johnny Mercer to write screen and stage scores. Wrote for *Almost Married, In the Navy, Keep 'em Flying, Moonlight In Hawaii, Ride 'em Cowboy, San Antonio Rose* 41/*Behind The Eight Ball, Get Hep To Love, Hellzapoppin', Pardon My Sarong, What's Cookin'?, When Johnny Comes Marching Home* 42/**Broadway Rhythm, I Dood It,** *Crazy House, Follow The Band, Hi Buddy, Hi' Ya Chum, Larceny With Music, What's Buzzin' Cousin?* 43/*Hi Good Lookin', Lost In A Harem, Night Club Girl, The Reckless Age, Weekend Pass*

44/*Wake Up And Dream, Song Of The Sarong* 45/ *A Date With Judy*, *A Song Is Born* 48/*Ichabod And Mr Toad, So Dear To My Heart* 49/*Alice In Wonderland* 51/*Seven Brides For Seven Brothers* 54/*You Can't Run Away From It* 56/ *Li'l Abner* 59.

De Sylva, B. G. 'Buddy'. Bn 27 January 1895, New York. Producer-composer-author-publisher (see also under De Sylva, Brown and Henderson). Sold his first-ever song to Al Jolson, and collaborated with Gershwin on Broadway shows. Came to Hollywood 1929 as writer and stayed to operate as producer for Fox. Executive producer at Paramount 1939–44, and founded Capitol Records 1942. Without his partners wrote songs for *Sally* 29/*Just Imagine* (also prod) 30/*My Weakness* 33/*Love Affair* 39. Wrote screenplay of **Born To Dance** 36/ and original story of **Dubarry Was A Lady** 43. Produced **Sunny Side Up** 29/**Take A Chance** (also wrote songs and screenplay) 33/*Captain January, The Little Colonel, The Littlest Rebel* 35/ **Sing Baby Sing**, *Poor Little Rich Girl, Stowaway* 36/*Merry-Go-Round Of 1938*/*You're A Sweetheart* 37/ **The Birth Of The Blues** 41/ **Lady In The Dark** 43/**Going My Way** 44/*The Stork Club* 46. Died 11 July 1950.

De Sylva, Brown and Henderson Songwriters. Buddy de Sylva (*q.v.*), Lew Brown (bn 10 December 1883, Odessa, Russia) and Ray Henderson (bn 1 December 1896, Buffalo, N.Y.) joined forces in 1925 to become one of the most powerful writing and publishing teams in show business. Film scores (many being adaptations of their Broadway hits): **The Singing Fool** 28/*Say It With Songs, Sunny Side Up* 29/*Good News* (remade 47), *Follow Through, Just Imagine, Follow The Leader, The Big Pond* 30/*Flying High* 31. Later they all worked with other writers, and Brown became a producer. He died on 5 February 1958, shortly after Fox had produced a lavish (and probably no more inaccurate than most) biography of the triumvirate, **The Best Things In Life Are Free** 55. Henderson died 31 December 1970.

Deutsch, Adolph. Bn 20 October 1897, London. Musical director. Wrote for Paul Whiteman and other bands before Hollywood called. Received credits as composer and/or arranger of background scores with Warner (37–46), and as musical director of MGM and other musicals from 1948 *Mr Dodd Takes The Air* 37/*A Cowboy From Brooklyn, Fools For Scandal, Swing Your Lady* 38/*Juke Girl* 42/ **Take Me Out To The Ball Game** 48/*Annie Get Your Gun* 49/*Pagan Love Song* 50/**Show Boat** 51/ **The Belle Of New York** 52/*Torch Song,*

The Band Wagon 53/**Deep In My Heart, Seven Brides For Seven Brothers** 54/ **Oklahoma!** 55/**Funny Face** 56/**Les Girls** 57/ **Some Like It Hot** 59.

De Wolfe, Billy. Bn *c.* 1905 as William Andrew Jones. Comedian, who could provide relief equally well as the hero's best friend or heroine's toothy, stuffy fiancé. Musicals *Dixie* 43/**Blue Skies** 46/*The Perils Of Pauline, Variety Girl* 47/*Isn't It Romantic?* 48/**Tea For Two** 50/ *Lullaby Of Broadway* 51/**Call Me Madam** 53.

Dieterle, William. Bn 1893, Rheinpfalz, Germany. Director. On stage in Germany and came to Hollywood in 1921 to star in and direct German versions of 1st National films. Started directing in earnest in 1926 and although he specialized in classical and historical dramas and biographies (Reuter, Zola, Pasteur, Ehrlich) did a few musicals *Her Majesty Love* 31/*Adorable* 33/*Fashions Of 1934* 34/*Syncopation* 42.

Dietz, Howard. Bn 8 September 1896, New York. Author-lyricist. Combined two careers, as lyricist with Arthur Schwartz (*q.v.*) from 1928–36, also with Jerome Kern, Vernon Duke and others on many Broadway shows and films, and as film executive. Director of advertising and promotion with Goldwyn in early twenties then with MGM from 1924 (it was Dietz who devised the MGM symbol of Leo and 'Ars Gratia Artis'), becoming vice-president of Loew's Inc. Wrote scores for *Lottery Bride* 30/*The Battle Of Paris* 31/*Hollywood Party, Operator 13* 34/*Under Your Spell* 36/*Three Daring Daughters* 47/*Dancing In The Dark* 49/ **The Band Wagon**, *Torch Song* 53.

Disney, Walt. Bn 5 December 1901, Chicago Producer, former cartoonist whose creation of Mickey Mouse heralded the beginning of the world's greatest organization devoted to the production of family entertainment. From the first of the series of 'Silly Symphony' cartoon shorts, music has always been an integral part of Disney films, even the later 'True Life Adventure' series depending on clever music scoring for a great part of their appeal. The majority of full-length cartoon features were, in essence, musicals, scored by Disney's own staff writers, and many famous singers lent their talents to the soundtracks. In the sixties the studio turned over to live action films, with the emphasis on whimsical comedy, but one or two full-scale musicals were attempted, with **Mary Poppins** as one of the best examples of the genre, combining live action, trick photography and cartoon techniques with a good score by Richard and Robert Sherman, who had taken over as resident composers. After Walt Disney's death in 1966 the production company carried on as a family concern, though it remains to be seen whether the spark of genius that Disney himself possessed has been handed on. Musicals *Snow White And The Seven Dwarfs* 37/*Pinocchio, Fantasia* 40/*Dumbo, The Reluctant Dragon* 41/ *Bambi* 42/*Saludos Amigos* 43/*The Three Caballeros* 45/*Song Of The South*, **Make Mine Music,** *Fun And Fancy Free* 46/*Melody Time* 48/ *Ichabod And Mr Toad, So Dear To My Heart* 49/ *Cinderella* 50/*Alice In Wonderland* 51/*Peter Pan* 53/*Lady And The Tramp* 55/*Sleeping Beauty* 59/ *101 Dalmatians* 60/*Babes In Toyland* 61/*The Magnificent Rebel* 62/*Summer Magic, In Search Of The Castaways* 63/*Born To Sing*, **Mary Poppins** 64/*The Waltz King* 65/*The Jungle Book*, **The Happiest Millionaire** 67/*The One And Only Genuine Original Family Band* 68/*The Aristocats* 70.

Dodd, Claire. Bn New England. Blonde singer-actress of early Warner musicals **Whoopee** 30/*Crooner* 32/**Footlight Parade** 33/**Roberta** 34/**The Singing Kid** 36/*Romance In The Dark* 38 /*In The Navy* 41/*Mississippi Gambler* 42.

Dolan, Robert Emmett. Bn 3 August 1906, Hartford, Conn. Composer-conductor-producer. Conducted theatre orchestras on Broadway before joining Paramount in 1941. Scored many films and conducted musicals **The Birth Of The Blues,** *Louisiana Purchase* 41/**Holiday Inn,** *Happy Go Lucky*, **Star Spangled Rhythm** 42/*Dixie, Let's Face It,* **Lady In The Dark** 43/*The Incendiary Blonde,* **Going My Way** 44/*The Bells Of St Mary's, Bring On The Girls*, **Here Come The Waves,** *Duffy's Tavern* 45/ *Cross My Heart, The Stork Club*, **Blue Skies,** *Monsieur Beaucaire* 46/*Welcome Stranger, The Perils Of Pauline, The Road To Rio* 47/*Let's Dance, Top o' The Morning* 49/*Aaron Slick From Punkin Crick* 51. Produced **White Christmas** 54/**Anything Goes** 56.

Donaldson, Walter. Bn 15 February 1893, Brooklyn. Composer-lyricist. One of the great popular song-writers of the twenties and thirties, often in partnership with Gus Kahn (*q.v.*). Seldom wrote complete scores, but his songs were featured in **The Jazz Singer** 27/ **Glorifying The American Girl** 29/**Whoopee** 30/*Big City Blues* 32/*Hollywood Party, Kid Millions, Million Dollar Ransom, Operator 13* 34/ *Reckless, Here Comes The Band* 35/**The Great Ziegfeld** 36/*Broadway Serenade* 39/*Two Girls On Broadway* 40/**Ziegfeld Girl** 41/*Give Out Sisters, Panama Hattie* 42/*What's Buzzin' Cousin?* 43/ *Beautiful But Broke, Kansas City Kitty, Follow The Boys* 44/*It's A Pleasure* 45/*The Daughter Of Rosie O'Grady* 50/**I'll See You In My Dreams** 51/*Everything I Have Is Yours* 52. Died 15 July 1947.

Donen, Stanley. Bn 1924. Director-choreographer. Former dancer who helped Gene Kelly with the choreography on **Cover Girl,** accompanied Kelly back to MGM and gave a new look to that company's musicals, initially as dance director, then director (although he still took a hand in the choreography, especially on Kelly's films). Moved to Warner as producer-director and eventually out of musicals. Musicals (as dance director) *Hey Rookie, Jam Session, Kansas City Kitty,* **Cover Girl, Anchors Aweigh** 44/*Holiday in Mexico, No Leave No Love* 45/*Living In A Big Way, This Time For Keeps* 46/*A Date With Judy, The Kissing Bandit, Big City*, **Take Me Out To The Ball Game** 48 (as director) **On The Town** 49/ **Royal Wedding** 51/**Singin' In The Rain** 52/ **Give A Girl A Break** 53/**Seven Brides For Seven Brothers, Deep In My Heart** 54/**It's Always Fair Weather** 55/**Funny Face** 56 (as producer-director) **The Pajama Game** 57/ **Damn Yankees** 58. Appeared briefly in *Best Foot Forward* 43.

Donohue, Jack. Bn 1904. Director-dance director. Former stage dancer in Ziegfeld Follies and Broadway shows with Marilyn Miller. Appeared in first screen version of *Sunny* 30, and stayed in Hollywood as dance director on Fox and MGM musicals **George White's Scandals,** *Lottery Lover, Music In The Air* 34/*The Littlest Rebel, The Little Colonel, Music Is Magic, Captain January, Curly Top* 35/ *Louisiana Purchase* 41/**The Fleet's In,** *The Powers Girl* 42/*Best Foot Forward*, **Girl Crazy, Broadway Rhythm** 43/*Bathing Beauty, Lost In A Harem* 44/*Easy To Wed, Two Sisters From Boston* 45/**It Happened In Brooklyn** 46/*On An Island With You* 47/*The Duchess Of Idaho* 50/**Calamity Jane** 53. Directed *Lucky Me* 54/*Babes In Toyland* 61. Still active on stage and as director of TV's 'The Lucy Show'.

Dorsey, Jimmy. Bn 29 February 1904, Shenandoah, Pa. Swing bandleader (brother of Tommy) who made guest appearances with his band and co-starred in their biopic, *The Fabulous Dorseys* 47. Other musicals **Lady Be Good** 41/**The Fleet's In** 42/*I Dood It* 43/ *Lost In A Harem, Four Jills In A Jeep, Hollywood Canteen* 44/*Make Believe Ballroom* 49. Died 12 June 1957.

Dorsey, Tommy. Bn 19 November 1905, Shenandoah, Pa. Trombone-playing bandleader known as 'The Sentimental Gentleman Of Swing' (brother of Jimmy), whose orchestra kept on cropping up in MGM wartime musicals *Las Vegas Nights* 41/*Ship Ahoy, Presenting Lily Mars* 42/**Broadway Rhythm, Dubarry Was A Lady, Girl Crazy,** *Swing Fever* (gag appearance) 43 / *The Thrill Of A Romance* 45 / *The Fabulous Dorseys* (biopic) 47 / *A Song Is Born* 48 / *Disc Jockey* 51. Died 26 November 1956.

Douglas, Gordon. Bn *c.* 1910. Director. Started as scenarist for Hal Roach and came to directing via 'B' features. Has worked regularly for Warners since the fifties. Musicals *Road Show* 40 / *The Girl Rush* 44 / *If You Knew Susie* 47 / *She's Back On Broadway* 52 / *So This Is Love* 53 / *Young At Heart* 54 / *Sincerely Yours* 55 / *Follow That Dream* 62 / *Robin And The Seven Hoods* 64.

Downey, Morton. Bn 14 November 1901, Wallingford, Conn. Singer. Radio star of the twenties and thirties, his tenor style paved the way for Kenny Baker and Dennis Day. Musicals *Syncopation, Lucky In Love, Mother's Boy* 29 / *Ghost Catchers* 44. Gave up show-business for a business career.

Downs, Johnny. Bn 10 October 1913. Song-and-dance man who started as a child in 'Our Gang' comedies, but never really worked his way out of 'B' films. Musicals *Babes In Toyland* 34 / *Coronado* 35 / *College Holiday, First Baby, Pigskin Parade* 36 / *The Thrill Of A Lifetime, Turn Off The Moon* 37 / *Swing Sister Swing, Hold That Co-ed* 38 / *Laugh It Off, Hawaiian Nights* 39 / *I Can't Give You Anything But Love, Sing Dance Plenty Hot, Melody And Moonlight* 40 / *Redhead, Moonlight In Hawaii, Sing Another Chorus, All-American Co-ed* 41 / *Behind The Eight Ball* 42 / *Harvest Melody* 43 / *Campus Rhythm, Trocadero, Twilight On The Prairie* 44 / *Rhapsody In Blue* 45 / *The Kid From Brooklyn* 46 / *Cruisin' Down The River* 53.

Dragon, Carmen. Bn 28 July 1914, Antioch, Calif. Composer-conductor. Wrote background scores for U-A, MGM and Columbia, scored and orchestrated *Cover Girl* 44, and was musical director of *The Kid From Brooklyn* 46 / *Lovely To Look At* 52. Left films for the concert field (conductor of Hollywood Bowl Symphony, etc.).

Dreifuss, Arthur. Bn 1908, Germany. Director. Also producer-composer-conductor-choreographer. Favourite director of producer Sam Katzman (*q.v.*) and an expert at turning out 'quickies' for Columbia, Monogram, Republic, etc. Musicals *Melody Parade, The Sultan's Daughter* 43 / *Campus Rhythm, Eadie Was A Lady, Ever Since Venus* 44 / *The Gay Senorita* 45 / *Betty Co-ed, Freddie Steps Out, High School Hero, Junior Prom* 46 / *Glamour Girl, Two Blondes And A Redhead, Little Miss Broadway, Sweet Genevieve* 47 / *Mary Lou, I Surrender Dear* 48 / *Manhattan Angel, There's A Girl In My Heart* (also prod) 49.

Dubin, Al. Bn 10 June 1891, Zurich,

Switzerland. Lyricist. Worked with many composers, but film work mainly for Warners with Joe Burke (*q.v.*) (29–32) and Harry Warren (*q.v.*) (32–8) with whom he wrote the memorable songs of the Busby Berkeley era. Other musicals *Down Argentine Way* 40 / *Stage Door Canteen* 43. Died 11 February 1945.

Duchin, Eddie. Bn 1 April 1909, Boston. Pianist-bandleader, a society favourite of the thirties. Musicals *Mr Broadway* 33 / *Coronado* 35 / *Hit Parade Of 1937* 36. Portrayed by Tyrone Power in his biopic, *The Eddie Duchin Story* 56. Died of leukaemia 1951.

Duke, Vernon. Bn 10 October 1903, Parafianovo, Russia, as Vladimir Dukelsky. Composer. In the classical field composed for Diaghilev's Ballet Russe, concertos, chamber music, etc., but also brought a touch of class to Broadway shows and Hollywood musicals with his songs and scores for *The Goldwyn Follies* 38 / *Cabin In The Sky* 42 / *Crazy House* 43 / *Hollywood Canteen* 44 / *She's Working Her Way Through College* 51 / *April In Paris* 52. Died 10 January 1969.

Duna, Steffi. Bn Budapest. Actress. Theatrical début at eleven in ballet at Budapest Opera House; later appeared in a few minor musicals *Hi Gaucho, The Dancing Pirate* 36 / *Rascals* 38 / *Way Down South* 39 / *The Girl From Havana* 40.

Dunbar, Dixie. Pert song-and-dance girl in Fox musicals of the thirties *George White's Scandals* 34/ *The King Of Burlesque* 35/ *Pigskin Parade, First Baby,* **Sing Baby Sing** 36/ *Sing And Be Happy, One In A Million* 37/ *Walking Down Broadway,* **Alexander's Ragtime Band,** *The Freshman Year, Rebecca Of Sunnybrook Farm* 38.

Dunne, Irene. Bn 20 December 1904, Louisville, Kentucky. Actress-singer. Ladylike heroine of many dramas, comedies and musicals. A graduate of the musical stage, she showed her soprano voice to advantage in *Leathernecking* 30/ **Sweet Adeline, Roberta** 34/ **Show Boat** 36/ **High, Wide And Handsome** 37/ *The Joy Of Living* 38/ *Love Affair* 39. Retired in the fifties.

Durante, Jimmy. Bn 19 February 1893. Comedian. A bare description that hardly does justice to a unique performer held in universal affection. Began as a Coney Island pianist, led one of the earliest jazz bands, opened his own club, appeared in vaudeville and on Broadway. At first made little impact in films owing to poor material, but later came into his own in MGM's wartime films. Musicals *Roadhouse Nights* (début) 29/ **Cuban Love Song** 31/ *Blondie Of The Follies, The Phantom President* 32/ *Broadway To Hollywood* 33/ **George White's Scandals,** *Hollywood Party, Palooka, Strictly*

Dynamite, Student Tour 34/ *Carnival* 35/ *Sally Irene And Mary, Start Cheering* 37/ *Little Miss Broadway* 38/ *Melody Ranch* 40/ **Two Girls And A Sailor,** *Music For Millions* 44/ *Two Sisters From Boston* 45/ *This Time For Keeps,* **It Happened In Brooklyn** 46/ *On An Island With You* 47/ *The Milkman* 50/ *Beau James* 57/ **Pepe** 60/ **Billy Rose's Jumbo** 62.

Durbin, Deanna. Bn 4 December 1922, Winnipeg, Canada, as Edna Mae Durbin. Singer-actress. Discovered by producer Joe Pasternak, who co-starred her with Judy Garland in an MGM short, then took her to Universal and created a series of musicals that put teenage sopranos on the map and sparked off a host of imitators, none of whom possessed the Durbin charm and durability. Musicals *Three Smart Girls* (début), **100 Men And A Girl** 36/ *Mad About Music, That Certain Age* 38/ *Three Smart Girls Grow Up, First Love* 39/ *It's A Date, Spring Parade* 40/ *Nice Girl* 41/ *It Started With Eve* 42/ *Hers To Hold, His Butler's Sister* 43/ *Christmas Holiday* 44/ *Because Of Him,* **Can't Help Singing,** *Lady On A Train* 45/ *I'll Be Yours, Something In The Wind* 47/ *Up In Central Park* 48. Retired to become a French farmer's wife.

Dwan, Allan. Bn 1895, Toronto, Canada. Director-producer-author. Started directing in silents and has worked prolifically if unmemorably. From 1942 produced many of his own films. Musicals *Song And Dance Man* 36/ *Heidi* 37/ *Josette, Rebecca Of Sunnybrook Farm* 38/ *The Three Musketeers* 39/ *Young People* 40/ *Rise And Shine* 41/ *Here We Go Again* 42/ *Around The World* 43/ *Calendar Girl, Northwest Outpost* 47/ *I Dream Of Jeannie* 52.

Ebsen, Buddy. Bn 2 April 1908 as Christian Rudolf Ebsen. Actor now best known as Jed Clampett in TV's 'Beverly Hillbillies'. Former Broadway star who danced in musicals of the thirties after 'Ziegfeld Follies' success. Musicals *Captain January,* **The Broadway Melody Of 1936** 35/ *Banjo On My Knee,* **Born To Dance** 36/ **The Broadway Melody Of 1938** 37/ *The Girl Of The Golden West, My Lucky Star* 38/ *They Met In Argentina* 41/ *Sing Your Worries Away* 42/ **Red Garters** 54/ *The One And Only Genuine Original Family Band* 68.

Eddy, Nelson. Bn 29 June 1901, Providence, Rhode Island. Singer, whose operetta-styled films with Jeanette MacDonald were hits of the thirties, but who contributed little to films when the novelty wore off and fashions changed. Musicals **Dancing Lady** (début), *Broadway To Hollywood* 33/ *Student Tour* 34/ **Naughty Marietta, Rose Marie** 35/ *Maytime,* **Rosalie** 37/ *The Girl Of The Golden West, Sweethearts* 38/ *Let Freedom Ring, Balalaika* 39/ *Bitter Sweet, New*

Buddy Ebsen with Judy Garland/The Broadway Melody of 1938

Moon 40 / *The Chocolate Soldier* 41 / *I Married An Angel* 42 / *The Phantom Of The Opera* 43 / *Knickerbocker Holiday* 44 / **Make Mine Music** (soundtrack) 46 / *Northwest Outpost* 47. Later developed a cabaret act and worked steadily until his death on 6 March 1967.

Edens, Roger. Bn 9 November 1905, Hillsboro', Texas. Composer-lyricist-arranger-producer. Since 1933 has been MGM's 'man about music', functioning in all capacities as right-hand man to producer Arthur Freed, later turning producer himself. Musical associate on the following, also wrote songs and special material for films marked * *The Broadway Melody Of 1936* 35 / **Born To Dance** 36 / *A Day At The Races,* **The Broadway Melody Of 1938*** 37 / *Everybody Sing, Listen Darling, Love Finds Andy Hardy* 38 / *At The Circus,* Ice Follies Of 1939, **Babes In Arms*** 39 / *The Broadway Melody Of 1940,* **Strike Up The Band*,** *Andy Hardy Meets Debutante, Go West*,* *Little Nellie Kelly*, Two Girls On Broadway* 40 / **Lady Be Good*, Babes On Broadway*,** **Ziegfeld Girl*** 41 / **For Me And My Gal,** **Cabin In The Sky,** *Panama Hattie*, Presenting Lily Mars** 42 / **Thousands Cheer*, Dubarry Was A Lady*, Girl Crazy** 43 / **The Ziegfeld Follies*, Meet Me In St Louis** 44 / **The Harvey Girls** (ass. prod) 45 / **Good News*** 47 / **Easter Parade, Take Me Out To The Ball Game*** 48 / **Annie Get Your Gun, On The Town*** 49 / *Pagan Love Song** 50 / **The Belle Of New York*** 52 / **The Band Wagon** (ass. prod) 53 / **Deep In My Heart** (prod) 54 / **Funny Face*** (prod) 56 / **Billy Rose's Jumbo*** (ass. prod) 62 / **The Unsinkable Molly Brown** (prod) 64 / **Hello Dolly!** (prod) 69. Died July 1970.

Edwards, Blake. Bn 1922. Director-producer-screenwriter, generally in comedy or crime. Occasional musicals *All Ashore, Rainbow Round My Shoulder* 52 / *Cruisin' Down The River* 53 / *Bring Your Smile Along, My Sister Eileen* 55 / *He Laughed Last* 56 / *High Time* 60 / *Darling Lili* 69.

Edwards, Cliff. Bn 1895, Hannibal, Mo. Singer-comedian from vaudeville, clubs, radio and records (known as 'Ukelele Ike') who had featured roles in early musicals **Hollywood Revue** (début), *Lord Byron Of Broadway, Marianne, So This Is College* 29 / *Dance Fools Dance, Good News, Montana Moon* 30 / *Stepping Out* 31 / **Take A Chance** 33 / **George White's Scandals** 34 / *George White's Scandals Of 1935* 35 / *The Girl Of The Golden West* 38 / *Pinocchio* (soundtrack) 40 / *Dumbo* (soundtrack) 41 / *Salute For Three* 43 / *Fun And Fancy Free* (soundtrack) 46. Later in bit parts.

Ellington, Duke (Edward Kennedy). Bn 29 April 1899, Washington, D.C. Negro band-

leader-composer, the greatest in the history of jazz. Also a good showman, he presented his band to good effect in *Check And Double Check* 30 / *The Belle Of The Nineties, Many Happy Returns, Murder At The Vanities* 34 / *Hit Parade Of 1937* 36 / *A Day At The Races* 37 / **Cabin In The Sky** 42 / *Reveille With Beverley* 43. Composed, and his orchestra played, soundtrack music for *Paris Blues* 61.

Enright, Ray. Bn 1896, Anderson, Indiana. Director. Began in silents with Mack Sennett; later joined Warners and directed many musicals with Dick Powell, Ruby Keeler *et al. Dancing Sweeties, The Golden Dawn, Song Of The West* 30 / **Dames, Twenty Million Sweethearts** 34 / *We're In The Money* 35 / **Sing Me A Love Song** 36 / *The Singing Marine,* **Ready, Willing And Able** 37 / *Swing Your Lady, Hard To Get, Going Places,* **The Golddiggers In Paris** 38 / **On Your Toes, Naughty But Nice** 39. Later did Westerns. Died 1965.

Etting, Ruth. Bn David City, Nebraska. Broadway and radio singer of the twenties whose career was documented in **Love Me Or Leave Me** 55, with Doris Day as Etting. Did little on screen apart from **Roman Scandals** (début), *Mr Broadway* 33 / *Gift Of Gab, Hips Hips Hooray* 34. Retired to the mid-West in 1938, returned to radio briefly in 1947.

Fain, Sammy. Bn 17 June 1902. Composer of many film themes and songs, mainly with lyrics by Irving Kahal (29–35), Ralph Freed (43–5) and Paul Francis Webster (53–8), for *It's A Great Life* 29 / *Young Man Of Manhattan, The Big Pond, Follow The Leader, Dangerous Nan McGrew* 30 / *Crooner* 32 / **Footlight Parade,** *Moonlight And Pretzels, College Coach* 33 / *Strictly Dynamite, Harold Teen,* **Dames,** *Fashions Of 1934, Happiness Ahead* 34 / *Sweet Music, Goin' To Town* 35 / *New*

Faces Of 1937, Vogues Of 1938 37 | **I Dood It,**
Meet The People, Swing Fever 43 | **Two Girls And
A Sailor,** Lost In A Harem, **Anchors Aweigh** 44 |
Holiday In Mexico, No Leave No Love, This Time
For Keeps, The Thrill Of A Romance, Two Sisters
From Boston, George White's Scandals of 1945, A
Weekend At The Waldorf 45 | Three Daring
Daughters 47 | The Milkman 50 | Call Me Mister,
Alice In Wonderland 51 | **The Jazz Singer** 52 |
Peter Pan, Two Sailors And A Girl, **Calamity
Jane** 53 | Lucky Me, **Young At Heart** 54 | April
Love 57 | Mardi Gras 58 | Sleeping Beauty 59.

Farnon, Robert. Bn 24 July 1917, Toronto,
Canada. Composer-conductor, renowned for his
contributions to British light music. Scored
many British films including the Wilcox-Neagle
series. Musical director of **Where's Charley?** 52 |
Invitation To The Dance 54 | **Gentlemen Marry
Brunettes** 55 | **The Road To Hong Kong** 62.

Faye, Alice. Bn 5 May 1912, New York as
Alice Leppert. Actress-singer, discovered by
Rudy Vallee on stage in 'George White's
Scandals' and taken to Hollywood for film
version. Starred regularly in Fox musicals and
retired at the height of her fame shortly after
marriage to Phil Harris (q.v.). Musicals **George
White's Scandals** (début), She Learned About
Sailors, 365 Nights In Hollywood 34 | Music Is
Magic, George White's Scandals Of 1935, **The King
Of Burlesque, Every Night At Eight** 35 | Sing
Baby Sing, Poor Little Rich Girl, Stowaway 36 |
Sally Irene And Mary, You're A Sweetheart, **On
The Avenue, Wake Up And Live, You Can't
Have Everything,** In Old Chicago 37 | **Alexander's
Ragtime Band** 38 | **Rose Of Washington
Square,** Tail Spin 39 | Little Old New York, **Tin
Pan Alley,** Lillian Russell 40 | The Great
American Broadcast, **A Weekend In Havana,
That Night In Rio** 41 | **The Gang's All Here,**
Hello, Frisco, Hello 43 | Four Jills In A Jeep 44 |
State Fair 62.

Feist, Felix E. Bn 1910. Director, mainly of
shorts and supporting features. Musicals All By
Myself, You're A Lucky Fellow Mr Smith 43 |
Pardon My Rhythm, The Reckless Age (also prod),
This Is The Life 44 | George White's Scandals Of 1945
45. Died 1965.

Felix, Seymour. Bn 23 October 1892, New
York. Dance director, formerly in vaudeville,
who worked on many important musicals as well
as routine assignments, including **Sunny Side
Up** 29 | Hollywood Party, The Cat And The Fiddle,
Kid Millions 34 | The Girl Friend 35 | **The Great
Ziegfeld** 36 | Vogues Of 1938, **On The Avenue**
37 | Alexander's Ragtime Band 38 | **Rose Of
Washington Square** 39 | **Tin Pan Alley** 40 |
Navy Blues 41 | **Yankee Doodle Dandy** 42 | Let's
Face It 43 | Atlantic City, **Cover Girl,** Greenwich

Village 44 | The Dolly Sisters, Three Little Girls In
Blue, Do You Love Me? 46 | **Mother Wore Tights**
47 | When My Baby Smiles At Me, Give My Regards
To Broadway 48 | Oh You Beautiful Doll 49 | Golden
Girl 51 | Down Among The Sheltering Palms 52 |
The 'I Don't Care' Girl 53. Died 1961.

Fields, Dorothy. Bn 15 July 1905, Allenhurst,
N.J. Author-lyricist. Sister of playwrights
Herbert and Joseph Fields, daughter of comic
Lew Fields. Wrote many Broadway shows (books
and lyrics) and wrote regularly for films with
Jimmy McHugh from 1930–5 (q.v. for credits),
also with Jerome Kern* **Roberta*** 34 | I Dream
Too Much*, In Person 35 | **Swing Time*,** The
King Steps Out, The Smartest Girl In Town 36 | When
You're In Love* 37 | The Joy Of Living* 38 | **One
Night In The Tropics*** 40 | Up In Central Park
48 | Excuse My Dust, Mr Imperium, Texas Carnival
51 | **The Farmer Takes A Wife, Lovely To
Look At*** 52 | **Sweet Charity** 68. Also wrote the
original stage books of Let's Face It 43 | **Annie
Get Your Gun** 49.

Fields, W. C. Bn 10 February 1879, Philadelphia as William Claude Dukinfield. Legendary
comedian whose own films were more impressive
than the musicals in which he was sometimes the
only asset. Stole scenes in Her Majesty Love 31 |
International House 33 | **Mississippi** 35 | Poppy 36 |
The Big Broadcast Of 1938 37 | Follow The
Boys, Sensations Of 1945, The Song Of The Open
Road 44. Died 24 December 1946.

Fine, Sylvia. Bn 29 August, New York.
Author-lyricist. Writes all special material and
songs for husband Danny Kaye, sometimes
taking a hand in stories and production. Probably
associated with all Kaye films, but received
credits for **Up In Arms** 44 | **The Kid From
Brooklyn** 46 | The Inspector General 49 | On The
Riviera 51 | Knock On Wood 54 | The Court Jester 56 |
The Five Pennies 59.

Fio Rito, Ted. Bn 20 December 1900,
Newark, N.J. Bandleader who made guest
appearances in musicals The Sweetheart Of Sigma
Chi (with Betty Grable as his vocalist) 33 |
Twenty Million Sweethearts 34 | **Every Night
At Eight** 35 | Rhythm Parade 42 | Melody Parade,
Silver Skates 43 | Out Of This World 45.

Fisher, Doris. Bn 2 May 1915, New York.
Composer-lyricist. Daughter of old-time song-
writer Fred Fisher. Former singer and recording
artist who became Columbia contract writer,
working exclusively in partnership with Allan
Roberts (q.v.). Musicals Betty Co-ed, Singin' In The
Corn, Sing While You Dance, The Thrill Of Brazil,
Talk About A Lady 46 | Two Blondes And A
Redhead, Sweet Genevieve, Little Miss Broadway,
Glamour Girl, Down To Earth, Cigarette Girl 47 | I
Surrender Dear, Mary Lou 48 | Holiday In Havana 49.

Fitzgerald, Ella. Bn 25 April 1918, Newport News, Va. Popular singer supreme, a great recording and concert artist who has made only occasional screen appearances. Musicals *Ride 'em Cowboy* 41 / **Pete Kelly's Blues** 55 / *St Louis Blues* 58.

Fleming, Victor. Bn February 1883, Pasadena, Calif. Director who started in very early silents as cameraman; later successful MGM and freelance director. Only musicals *Reckless* 35 / *The Great Waltz* 38 / **The Wizard Of Oz** 39. Died 1949.

Foran, Dick. Bn 18 June 1910, Flemington, N.J. Actor-singer, often typed as cowboy or cop. A consistent second lead in musicals *Lottery Lover, Stand Up And Cheer* 34 / *Shipmates Forever* 35 / *A Cowboy From Brooklyn* 38 / *In The Navy, Keep 'em Flying, Ride 'em Cowboy* 41 / *Private Buckaroo, Behind The Eight Ball* 42 / *Hi Buddy* 43.

Forbstein, Leo F. Musical director of Warner Brothers, receiving credit on all films (although in fact people like Max Steiner and Ray Heindorf actually did the donkey work) from the first sound films till his death in 1948.

Fosse, Robert (Bob). Bn 23 June 1927, Chicago. Dancer-choreographer-director. Broadway star who came to the screen as a dancer, and has since succeeded Donen as creator of energetic dance routines. Musicals (as actor-dancer (a), choreographer (c), director (d)) **Kiss Me Kate** (a–c), *The Affairs Of Dobie Gillis* (a), **Give A Girl A Break** (a) 53 / *My Sister Eileen* (a–c) 55 / **The Pajama Game** (c) 57 / **Damn Yankees** (a–c) 58 / **How To Succeed In Business Without Really Trying** (c) 66 / **Sweet Charity** (c–d) 68.

Foster, Susanna. Bn 6 December 1924, Chicago, as Suzan Larsen. One of the many teenage sopranos who followed in the Durbin wake. After progressively unimportant films, married singer Wilbur Evans and retired. Musicals *The Great Victor Herbert* (début), *The Star Maker* 39 / *There's Magic In Music, Glamour Boy* 41 / *Top Man, The Phantom Of The Opera* 43 / *Bowery To Broadway, The Climax, Follow The Boys, This Is The Life* 44 / *That Night With You, Frisco Sal* 45.

Foy, Eddie, Jnr. Bn 1908. Comedian. The only one of the Seven Little Foys to emulate his famous father, although his many screen roles have hardly been stellar. A good secondary player and comedy relief in musicals *Leathernecking* (début), *The Queen Of The Nightclubs* 30 / *Broadway Thru' A Keyhole* 33 / *Scatterbrain, Lillian Russell* 40 / *Puddin' Head, Rookies On Parade, Four Jacks And A Jill* 41 / **Yankee Doodle Dandy** (as his father), *Joan Of Ozark,*

Yokel Boy, Moonlight Masquerade 42 / *Dixie* 43 / **And The Angels Sing** 44 / *Honey Chile* 51 / **The Farmer Takes A Wife** 52 / *Lucky Me* 54 / **The Pajama Game** 57 / **Bells Are Ringing** 60.

Francis, Connie. Bn 12 December 1938, Newark, N.J., as Constance Franconero. Singer from the rock and roll era who developed into a 'standard' performer and business woman. First heard dubbing songs for Tuesday Weld in *Rock, Rock, Rock* 55, went on to appear in teenage musicals, *Jamboree* 57 / *Where The Boys Are* 60 / *Follow The Boys* 63 / *Looking For Love* 64 / *Girl Crazy* 66.

Frank, Melvin. Bn *c.* 1917. Director-producer-screenwriter, in partnership with Norman Panama (*q.v.* for credits) for twenty-five years. Additional credit *A Funny Thing Happened On The Way To The Forum* 67, as producer-writer.

Frazee, Jane. Bn 18 July 1919, Duluth, Minn., as Mary Jane Frahse. Singer-actress of stage, cabaret and radio who became the Queen of Universal 'B' musicals during the war years *Melody And Moonlight* 40 / *Moonlight In Hawaii, San Antonio Rose, Sing Another Chorus, Don't Get Personal, Buck Privates, Angels With Broken Wings, Almost Married* 41 / *Get Hep To Love, Moonlight In Havana, Hellzapoppin', Moonlight Masquerade, What's Cookin'?, When Johnny Comes Marching Home* 42 / *Hi' Ya Chum, Rhythm Of The Islands* 43 / *Rosie The Riveter, She's A Sweetheart, Kansas City Kitty, Beautiful But Broke, Swing In The Saddle* 44 / *Swingin' On A Rainbow* 45 / *Calendar Girl* 47 / *Rhythm Inn* 51.

Freed, Arthur. Bn 9 September 1894, Charleston, S. Carolina. Producer-lyricist. President of Academy of Motion Picture Arts and Sciences. Came to Hollywood to write score for **The Broadway Melody** in 1929 and stayed at MGM permanently. The most prolific, and best, producer of musicals, he succeeded by surrounding himself with the best talent, both technical and artistic, and allowing everyone to operate on their own level with minimum front office interference. If he did not discover them, at least he encouraged such as Gene Kelly, Stanley Donen, Vincente Minnelli, André Previn, Michael Kidd *et al* to put their best work into his films. Produced **Babes In Arms, The Wizard Of Oz** 39 / *Andy Hardy Meets Debutante,* **Strike Up The Band,** *Little Nellie Kelly* 40 / **Babes On Broadway, Lady Be Good** 41 / **Cabin In The Sky, For Me And My Gal,** *Born To Sing, Cairo, Panama Hattie* 42 / **Dubarry Was A Lady, Girl Crazy,** *Best Foot Forward* 43 / **Meet Me In St Louis, The Ziegfeld Follies** 44 / **Yolanda And The Thief, The Harvey Girls** 45 / **Summer Holiday, Till The Clouds Roll By** 46 / *Good News,* **The Pirate** 47 / **The Barkleys Of Broadway, Easter Parade, Take Me Out To The Ball Game, Words And Music** 48 / **Annie Get Your Gun, On The Town** 49 / *Pagan Love Song* 50 / **An American In Paris, Show Boat, Royal Wedding** 51 / **The Belle Of New York, Lovely To Look At, Singin' In The Rain** 52 / **The Band Wagon** 53 / **Brigadoon, Kismet, It's Always Fair Weather** 55 / **Silk Stockings** 56 / **Gigi** 58 / **Bells Are Ringing** 60. As a song-writer, Freed's partnership with composer Nacio Herb Brown (*q.v.*) gave the world some of its greatest songs, constantly revived in MGM films of later years, and Freed's songs (with Brown and other writers) were heard in **The Broadway Melody, Hollywood Revue,** *Lord Byron Of Broadway, Marianne* 29 / *Good News, Montana Moon, Pagan Love Song* 30 / **The Big Broadcast,** *Blondie Of The Follies* 32 / *Dancing Lady, Going Hollywood* 33 / *Student Tour, Hollywood Party* 34 / **The Broadway Melody Of 1936** / *A Night At The Opera* 35 / *San Francisco* 36 / **The Broadway Melody Of 1938** / *Thoroughbreds Don't Cry* 38 / **Babes In Arms,** *Ice Follies Of 1939* 39 / *Little Nellie Kelly, Two Girls On Broadway, Andy Hardy Meets Debutante,* **Strike Up The Band** 40 / **Lady Be Good** 41 / *Born To Sing* 42 / **The Ziegfeld Follies** 44 / **Yolanda And The Thief** 45 / **Singin' In The Rain** 52.

Freed, Ralph. Bn 1 May 1907, Vancouver. Composer-lyricist. Wrote songs for pre-war Paramount and Universal films before joining brother Arthur Freed at MGM, where he did some of his best work. Contributed to *Careless Lady* 32 / *College Holiday* 36 / *Hideaway Girl, Swing High Swing Low* 37 / *You And Me, Double or Nothing, The Cocoanut Grove, Her Jungle Love* 38 / *First Love* 39 / *Spring Parade, A Little Bit Of Heaven, It's A Date* 40 / **Ziegfeld Girl, Babes On Broadway** 41 / *Presenting Lily Mars,*

Seven Sweethearts 42 / **Dubarry Was A Lady, Thousands Cheer, I Dood It,** *Meet The People, Swing Fever* 43 / **Anchors Aweigh,** *Lost In A Harem,* **Two Girls And A Sailor** 44 / *Holiday In Mexico, The Thrill Of A Romance, Two Sisters From Boston, No Leave No Love* 45 / *This Time For Keeps* 46. Later a TV producer.

Freeland, Thornton. Bn 10 February 1898, Hope, N. Dakota. Director. Ex-actor in stock companies who came to Hollywood 1916. Worked his way up to director from assistant cameraman with Griffith, Lubitsch and others. Output mainly lightweight, including musicals **Whoopee,** *Be Yourself* 30 / **Flying Down To Rio** 33 / *George White's Scandals* 34 / *Too Many Blondes* 41.

Friml, Rudolph. Bn 7 December 1879, Prague. Composer who studied with Dvorak. Wrote his first operetta (*The Firefly*) in 1912. Several were filmed, and he also wrote specially for films **The Vagabond King** 30 and 56 / *Lottery Bride* 30 / **Rose Marie** 35 and 54 / *Music For Madame, The Firefly* 37 / *Northwest Outpost* 47.

Froman, Jane. Bn *c.* 1910, St Louis, Missouri, as Jane Frohman. Singer (ex-Paul Whiteman and 'Ziegfeld Follies') who appeared in **Stars Over Broadway** 35 / *Radio City Revels* 37. Severely injured in 1943 plane crash, she went on to sing to US forces in the war zones, and was later rewarded with a Hollywood biopic, **With A Song In My Heart** 52, in which Susan Hayward mimed to the Froman voice.

Funicello, Annette. Bn 22 October 1942, Ulica, N.Y. Disney starlet, formerly known as Annette, who later found a niche in teen-appeal musicals of the sixties *Babes In Toyland* 61 / *Beach Party, Bikini Beach* 64 / *Fireball 500* 66.

Garland, Judy. Bn 10 June 1922, Grand Rapids, Minn., as Frances Gumm. One of the few child stars to move naturally into adult stardom, doing so via a series of superb musicals which showed every facet of a marvellous talent *Pigskin Parade* (début) 36 / *Thoroughbreds Don't Cry,* **The Broadway Melody Of 1938** 37 / *Everybody Sing, Listen Darling, Love Finds Andy Hardy* 38 / **Babes In Arms, The Wizard Of Oz** 39 / **Strike Up The Band,** *Andy Hardy Meets Debutante, Little Nellie Kelly* 40 / **Babes On Broadway, Ziegfeld Girl** 41 / *Presenting Lily Mars, For Me And My Gal* 42 / **Girl Crazy, Thousands Cheer** 43 / **Meet Me In St Louis, The Ziegfeld Follies** 44 / **The Harvey Girls** 45 / **Till The Clouds Roll By** 46 / **The Pirate** 47 / **Easter Parade, Words And Music** 48 / *In The Good Old Summertime* 49 / **Summer Stock** 50 / **A Star Is Born** 54 / *Pepe* (soundtrack) 60 / **Gay Purr-ee** (soundtrack) 62 / *I Could Go On Singing* 63. Toured extensively for many years but not always with happy results, owing to personal problems, and died in London 22 June 1969.

Garrett, Betty. Bn 23 May 1919, St Joseph, Missouri. Actress-singer who went from night-clubs via Broadway to MGM. Musicals *Big City* (début), *Words And Music, Take Me Out To The Ball Game* 48 / *Neptune's Daughter, On The Town* 49 / *My Sister Eileen* 55. Later did live appearances with husband Larry Parks (*q.v.*).

Gaxton, William. Bn 1894. Veteran comedy actor, mostly in Broadway shows. Appeared in musicals *Best Foot Forward, The Heat's On, Something To Shout About* 43 / *Billy Rose's Diamond Horseshoe* 45. Died 1963.

Gaynor, Mitzi. Bn 4 September 1930, Chicago, as Francesca Mitzi Marlene de Charney von Gerber. Talented song-and-dance girl who brought a breath of fresh air to musicals of the fifties *My Blue Heaven* (début) 50 / *Golden Girl* 51 / *Down Among The Sheltering Palms* 52 / *The 'I Don't Care' Girl* 53 / *There's No Business Like Show Business* 54 / *Anything Goes* 56 / *The Joker Is Wild, Les Girls* 57 / *South Pacific* 58.

Gershenson, Joseph. Bn 1904, Russia. Conductor. After long experience in films joined Universal music department, and received musical director credit on most of the studio's films from 1952 to date. Musicals *Has Anybody Seen My Gal, Here Come The Nelsons, Meet Danny Wilson, Meet Me At The Fair* 52 / *So This Is Paris, The Glenn Miller Story* 54 / *The Benny Goodman Story* 55 / *The Big Beat, Rock Pretty Baby* 57 / *I'd Rather Be Rich* 64 / *Thoroughly Modern Millie* 67.

Gershwin, George. Bn 26 September 1898, Brooklyn. (Family name originally Gershovitz, then Gershvin.) Composer who wrote his first Broadway show at twenty and never looked back. Probably the greatest influence on modern popular music (and on film writers – most 'big city' theme music stems from Gershwin) who in the thirty-nine years of his life contributed more to American music than most others did in a lifetime. Film scores, all with lyrics by brother Ira (*q.v.*) *Delicious* 31 / *Girl Crazy* 32 / *Shall We Dance?, A Damsel In Distress* 37 / *The Goldwyn Follies* 38. While working on the latter suffered from a brain tumour, and died 11 June 1937, the score being completed by Vernon Duke and Kurt Weill. Later films with Gershwin scores or songs were *Lady Be Good* 41 / *Girl Crazy* 43 / *The Shocking Miss Pilgrim* 47 / *The Barkleys Of Broadway* 48 / *An American In Paris* 51 / *Funny Face* 56 / *Porgy And Bess* 59 / *Kiss Me Stupid* 65 / *Girl Crazy* 66, and a marvellously inaccurate but equally entertaining biopic *Rhapsody In Blue* 45.

Gershwin, Ira. Bn 6 December 1896, Brooklyn. Lyricist. Brother of George Gershwin, with whom he worked exclusively until George's death in 1937. He then wrote with composers like Kurt Weill, Jerome Kern, Harold Arlen and Harry Warren in discriminatingly chosen screen musicals *The Goldwyn Follies* 38 / *Lady In The Dark* 43 / *Cover Girl* 44 / *Where Do We Go From Here?* 45 / *The Barkleys Of Broadway* 48 / *Give A Girl A Break* 53 / *The Country Girl, A Star Is Born* 54.

Gibson, Virginia. Bn 9 April, St Louis, Missouri. Singer-actress in supporting roles. Musicals *Tea For Two* 50 / *Painting The Clouds With Sunshine, Starlift* 51 / *She's Back On Broadway, About Face* 52 / *Seven Brides For Seven Brothers, Athena* 54.

Gingold, Hermione. Bn 9 December 1897, London. Veteran comedienne of British films and intimate revue who suddenly emerged in uncharacteristic but effective roles in musicals *Gigi* 58 / *Gay Purr-ee, The Music Man* 62 / *I'd Rather Be Rich* 64.

Goldwyn, Samuel. Bn 22 August 1884, Warsaw, Poland, as Samuel Goldfish. Producer. Founded the Lasky Company in 1913, and three years later merged with Famous Players. He then formed the Goldwyn Pictures Corporation with Archie and Edgar Selwyn, and sold his interest to Metro in 1922 (the start of the Metro-Goldwyn-Mayer empire). Later Goldwyn productions were distributed by United Artists and RKO Radio. Probably the most com-mercially (and often artistically) successful producer in Hollywood, Goldwyn always operated on a large scale, in musicals no less than in the dramatic and spectacular field, and was responsible for the initial screen successes of Eddie Cantor and Danny Kaye. Musicals *One Heavenly Night, Whoopee* 30 / *Palmy Days* 31 / *The Kid From Spain* 32 / *Roman Scandals* 33 / *Kid Millions* 34 / *Strike Me Pink* 35 / *The Goldwyn Follies* 38 / *Up In Arms* 44 / *Wonder Man* 45 / *The Kid From Brooklyn* 46 / *A Song Is Born* 48 / *Hans Christian Andersen* 52 / *Guys And Dolls* 55 / *Porgy And Bess* 59.

Goodman, Benny. Bn 30 May 1909, Chicago. Clarinet-playing bandleader known in the thirties as 'The King Of Swing'. Appeared with his band in several musicals, and played on the soundtrack of his biopic, *The Benny Goodman Story* 55 (Steve Allen played Goodman). Musicals *The Big Broadcast Of 1937* 36 / *Hollywood Hotel* 37 / *The Powers Girl, Syncopation* 42 / *The Gang's All Here, Stage Door Canteen* 43 / *Sweet And Lowdown* 44 / *Make Mine Music* (soundtrack) 46 / *A Song Is Born* 48.

Goodwins, Leslie. Bn 1899, England. Director of 'B' features for RKO, Universal and

independents. Musicals *With Love And Kisses* 36 /
Swing While You're Able 37 / *The Girl From
Mexico* 39 / *Let's Make Music* 40 / *They Met In
Argentina* 41 / *Gals Incorporated, Silver Skates* 43 /
*Casanova In Burlesque, Hi Beautiful, Murder In The
Blue Room, The Singing Sheriff* 44 / *Radio Stars On
Parade, I'll Tell The World* 45 / *Vacation In Reno,
Riverboat Rhythm* 46.

Gordon, Mack. Bn 21 June 1904, Warsaw,
Poland. Lyricist responsible for some of the best
songs ever heard in films, most of his work
being done for Fox in collaboration with Harry
Revel (33–9), Harry Warren (40–4), James V.
Monaco (44–6) and Josef Myrow (46–56) (*q.v.*
for credits). Other musicals *Song Of Love* 29 /
Swing High 30 / *Lillian Russell, Little Old New
York* 40. Died 1 March 1959.

Gorney, Jay. Bn 12 December 1896,
Bialystock, Russia. Composer-author-producer.
Practised as a lawyer in Detroit before going into
films, then wrote songs for *Roadhouse Nights,
Applause,* **Glorifying The American Girl** 29 /
The Battle Of Paris 31 / *Jimmy And Sally, Moonlight
And Pretzels* 33 / *Lottery Lover, Romance In The
Rain, Stand Up And Cheer, Marie Galante* 34 /
Redheads On Parade, Spring Tonic 35 / *Romance In
The Dark* 38 / *The Heat's On* 43 / *Hey Rookie* 44.
Produced *The Gay Senorita* 45.

Gould, Dave. Bn 1905, Hungary. Dance
director *Melody Cruise,* **Flying Down To Rio**
33 / *The Gay Divorcee, Hollywood Party, Hips
Hips Hooray* 34 / **The Broadway Melody Of
1936** / *Folies Bergère* 35 / **Born To Dance** 36 /
The Broadway Melody Of 1938, *A Day at the
Races* 37 / *Breaking The Ice* 38 / *Everything's On Ice*
39 / **The Boys From Syracuse** 40 / *Rhythm Parade*
(dir) 42 / *My Best Gal* 43 / *Rosie The Riveter, Lady
Let's Dance, Casanova In Burlesque* 44.

Goulet, Robert. Bn 26 November 1933,
Lawrence, Mass. Robust Broadway baritone,
who, despite good looks and virile personality,
has not yet made any impact in films. Provided
soundtrack singing in the cartoon **Gay Purr-ee**
62, and little in *I'd Rather Be Rich* 64.

Grable, Betty. Bn 18 December 1916, St Louis.
Blonde dancing star who, during the war years,
kept the Fox front office happy with her films
and the serving men delirious with her pin-up
pictures. Came into films at fourteen and is still
active. Musicals *Let's Go Places* (début),
New Fox Follies of 1931, Happy Days,
Whoopee 30 / **Palmy Days** 31 / **The Kid From
Spain** 32 / *The Sweetheart Of Sigma Chi,* **Roman
Scandals** 33 / *Student Tour, Hips Hips Hooray,*
The Gay Divorcee 34 / *Old Man Rhythm,
Nitwits,* **Follow The Fleet** 35 / *Collegiate,
Pigskin Parade* 36 / *This Way Please, The Thrill Of A
Lifetime* 37 / *College Swing, Give Me A Sailor* 38 /
Man About Town 39 / **Down Argentine Way,
Tin Pan Alley** 40 / **Moon Over Miami** 41 /
Footlight Serenade, Song Of The Islands, **Springtime
In The Rockies** 42 / *Sweet Rosie O'Grady, Coney
Island* 43 / *Four Jills In A Jeep, Pin-Up Girl* 44 /
Billy Rose's Diamond Horseshoe 45 / *The
Dolly Sisters* 46 / **Mother Wore Tights,** *The
Shocking Miss Pilgrim* 47 / *That Lady In Ermine,
When My Baby Smiles At Me* 48 / *The Beautiful
Blonde From Bashful Bend* 49 / *Wabash Avenue, My
Blue Heaven* 50 / *Meet Me After The Show, Call Me
Mister* 51 / **The Farmer Takes A Wife** 52 /
Three For The Show 55.

Grant, Kirby. Bn 24 November 1914, Butte,
Montana, as Kirby Grant Horn. Actor, with a
pleasant singing voice, who made a living out of
Westerns and minor musicals *Blondie Goes Latin*
41 / *Babes On Swing Street* 43 / *Ghost Catchers,
Hi Good Lookin', In Society* 44 / *Penthouse
Rhythm, Easy To Look At* 45 / *Coming Round
The Mountain, Rhythm Inn* 51. Subsequently
worked variously in TV, circus, cabaret and
stock companies.

Gray, Alexander. Leading man of early
musicals *Sally* 29 / *No, No, Nanette, The Song Of
The Flame, Spring Is Here, Viennese Nights* 30.

Gray, Dolores. Bn 7 June 1926, Chicago.
Singer-actress, formerly on radio with Rudy
Vallee and Milton Berle. Took London by storm
in 'Annie Get Your Gun' but had to wait eight
years for a film break. Musicals **It's Always
Fair Weather, Kismet** 55 / *The Opposite Sex* 56.
Returned to stage.

Gray, Lawrence. Bn 28 July 1898, San
Francisco. Leading man of early talkies who also
functioned as supervising producer for Famous
Players-Lasky. Musicals *It's A Great Life,
Marianne* 29 / *Spring Is Here, Sunny* 30 / *The Old
Homestead, Dizzy Dames* 35.

Grayson, Kathryn. Bn 9 February 1922,
Winston-Salem, N. Carolina, as Zelma Kathryn
Elizabeth Hedrick. Singer who starred in many
MGM musicals of the forties and fifties (trilling
the inevitable aria among the popular ballads)

Betty Grable with Jack Lemmon / Three for the Show

but failed to make the sixties. Musicals *Rio Rita, Seven Sweethearts* 42 / *Thousands Cheer* 43 / *Anchors Aweigh, The Ziegfeld Follies* 44 / *Two Sisters From Boston* 45 / *It Happened In Brooklyn, Till The Clouds Roll By* 46 / *The Kissing Bandit* 48 / *That Midnight Kiss* 49 / *The Toast Of New Orleans* 50 / *Show Boat* 51 / *The Desert Song, Lovely To Look At* 52 / *So This Is Love, Kiss Me Kate* 53 / *The Vagabond King* 56.

Green, Alfred E. Bn 1889, Perris, Calif. Director who started in silents with the Selig Polyscope Co. (1912). Under contract to Warner/1st National as sound arrived, but later freelanced. Musicals *Sweet Music, Here's To Romance* 35 / *Colleen* 36 / *Mr Dodd Takes The Air, Thoroughbreds Don't Cry* 37 / *The Mayor Of 44th Street* 42 / *The Jolson Story, Tars And Spars* 46 / *The Fabulous Dorseys, Copacabana* 47 / *Two Gals And A Guy* 51 / *The Eddie Cantor Story* 53. Died 1960.

Green, John (Johnny). Bn 10 October 1908, New York. Composer-conductor. With Paramount as arranger (30–3), then radio and recording bandleader till 1940, when he joined MGM's musical staff. Took over as General Music Director 1949 till 1958. Has since conducted major symphony orchestras and produced for TV. Wrote songs for *Dude Ranch* 31 / *Cockeyed Cavaliers* 34 / *Start Cheering* (also appeared with his band) 37 / *Beat The Band* 46 / *Everything I Have Is Yours* 52. Musical director of *The Big Pond, Follow The Leader, Animal Crackers, Heads Up, Queen High* 30 / *Broadway Rhythm* 43 / *Bathing Beauty* 44 / *A Weekend At The Waldorf, Easy To Wed* 45 / *Fiesta, It Happened In Brooklyn* 46 / *Something In The Wind* 47 / *Up In Central Park, Easter Parade* 48 / *The Inspector General* 49 / *Summer Stock, The Toast Of New Orleans* 50 / *An American In Paris, Royal Wedding, The Great Caruso* 51 / *Because You're Mine* 52 / *Lili* 53 / *Invitation To The Dance* 54 / *Brigadoon, I'll Cry Tomorrow* 55 / *High Society, Meet Me In Las Vegas, Silk Stockings* 56 / *Pepe* 60 / *West Side Story* 61 / *Bye Bye Birdie* 63 / *Oliver!* 68.

Greenwood, Charlotte. Bn 25 June 1893, Philadelphia. Leggy comedienne who entered silents 1918 and became a star ten years later in the 'Letty' series. Contributed telling cameos as the comic matron in musicals *So Long Letty, She Couldn't Say No* 30 / *Palmy Days, Flying High, Stepping Out* 31 / *Down Argentine Way, Young People* 40 / *Moon Over Miami, Tall Dark And Handsome* 41 / *Springtime In The Rockies* 42 / *The Gang's All Here* 43 / *Wake Up And Dream* 46 / *Oh You Beautiful Doll* 49 / *Dangerous When Wet* 53 / *Oklahoma!* 55 / *The Opposite Sex* 56.

Guinan, Texas. Night-club owner and performer of the twenties whose gay yet tragic career was portrayed by Betty Hutton in *The Incendiary Blonde* 44. Appeared more or less as herself in *The Queen Of The Nightclubs* 29 / *Broadway Thru' A Keyhole* 33, but died before her screen potential could be realized.

Guizar, Tito. Bn 1909, Mexico. Singer, Studied in Italy, sang opera in Mexico and was discovered and promoted in New York by Texas Guinan. Brought Latin vocal charm to musicals *The Big Broadcast Of 1938* 37 / *Tropic Holiday* 38 / *St Louis Blues* 39 / *Blondie Goes Latin* 41 / *Brazil* 44 / *Mexicana* 45 / *The Thrill Of Brazil* 46.

Haley, Jack. Bn 1902, Boston. Comedy actor who came from vaudeville to screen in Vitaphone shorts and went on to become a useful second lead (and sometimes star) in musicals *Follow Through* 30 / *Sitting Pretty, Mr Broadway* 33 / *Coronado, The Girl Friend, Redheads On Parade, Spring Tonic* 35 / *Pigskin Parade, Poor Little Rich Girl* 36 / *Wake Up And Live, Pick A Star* 37 / *Thanks For Everything, Hold That Co-ed, Alexander's Ragtime Band, Rebecca Of Sunnybrook Farm* 38 / *The Wizard Of Oz* 39 / *Navy Blues, Moon Over Miami* 41 / *Beyond The Blue Horizon* 42 / *Higher And Higher* 43 / *Take It Big* 44 / *Sing Your Way Home, People Are Funny, George White's Scandals Of 1945* 45 / *Vacation In Reno* 46 / *Make Mine Laughs* 49. Seen occasionally on TV.

Hammerstein, Oscar II. Bn 12 July 1895, New York. Lyricist-librettist. One of the great figures of American stage and film music, via his early collaborations with Romberg, Kern and Friml, but best known for the partnership with Richard Rodgers that lasted from 1943 till Hammerstein's death on 23 August 1960. Some films were written for the screen, others

adaptations of Hammerstein's own stage book, from which he often did the screenplay. *The Desert Song* 29, 43, 52 | *The New Moon* 30, 40 | *Sunny* 30, 41 | *The Golden Dawn, Song Of The West, Three Sisters, Viennese Nights* 30 | *Children Of Dreams* 31 | *Rose Marie* 34, 54 | *Music In The Air, The Night Is Young,* **Sweet Adeline** 34 | *Show Boat* 36, 51 | *Give Us This Night* 36 | **High, Wide And Handsome,** *I'll Take Romance* 37 | *The Great Waltz, The Lady Objects* 38 | **One Night In The Tropics** 40 | **State Fair** 45, 62 | **Till The Clouds Roll By, Centennial Summer** 46 | **Carmen Jones** 54 | **Oklahoma!** 55 | *The King And I,* **Carousel** 56 | **South Pacific** 58 | **The Flower Drum Song** 61 | **The Sound Of Music** 65.

Harburg, E. Y. ('Yip'). Bn 8 April 1898, New York. Lyricist-librettist of many Broadway shows and films, working with such composers as Harold Arlen, Jerome Kern and Burton Lane. Wrote individual songs or full scores for *Roadhouse Nights, Applause,* **Glorifying The American Girl** 29 | *Queen High, Follow The Leader* 30 | *Moonlight And Pretzels,* **Take A Chance** 33 | *The Singing Kid,* **Stage Struck** 36 | **The Golddiggers Of 1937** 37 | **At The Circus, Babes In Arms, The Wizard Of Oz** 39 | *Andy Hardy Meets Debutante* 40 | **Babes On Broadway** 41 | *Cairo, Panama Hattie,* **Rio Rita, Cabin In The Sky,** *Presenting Lily Mars, Ship Ahoy* 42 | **Thousands Cheer,** *Meet The People* (also prod), **Dubarry Was A Lady** 43 | *Hollywood Canteen* 44 | **Can't Help Singing** 45 | **California** 46 | **April In Paris** 52 | **Gay Purr-ee,** *I Could Go On Singing* 62 | **Finian's Rainbow** 68.

Harline, Leigh. Bn 26 March 1907, Salt Lake City. Composer-conductor. Ex-radio arranger on Disney music staff 1932–41, then with RKO till 1952, with time out for assignments for Paramount, Columbia, etc. Musical director of *Snow White And The Seven Dwarfs* 37 | *Pinocchio* 40 | *Blondie Goes Latin* 41 | **You Were Never Lovelier** 42 | **The Sky's The Limit** 43 | *Follow The Boys, Music In Manhattan* 44 | *George White's Scandals Of 1945, The Road To Utopia* 45 | *Beat The Band* 46 | *My Friend Irma Goes West* 50 | *Call Me Mister, Double Dynamite, That's My Boy* 51 | *The Las Vegas Story* 52 | *The Wonderful World Of The Brothers Grimm* 62. Died December 1969.

Harling, W. Franke. Bn 18 January 1887, London. Composer who combined classical and religious composition with film scoring. Musical director of **Monte Carlo,** *Honey* (also wrote songs for both) 30 | **One Hour With You** 32 | *You And Me* 38. Died 22 November 1958.

Harris, Phil. Bn 24 June 1906, Linton, Indiana. Bandleader-comedian who came to prominence with Jack Benny. A rumbustious character who comes over better on TV and records than in most of his screen vehicles *Melody Cruise* 33 | *Turn Off The Moon* 37 | *Man About Town* 39 | *Dreaming Out Loud, Buck Benny Rides Again* 40 | *I Love A Bandleader* 46 | *Wabash Avenue* 50 | *Starlift, Here Comes The Groom* 51 | *Anything Goes* 56 | *The Jungle Book* (soundtrack) 67 | *The Aristocats* (soundtrack) 70. Married to Alice Faye since 1940.

Hart, Lorenz. Bn 2 May 1895, New York. Author of some of the wittiest and most sophisticated lyrics in popular music. A qualified linguist, he was translating German plays for the Shubert management when he met Richard Rodgers. They collaborated on college shows and remained together till Hart's death on 22 November 1943. (See Rodgers for credits.)

Haver, June. Bn 10 January 1926, Rock Island, Illinois as June Stovenour. Singer-actress, a talented Fox star of the forties who became disillusioned with movies and entered a convent. Re-emerged, married Fred MacMurray and stayed in retirement. Musicals *Irish Eyes Are Smiling, Sweet And Lowdown* 44 | **Where Do We Go From Here?** 45 | *The Dolly Sisters, Three Little Girls In Blue, Wake Up And Dream* 46 | *I Wonder Who's Kissing Her Now* 47 | **Look For The Silver Lining,** *Oh You Beautiful Doll* 49 | *The Daughter Of Rosie O'Grady, I'll Get By* 50.

Havoc, June. Bn 8 November 1916 as June Hovick. Actress. Sister of Gypsy Rose Lee, on stage at three as 'Baby June', made Broadway début 1936. Mainly a stage performer but appeared in musicals *Four Jacks And A Jill* 41 | *Sing Your Worries Away* 42 | **Hello, Frisco, Hello** 43 | *Casanova In Burlesque* 44 | *When My Baby Smiles At Me* 48 | *Red Hot And Blue* 49.

Hawks, Howard. Bn 30 May 1896, Goshen, Indiana. Producer-director of dramas, comedies

Lennie Hayton with Lena Horne

and adventure stories: entered films at twenty-two as scenarist. His only musicals, *A Song Is Born* 48, and *Gentlemen Prefer Blondes* 53, showed no lessening of his usual quality.

Haydn, Richard. Bn 10 March 1905, London. Actor-director-author, well known as the adenoidal 'Edwin Carp', a role he repeated with variations in many comedies and musicals *Tonight And Every Night* 44 / *The Emperor Waltz* 48 / *Mr Music* (also dir) 50 / *Alice In Wonderland* (soundtrack) 51 / *The Merry Widow* 52 / *Jupiter's Darling* 55 / *The Sound of Music* 65.

Haymes, Dick. Bn 13 September 1918, Buenos Aires of Irish parents. Former stunt man who became a top singer of the forties with Harry James and Tommy Dorsey bands. Starred in Fox musicals until personal problems removed him from the limelight *Dubarry Was A Lady* (as Dorsey band vocalist) 43 / *Four Jills In A Jeep, Irish Eyes Are Smiling* 44 / *Billy Rose's Diamond Horseshoe, State Fair* 45 / *Do You Love Me* 46 / *Carnival In Costa Rica, The Shocking Miss Pilgrim* 47 / *One Touch Of Venus, Up In Central Park* 48 / *All Ashore* 52 / *Cruisin' Down The River* 53. Now lives in London, writing and occasionally singing.

Hayton, Lennie. Bn 13 February 1908, New York. Conductor-arranger. Former Paul Whiteman pianist-arranger, with MGM 1940–53. Musical director of some of the great musicals *Born To Sing* 42 / *Best Foot Forward, Meet The People* 43 / *The Ziegfeld Follies* 44 / *The Harvey Girls, Yolanda And The Thief* 45 / *Living In A Big Way, Summer Holiday, Till The Clouds Roll By* 46 / *Good News, The Pirate* 47 / *The Barkleys Of Broadway, Words And Music* 48 / *On The Town* 49 / *Singin' In The Rain* 52 / *Easy To Love* 53 / *Star!* 68 / *Hello Dolly!* 69. Recently mainly musical director for his wife Lena Horne. Died 24 April 1971.

Hayworth, Rita. Bn 17 October 1918, New York, as Margarita Carmen Cansino. Dancer-actress whose red hair and lithe dancing made her a star and pin-up girl of World War II. Later made more headlines for her private life, but is still fondly remembered for some lively musicals (always with someone else's singing voice) *Paddy O'Day* (made under her real name) 35 / *Music In My Heart* 40 / *You'll Never Get Rich, The Strawberry Blonde* 41 / *You Were Never Lovelier, My Gal Sal* 42 / *Cover Girl, Tonight And Every Night* 44 / *Down To Earth* 47 / *Pal Joey* 57.

Heindorf, Ray. Bn 25 August 1908, Haverstraw, N.Y. Composer-conductor with Warners since 1935, first as composer-arranger under music chief Leo F. Forbstein (*q.v.*). Became chief musical director on Forbstein's death in 1948

and has received musical director credit on most Warner films since then.

Henie, Sonja. Bn 8 April 1913, Oslo, Norway. Olympic skating champion who was signed to a Fox contract by Darryl F. Zanuck and made a series of enjoyable if undemanding musicals over the next decade *One In A Million, Thin Ice* 37 / *My Lucky Star, Happy Landing* 38 / *Second Fiddle* 39 / *Sun Valley Serenade* 41 / *Iceland* 42 / *Wintertime* 43 / *It's A Pleasure* 45 / *The Countess Of Monte Cristo* 46. Retired from the screen, but promoted and starred in ice spectaculars, becoming allegedly a millionairess. Died of leukaemia 19 October 1969.

Herbert, Victor. Bn 1 February 1859, Dublin. Composer who was probably the father of the American musical as we know it. Symphony orchestra 'cellist, led his own orchestra prior to writing many successful operettas, some of which were filmed *Babes In Toyland* 34, 61 / *Naughty Marietta* 35 / *Sweethearts* 38. Walter Connolly played the composer in his biopic, *The Great Victor Herbert* 39. Died 26 May 1924.

Herman, Jerry. Bn 10 July 1933, New York. Composer-lyricist. One of the new generation of Broadway show writers. Only film to date *Hello Dolly!* 69.

Herman, Woody (Woodrow Wilson). Bn 16 May 1913, Milwaukee. Swing bandleader (started out in 1931 in a band with Ginny Simms and Tony Martin) who made guest appearances in musicals *What's Cookin'?* 42 / *Wintertime* 43 / *Sensations Of 1945* 44 / *Earl Carroll's Vanities* 45 / *Hit Parade Of 1947* 46 / *New Orleans* 47.

Hilliard, Harriett. Bn 1914. Singer-actress. Made a good start in films, but after marriage to bandleader Ozzie Nelson (*q.v.*) concentrated on appearing with his band, which involved a number of second-string musicals *Follow The*

Fleet 35 / *The Life Of The Party, New Faces Of 1937* 37 / *The Cocoanut Grove* 38 / *Sweetheart Of The Campus* 41 / *Juke Box Jennie, Strictly In The Groove* 42 / *Swingtime Johnny, Gals Incorporated, Hi Buddy, Honeymoon Lodge* 43 / *Hi Good Lookin', Take It Big* 44 / *Here Come The Nelsons* 52. Latterly on TV.

Hollander, Frederick. Bn 18 October 1896, London. Composer-conductor. Associate conductor of the Prague Opera at eighteen, later worked with Max Reinhardt in Berlin. First noted for scoring of *The Blue Angel* and brought to Hollywood. With Paramount (35–40), Warner (40–8) and RKO (49–50). Wrote scores and/or songs for musicals *I Am Suzanne* 34 / *It's A Great Life, Poppy, Rhythm On The Range,* **Anything Goes, 100 Men And A Girl** 36 / *Artists And Models, Champagne Waltz, This Way Please, The Thrill Of A Lifetime* 37 / *You And Me, The Cocoanut Grove, Her Jungle Love* 38 / *Man About Town* 39 / *A Night At Earl Carroll's, Moon Over Burma* 40 / *There's Magic In Music* 41 / *Cinderella Jones, Two Guys From Milwaukee,* **The Time, The Place And The Girl** 46 / *That Lady In Ermine* 48.

Holloway, Sterling. Bn 1905. Rusty-voiced, straw-haired comic, essentially a featured player but often seen in musicals *International House* 33 / *Lottery Lover,* **The Merry Widow,** *Strictly Dynamite, Down To Their Last Yacht* 34 / *Palm Springs* 36 / **The Varsity Show** 37 / *Doctor Rhythm* 38 / *The Bluebird* 39 / **Star Spangled Rhythm** 42 / *The Beautiful Blonde From Bashful Bend* 49. Does many Disney records and soundtracks (he's the voice of Winnie The Pooh) including **Make Mine Music** 46 / *Alice In Wonderland* 51 / *Jungle Book* 67 / *The Aristocats* 70.

Holm, Celeste. Bn 29 April 1919, New York. Actress-singer. Cool blonde, self-reliant and cynical, but usually warm-hearted under it all.

Though a musical comedy star on Broadway ('Oklahoma!', etc.), did surprisingly few musicals *Three Little Girls In Blue* 46 / *Carnival In Costa Rica* 47 / **High Society** 56.

Hope, Bob. Bn 29 May 1903, Eltham, England, as Lester Townes Hope. Comedian, one of the earliest and best of the stand-up, wisecracking variety, he carried the technique through dozens of Paramount comedies, often having to transcend his material. Best known for his association with Crosby and Lamour in the 'Road' series. Musicals **The Big Broadcast Of 1938** (début) 37 / *Thanks For The Memory, College Swing, Give Me A Sailor* 38 / *Never Say Die, Some Like It Hot* 39 / **The Road To Singapore** 40 / *Louisiana Purchase, The Road To Zanzibar* 41 / **Star Spangled Rhythm,** *The Road To Morocco* 42 / *Let's Face It* 43 / *The Road To Utopia, Duffy's Tavern* 45 / *Monsieur Beaucaire* 46 / *The Road To Rio, Variety Girl* 47 / *The Paleface* 48 / *Fancy Pants* 50 / *The Road To Bali, Here Come The Girls, Son Of Paleface* 52 / *The Seven Little Foys* 55 / *Beau James* 57 / **The Road To Hong Kong** 62 / *I'll Take Sweden* 65.

Horne, Lena. Bn 30 June 1917, Brooklyn. Singer, who apart from one strong acting role* was a victim of Hollywood limitations on coloured artists. Did guest numbers in elegant poses in musicals *Duke Is Tops* (début) 38 / *Panama Hattie,* **Cabin In The Sky*** 42 / *I Dood It, Swing Fever,* **Broadway Rhythm,** *Stormy Weather, Thousands Cheer* 43 / *Two Girls And A Sailor, The Ziegfeld Follies* 44 / *Till The Clouds Roll By* 46 / **Words And Music** 48 / *The Duchess Of Idaho* 50 / **Meet Me In Las Vegas** 56. Still a top TV and cabaret star.

Horton, Edward Everett. Bn 18 March 1886, Brooklyn. Comedian, alternately dithering, flustered or shocked, a master of the double-take. On stage in Gilbert and Sullivan operas when they were new, he was already well established and middle-aged when co-starring in the Astaire-Rogers films. Musicals *Reaching For The Moon* 31 / *The Way To Love, Bedtime Story* 33 / **The Gay Divorcee,** *The Night Is Young,* **The Merry Widow** 34 / *Going Highbrow, All The King's Horses,* **Top Hat, In Caliente** 35 / *Hearts Divided,* **The Singing Kid** 36 / *The King And The Chorus Girl, Hitting A New High,* **Shall We Dance?** 37 / *College Swing* 38 / *Paris Honeymoon, That's Right You're Wrong* 39 / *Sunny, You're The One, Ziegfeld Girl* 41 / *I Married An Angel,* **Springtime In The Rockies** 42 / **Thank Your Lucky Stars, The Gang's All Here** 43 / *Brazil* 44 / *Lady On A Train* 45 / *Earl Carroll's Sketchbook, Cinderella Jones* 46 / *Down To Earth* 47. Was very active professionally in his eighties until his death in September 1970.

Celeste Holm

Humberstone, Bruce. Bn 1903, Buffalo, N.Y. Director, many years with Fox, and later with Warner, responsible for some colourful and enjoyable musicals *Rascals* 38 / *Pack Up Your Troubles* 39 / *Sun Valley Serenade, Tall Dark And Handsome* 41 / *Iceland* 42 / *Hello, Frisco, Hello* 43 / *Pin-Up Girl* 44 / *Wonder Man* 45 / *Three Little Girls In Blue* 46 / *Happy Go Lovely, She's Working Her Way Through College* 51 / *The Desert Song* 52.

Hunter, Ross. Bn *c.* 1916 as Martin Fuss. Producer-ex-actor. Promised well as a leading man in forties' musicals *Ever Since Venus, Louisiana Hayride, She's A Sweetheart* 44 / *The Sweetheart Of Sigma Chi, Hit The Hay* 46. Illness cut short his career and he went back to teaching. In the mid-fifties came back as producer of mass-market comedies and dramas with Universal, including musicals *The Flower Drum Song* 61 / *I'd Rather Be Rich* 64 / *Thoroughly Modern Millie* 67.

Hutton, Betty. Bn 26 February 1921, Battle Creek, Mich., as Betty Jane Thornberg. Blonde comedienne-singer, formerly Vincent Lopez vocalist. Discovered by B. G. De Sylva in 'Panama Hattie' on Broadway and bounced through many Paramount comedies and musicals of the forties *The Fleet's In* (début), *Happy Go Lucky, Star Spangled Rhythm* 42 / *Let's Face It* 43 / *The Incendiary Blonde, And The Angels Sing* 44 / *Duffy's Tavern, Here Come The Waves* 45 / *The Stork Club, Cross My Heart* 46 / *The Perils Of Pauline* 47 / *Dream Girl* 48 / *Annie Get Your Gun, Let's Dance, Red Hot And Blue* 49 / *Somebody Loves Me* 52.

Hutton, Marion. Bn 1919, Battle Creek, Mich., as Marion Thornberg. Started before sister Betty, as Glenn Miller vocalist, and went on into films but suffered because of the resemblance. Musicals *Sun Valley Serenade* 41 / *Orchestra Wives* 42 / *Babes On Swing Street, Crazy House* 43 / *In Society* 44 / *Love Happy* 48.

Iturbi, Jose. Bn 28 November 1895, Valencia, Spain. Classical pianist. Musicals *Thousands Cheer* 43 / *Music For Millions, Anchors Aweigh, Two Girls And A Sailor* 44 / *Holiday In Mexico, A Song To Remember* (dubbed sound-track for Cornel Wilde as Chopin) 45 / *Three Daring Daughters* 47 / *That Midnight Kiss* 49.

Ives, Burl. Bn 14 June 1909, Huntingdon, Illinois, as Burl Icle Ivanhoe. Actor-singer. Former folk artist ('The Wayfaring Stranger') with great Broadway musicals experience. Many dramatic films but, strangely enough for a singer, only two musicals *So Dear To My Heart* (soundtrack) 49 / *Summer Magic* 63.

James, Harry. Bn 15 March 1916, Albany, Ga. Trumpet-playing bandleader who, by frequent appearances on screen, developed more as an actor and personality than any of his contemporaries. Musicals *Hollywood Hotel* (as member of Benny Goodman band) 37 / *Private Buckaroo, Springtime In The Rockies, Syncopation* 42 / *Swing Fever* (gag appearance), *Best Foot Forward* 43 / *Two Girls And A Sailor, Bathing Beauty* 44 / *Do You Love Me?, If I'm Lucky* 46 / *Young Man With A Horn* (dubbed soundtrack for Kirk Douglas), *I'll Get By, Carnegie Hall* 50 / *The Benny Goodman Story* 55 / *The Opposite Sex* 56 / *The Big Beat* 57 / *Ladies' Man* 61.

James, Inez. Bn 15 December 1919, New York. Composer, with lyrics by Buddy Pepper and Sidney Miller, of many songs for Universal 'B' musicals of the war years *When Johnny Comes Marching Home* 42 / *Top Man, Rhythm Of The Islands, All By Myself, Babes On Swing Street, How's About It, Mr Big, Moonlight In Vermont* 43 / *On Stage Everybody, A Chip Off The Old Block, This Is The Life, The Singing Sheriff, Sing A Jingle, Follow The Boys, Hi Good Lookin'* 44 / *That's The Spirit* 45 / *Are You With It?, Big City* 48.

Jean, Gloria. Bn 14 April 1928, Buffalo, N.Y., as Gloria Jean Schoonover. Teenage soprano nursed through 'B' pictures by Universal possibly as insurance against any defection by Deanna Durbin. The day never came. Musicals *The Underpup* (début) 39 / *If I Had My Way, A Little Bit Of Heaven* 40 / *Get Hep To Love, What's Cookin'?, When Johnny Comes Marching Home* 42 / *Moonlight In Vermont, Mr Big* 43 / *Cinderella Swings It, Follow The Boys, Ghost Catchers, Pardon My Rhythm, The Reckless Age* 44 / *Easy To Look At* 45 / *Copacabana* 47 / *I Surrender Dear* 48 / *Manhattan Angel, There's A Girl In My Heart* 49 / *Ladies' Man* 61.

Jeanmaire. Bn 29 April 1924, Paris, as Renée Jeanmaire (later known as Zizi Jeanmaire). Ballet dancer with company of husband Roland Petit with whom she appeared in *Black Tights* 60. Other musicals *Hans Christian Andersen* 52 / *Anything Goes* 56.

Jeffreys, Anne. Bn 26 January 1923, Golds-borough, N. Carolina. Singer-actress who played in several musicals of the forties *I Married An Angel* (début), *Joan Of Ozark* 42 / *Chatterbox* 43 / *Step Lively* 44 / *Sing Your Way Home* 45 / *Vacation In Reno* 46 / *Melody Maker* 47. Left screen for opera and stage musical comedy, then starred with husband Robert Sterling in TV's 'Topper' and a night-club act.

Jessel, George. Bn 3 April 1898, New York. Comedian-author-producer, who started in a Gus Edwards troupe and became America's best-loved raconteur. Appeared in early talkies with little success and thereafter did guest spots before becoming a Fox producer. Musicals *Lucky Boy* 29 / *Happy Days* 30 / *Stage Door Canteen* 43 / *Four Jills In A Jeep* 44 / *Beau James* 57. Wrote songs for *The Life Of The Party* 37 / *Cinderella Swings It* 44. Produced *Do You Love Me,* *The Dolly Sisters* 46 / *I Wonder Who's Kissing Her Now* 47 / *When My Baby Smiles At Me* 48 / *Dancing In The Dark, Oh You Beautiful Doll* 49 / *Golden Girl, Meet Me After The Show* 51 / *Tonight We Sing, The 'I Don't Care' Girl* (also appeared) 53.

Johnson, Van. Bn 25 August 1916, Newport, Rhode Island. Freckled, eternally youthful actor first discovered singing and dancing in chorus of 'Pal Joey' on Broadway. Has worked steadily for MGM, including musicals *Too Many Girls* (in chorus) 40 / **Two Girls And A Sailor** 44 / *A Weekend At The Waldorf, Easy To Wed, No Leave No Love, The Thrill Of A Romance* 45 / **Till The Clouds Roll By** 46 / *In The Good Old Summertime* 49 / *The Duchess Of Idaho* 50 / *Easy To Love* 53 / **Brigadoon** 55.

Johnston, Arthur. Bn 10 January 1898, New York. Composer. Former musical director for Irving Berlin stage shows, came to Hollywood 1930 to write film scores, usually in collaboration with Sam Coslow* and/or Johnny Burke** *Melody Man* 30 / *College Coach*, College Humour*, Duck Soup* (musical director), *Hello Everybody*, The Way To Love*, Too Much Harmony** 33 / *Many Happy Returns*, The Belle Of The Nineties*, Murder At The Vanities*/** 34 / *The Girl Friend,* **Thanks A Million** 35 / *Pennies From Heaven**, Go West Young Man** 36 / *Double Or Nothing*/**, Song Of The South* 46. Died 1 May 1954.

Johnston, Johnny. Bn 1 December 1915, St Louis, Mo. Singer-actor. Former radio star who came to films 1942 but failed to maintain early promise. Musicals *Sweater Girl* (début), **Star Spangled Rhythm,** *Priorities On Parade* 42 / *You Can't Ration Love, Rainbow Island* 44 / **Till The Clouds Roll By,** *This Time For Keeps* 46 / *Rock Around The Clock* 56. Married Kathryn Grayson, with whom he did a stage act, but later divorced.

Jolson, Al. Bn 26 May 1880, Leningrad, as Asa Yoelson. Former Dockstader Minstrel who became the first voice in talking pictures. After early success his career slumped and he turned to radio. Even this had faded when Columbia did his biopic in 1946, which put him right back on top. His florid, hammy style now seems outdated, but the basic talent and showmanship mark him as one of *the* great entertainers of all time. Musicals *The Jazz Singer* 27 / **The Singing Fool** 28 / *Say It With Songs* 29 / **Mammy, Big Boy** 30 / *Hallelujah I'm A Bum* 33 / **Wonder Bar** 34 / *Go Into Your Dance* 35 / **The Singing Kid** 36 / *Swanee River,* **Rose Of Washington Square** 39 / *Rhapsody In Blue* 45. Dubbed soundtrack singing for Larry Parks in **The Jolson Story** 46, and **Jolson Sings Again** 49. Died 23 October 1950.

Jones, Allan. Born 1907, Scranton, Pa. Singer, of Welsh descent, who started in opera, then brought his big tenor voice to prominent musicals of the thirties. After joining Universal became featured in increasingly unimportant films. Musicals *A Night At The Opera,* **Rose Marie,** *Reckless* 35 / **Show Boat** 36 / *A Day At The Races, The Firefly* 37 / *Everybody Sing* 38 / *The Great Victor Herbert* 39 / **The Boys From Syracuse, One Night In The Tropics** 40 / *There's Magic In Music* 41 / *True To The Army, When Johnny Comes Marching Home, Moonlight In Havana* 42 / *You're A Lucky Fellow Mr Smith, Crazy House, Larceny With Music, Rhythm Of The Islands* 43 / *Sing A Jingle* 44 / *The Senorita From The West, Honeymoon Ahead* 45. Still works in clubs and on records, but is best known nowadays as the father of singer Jack Jones.

Jones, Shirley. Bn 31 March 1934, Smithton, Pa. Singing star of latter-day musicals **Oklahoma!** (début) 55 / **Carousel** 56 / *April Love* 57 / **Pepe** 60 / **The Music Man** 62. Sopranos being out of fashion she now acts 'straight', also does singing act with husband Jack Cassidy.

Jones, Spike. Bn 14 December 1911, Long Beach, Calif., as Lindley A. Jones. Comedy bandleader. Ex-studio drummer whose City Slickers were a popular record act in the forties and did guest spots in musicals *Meet The People,* **Thank Your Lucky Stars** 43 / *Bring On The Girls* 45 / *Breakfast In Hollywood* 46 / *Ladies' Man, Variety Girl* 47. Died 1 May 1964.

Kahal, Irving. Bn 5 March 1903, Houtzdale, Pa. Lyricist. Former minstrel singer who began song-writing at eighteen and was in at the birth of talkies, under contract to Paramount, then Warner, with his one and only composer-partner Sammy Fain (q.v. for credits). Died 2 July 1942.

Kahn, Gus. Bn 6 November 1886, Coblenz, Germany. Lyricist who worked with all the top writers of the twenties and thirties, producing songs for musicals *The Jazz Singer* 27 / **Whoopee** 30 / *Big City Blues* 32 / **Flying Down To Rio** 33 / *Bottoms Up, Cockeyed Cavaliers,* **One Night Of Love, The Merry Widow,** *Hollywood Party, Kid Millions, Operator 13* 34 / *Reckless, Love Me Forever,* **Naughty Marietta,** *The Girl Friend,* **Rose Marie, Thanks A Million** 35 / *Let's Sing Again, San Francisco, Three Smart Girls* 36 / *Music For Madame, A Day At The Races* 37 / *Everybody Sing, The Girl Of The Golden West* 38 /

*Let Freedom Ring, Honolulu, Broadway Serenade,
Balalaika* 39 / *Bitter Sweet, Go West, Lillian
Russell, Spring Parade, Two Girls On Broadway*
40 / **Ziegfeld Girl,** *The Chocolate Soldier* 41 /
Show Business 44 / **I'll See You In My Dreams**
(biopic with Danny Thomas as Kahn) 51. Died
8 October 1941.

Kalmar and Ruby. Song-writers-librettists-
screenwriters. Bert Kalmar (bn 16 February
1884) met Harry Ruby (bn 27 January 1895), a
fellow New Yorker, in 1917, when Ruby joined
his music publishing company. They wrote
many Broadway shows together, and went to
Hollywood 1930 to write songs and screenplays
for the Marx Brothers. Together they wrote
musicals *Animal Crackers, Top Speed, Cuckoos,
Check And Double Check* 30 / *Horse Feathers,
Manhattan Parade,* **The Kid From Spain** 32 /
Duck Soup 33 / *Happiness Ahead, Hips Hips
Hooray* 34 / *Bright Lights* 35 / *Walking On Air* 36 /
Everybody Sing 38. After Kalmar's death on 18
September 1947 Ruby worked with others on
songs for *Do You Love Me?, Wake Up And Dream*
46 / *Carnival In Costa Rica* 47, wrote the original
story for **Look For The Silver Lining** 49, and
the screenplay of **Lovely To Look At** 52. Some
of the duo's best songs were used in their
biopic **Three Little Words** 50, with Fred
Astaire as Kalmar and Red Skelton as Ruby.

Kane, Helen. Bn 4 August 1908, New York.
Comedienne-singer who started on stage with
the Marx Brothers. Her tiny voice earned her
the name of the Boop-a-doop Girl and was
heard in musicals *Sweetie* 29 / **Paramount On
Parade,** *Heads Up, Dangerous Nan McGrew* 30.
Dubbed singing for Debbie Reynolds (as Kane)
in **Three Little Words** 50. Died 1966.

Kaper, Bronislaw. Bn 5 February 1902,
Warsaw. Composer-conductor, a noted
European writer of concert music. With MGM
1935–53, writing songs, background scores, and
conducting. Songs for musicals *A Night At
The Opera* 35 / *San Francisco, Three Smart Girls* 36 /
A Day At The Races 37 / *Everybody Sing* 38 /
Balalaika 39 / *Go West, Lillian Russell* 40 / *The
Chocolate Soldier* 41 / **That Midnight Kiss** 49 /
Lili 53. Musical director of *The Chocolate
Soldier* 41 / *Song Of Love* 47 / *Mr Imperium* 51 /
Lili 53.

Karger, Fred. Composer-conductor.
Formerly on staff at Columbia as vocal coach,
song-writer, etc. Has since freelanced, often for
MGM and on latter-day Sam Katzman
productions. Wrote songs for *Little Miss
Broadway* 47 / *Holiday In Havana* 49. Musical
director of *Cruisin' Down The River* 53 / *He
Laughed Last* 56 / *Don't Knock The Twist, Twist
Around The Clock* 62 / *Kissin' Cousins* 64 / *Your*

Cheatin' Heart, Harem Scarum 65 / *Girl Crazy,
Frankie and Johnny, Hold On, Spinout, The
Swingin' Set* 66 / *The Fastest Guitar Alive* 68.

Katzman, Sam. Bn 7 July 1900, New York.
Producer of quickies, always over the years
keeping his finger on the teenage pulse with
changes in popular music fashions *Zis Boom Bah*
41 / *Spotlight Scandals* 43 / *Betty Co-ed, Freddie
Steps Out, High School Hero, Junior Prom* 46 /
*Little Miss Broadway, Sweet Genevieve, Glamour
Girl, Two Blondes And A Redhead* 47 / *I Surrender
Dear, Mary Lou* 48 / *Manhattan Angel* 49 / *Purple
Heart Diary* 51 / *Rock Around The Clock* 56 / *Don't
Knock The Rock* 57 / *Don't Knock The Twist,
Twist Around The Clock* 62 / *Kissin' Cousins* 64 /
Harem Scarum, Your Cheatin' Heart 65 / *The
Swingin' Set, Girl Crazy, Hold On* 66 / *The Fastest
Guitar Alive* 68.

Kaye, Danny. Bn 18 January 1913, Brooklyn,
as Daniel Kominsky. Supreme entertainer of the
last quarter century, a miraculous stage
performer whose early films for Goldwyn gave
more rein to his inspired lunacy than later
efforts, although he has developed as an actor.
Was greatly helped by the writing and songs of
his wife Sylvia Fine (*q.v.*). Musicals **Up In Arms**
(début) 44 / *Wonder Man* 45 / *The Kid From
Brooklyn* 46 / **A Song Is Born** 48 / *The Inspector
General, It's A Great Feeling* 49 / *On The Riviera*
51 / **Hans Christian Andersen** 52 / *Knock On
Wood,* **White Christmas** 54 / *The Court Jester* 56 /
Merry Andrew 58 / **The Five Pennies** 59.

Keel, Howard. Bn 13 April 1919, Gillespie,
Illinois. Massive baritone who strode robustly
through many of MGM's top musicals, but
languished into Westerns and comparatively
minor dramas when big voices went out of
style. Musicals **Annie Get Your Gun** 49 / *Pagan
Love Song* 50 / **Show Boat,** *Texas Carnival* 51 /

Lovely To Look At 52 / *Calamity Jane, Kiss Me Kate, I Love Melvin* (guest) 53 / *Deep In My Heart, Rose Marie, Seven Brides For Seven Brothers* 54 / *Jupiter's Darling, Kismet* 55.

Keeler, Ruby. Bn 25 August 1909, Halifax, Nova Scotia. Inevitable dancing star of Warner musicals of the Busby Berkeley era *Footlight Parade, The Golddiggers Of 1933, Forty-Second Street* 33 / *Dames, Flirtation Walk* 34 / *Go Into Your Dance, Shipmates Forever* 35 / *Colleen* 36 / *Ready, Willing And Able* 37 / *Sweetheart Of The Campus* 41. Made a Broadway comeback in 1971 after thirty years in retirement.

Kelly, Gene. Bn 23 August 1912, Pittsburgh. Dancer-actor-choreographer-singer-director-producer. The only serious opposition Astaire ever had, though a different, more balletic type of dancer, who carried his ideas into his (sometimes uncredited) work as choreographer and director. (Also directed non-musicals.) With Stanley Donen created some of the classic MGM musicals *For Me And My Gal* (début) 42 / *Dubarry Was A Lady, Thousands Cheer* 43 / *The Ziegfeld Follies, Christmas Holiday, Anchors Aweigh, Cover Girl* 44 / *Living In A Big Way* 46 / *The Pirate* 47 / *Words And Music, Take Me Out To The Ball Game* 48 / *On The Town* (also dir) 49 / *Summer Stock* 50 / *An American In Paris* 51 / *Singin' In The Rain* (also dir) 52 / *Invitation To The Dance* (also prod/dir), *Deep In My Heart* 54 / *Brigadoon, It's Always Fair Weather* (also dir) 55 / *Les Girls* 57 / *Let's Make Love* 60 / *The Young Girls Of Rochefort* 67 / *Hello Dolly!* (dir only) 69.

Kenton, Erle C. Bn 1 August 1895, Missouri. Director from silent days. Output fairly routine; did a few minor musicals for Universal in the forties *Song Of Love* 29 / *Everything's On Ice* 39 / *Pardon My Sarong* 42 / *Crazy House* (prod), *Always A Bridesmaid, How's About It, It Ain't Hay* 43.

Kern, Jerome. Bn 27 January 1885, New York. Composer who bridged the gap from operetta to modern musicals and was revered by Gershwin, Berlin *et al.* Many of his stage shows were filmed, and he willingly adopted the screen musical as a new vehicle for his enormous talent, working mostly with Oscar Hammerstein, Johnny Mercer, Yip Harburg, Dorothy Fields and Ira Gershwin *Sally* 29 / *Sunny* 30, 41 / *Three Sisters* 30 / *Sweet Adeline, Roberta, The Cat And The Fiddle, Music In The Air* 34 / *I Dream Too Much* 35 / *Show Boat* 36, 51 / *Swing Time* 36 / *When You're In Love, High, Wide And Handsome* 37 / *The Joy Of Living* 38 / *One Night In The Tropics* 40 / *You Were Never Lovelier* 42 / *Cover Girl* 44 / *Can't Help Singing* 45 / *Centennial Summer* 46. Died 11 November

1945; as a tribute MGM produced all-star biopic *Till The Clouds Roll By* 46. Kern songs were featured in *Reckless* 35 / *Look For The Silver Lining* 49 / *Lovely To Look At* (remake of *Roberta*) 52 and *The Helen Morgan Story* 57.

Kidd, Michael. Bn 1919. Ballet dancer who came to dance in films and ended up choreographing, with vitality and imagination, musicals of the fifties and sixties *Where's Charley?* 52 / *The Band Wagon* 53 / *Seven Brides For Seven Brothers* 54 / *It's Always Fair Weather* (also appeared), *Guys And Dolls* 55 / *Merry Andrew* (also dir) 58 / *Li'l Abner* 59 / *Star!* 68 / *Hello Dolly!* 69.

Kiepura, Jan. Bn 16 May 1902, Sosnoweece, Poland. Operatic tenor who made several films in London, and came to Hollywood for *Give Us This Night* 36. Despite good looks was not a success and returned to the stage, touring with his wife, singer Marta Eggerth. Died 12 August 1966.

King, Charles. Bn 31 October 1892. Singer-actor, ex-vaudeville and Ziegfeld Follies, who was a star of two-reelers before full stardom in *The Broadway Melody* 29. Other musicals *Hollywood Revue* 29 / *Chasing Rainbows, Oh Sailor Behave* 30, then into Westerns. Died 11 January 1944.

King, Dennis. Bn 2 November 1897, Coventry, England. Distinguished stage singer-actor who starred briefly in films before returning to the stage. Musicals *Paramount On Parade, The Vagabond King* 30. Died 1971.

King, Henry. Born 1892, Christiansburg, Va. Director (ex-silents) for many years with Fox. Musicals *Marie Galante* 34 / *In Old Chicago* 37 / *Alexander's Ragtime Band* 38 / *Little Old New York* 40 / *Margie* 46 / *Carousel* 56.

Kitt, Eartha. Bn 26 January 1930, Columbia, S. Carolina. Negro performer of point songs, definitely an acquired taste. Has done little on screen. Musicals *New Faces* 54 / *St Louis Blues* 58.

Koehler, Ted. Bn 14 July 1894, Washington, D.C. Lyricist. Former cinema pianist in silent days, then night-club producer. Best known for his collaboration with Harold Arlen*, wrote songs for musicals *Manhattan Parade*, *The Big Broadcast* 32 / *Let's Fall In Love* 34 / *Curly Top, The King Of Burlesque* 35 / *Dimples, Happy Go Lucky* 36 / *Artists And Models*, *The King And The Chorus Girl, Start Cheering, 23½ Hours Leave* 37 / *Love Affair* 39 / *Hullabaloo* 40 / *Stormy Weather* (also screenplay) 43 / *Rainbow Island, Hollywood Canteen, Up In Arms* 44 / *A Weekend At The Waldorf* 45 / *My Wild Irish Rose* 47.

Korjus, Miliza. Bn 1912, Warsaw. Continental

opera star who starred in *The Great Waltz* 38, but, like many others, found no future in films. Still active vocally.

Kostal, Irwin. Bn *c.* 1915. Conductor-composer-arranger, long a backroom boy of the music business, who came to the fore in the sixties as musical director of *West Side Story* 61 / *Mary Poppins* 64 / *The Sound Of Music* 65 / *Half A Sixpence* 67 / *Chitty Chitty Bang Bang* 68.

Koster, Henry. Bn 1 May 1905, Berlin, as Hermann Kosterlitz. Director of escapist movies, comedies and musicals, often working for Joe Pasternak (*q.v.*) *Three Smart Girls,* **100 Men And A Girl** 36 / *First Love, Three Smart Girls Grow Up* 39 / *Spring Parade* 40 / *It Started With Eve* 42 / *Music For Millions* 44 / *Two Sisters From Boston* 45 / *The Unfinished Dance* 47 / *The Inspector General* 49 / *My Blue Heaven, Wabash Avenue* 50 / *Stars And Stripes Forever* 52 / *The Flower Drum Song* 61 / *The Singing Nun* 66.

Krupa, Gene. Bn 15 January 1909, Chicago. Showman drummer-bandleader who was one of the few jazzmen accorded a biopic, *The Gene Krupa Story* 60, in which he played on the sound-track for Sal Mineo. Other musicals *Some Like It Hot* 39 / *Syncopation* 42 / *George White's Scandals Of 1945* 45 / *Beat The Band* 46 / *Glamour Girl* 47 / *Make Believe Ballroom* 49 / *The Glenn Miller Story* 54 / *The Benny Goodman Story* 55.

Kyser, Kay. Bn 18 June 1906 (has also been given as 1897), Rocky Mount, N. Carolina. Bandleader-entertainer whose Kollege of Musical Knowledge was transferred from radio to the screen in a series of surprisingly good if minor films *That's Right You're Wrong* 39 / *You'll Find Out* 40 / *Playmates* 41 / *My Favourite Spy* 42 / *Swing Fever,* **Thousands Cheer,** *Around The World, Stage Door Canteen* 43 / *Carolina Blues* 44. Retired mid-fifties.

Lahr, Bert. Bn 13 August 1895, Yorkville, N.Y., as Irving Lahrheim. Rubber-faced revue comedian who came to films 1931, but whose style was more suited to the auditorium than the camera. Musicals *Flying High* 31 / *Mr Broadway* 33 / *Merry-Go-Round Of 1938,* **Love And Hisses** 37 / *Josette, Just Around The Corner* 38 / *The Wizard Of Oz* 39 / *Ship Ahoy, Sing Your Worries Away* 42 / *Meet The People* 43 / *Always Leave Them Laughing* 49 / *Rose Marie* 54 / *The Night They Raided Minsky's* 68. Died 1967.

Laine, Frankie. Bn 30 March 1913, Chicago, as Frank Paul Lo Vecchio. Lusty singer of popular songs who made several Columbia 'B' musicals *Make Believe Ballroom* (début) 49 / *The Sunny Side Of The Street* 51 / *Rainbow Round My Shoulder* 52 / *When You're Smiling* 53 / *Bring Your Smile Along* 55 / *He Laughed Last,* **Meet Me In Las**

Vegas 56. Still records and does dramatic spots in TV series.

Lamas, Fernando. Bn 9 January 1917, Buenos Aires. Actor. Former Olympic swimmer who rose to stardom in Argentina before coming to Hollywood in 1950. Musicals *Rich Young And Pretty* 51 / *The Merry Widow* 52 / *Dangerous When Wet* 53 / *Rose Marie* 54 / *The Girl Rush* 55.

Lamb, Gil. Born 1906. Doleful acrobatic comedian who did guest spots in Paramount and other musicals *The Fleet's In, Star Spangled Rhythm* 42 / *Riding High* 43 / *Rainbow Island* 44 / *Hit Parade Of 1947* 46 / *Make Mine Laughs* 49. Still does bits.

Lamont, Charles. Bn 1898, San Francisco. Director, ex-silents, of many Universal wartime musicals *Oh Johnny* 40 / *San Antonio Rose, Sing Another Chorus, Moonlight In Hawaii, Almost Married, Don't Get Personal, Melody Lane* 41 / *Get Hep To Love, When Johnny Comes Marching Home* 42 / *Top Man, Mr Big, Hit The Ice* 43 / *A Chip Off The Old Block, Bowery To Broadway, The Merry Monahans* 44 / *That's The Spirit* 45 / *Coming Round The Mountain* 51. Subsequently directing on TV.

Lamour, Dorothy. Bn 10 December 1914, New Orleans, as Dorothy Kaumeyer. Singer-actress who eventually shook off the 'Sarong girl' tag to become one of Paramount's most dependable stars (and an essential part of the 'Road' series with Hope and Crosby). Musicals *College Holiday* 36 / *The Thrill Of A Lifetime,* **The Big Broadcast Of 1938,** *Swing High Swing Low,* **High, Wide And Handsome** 37 / *Her Jungle Love, Tropic Holiday* 38 / *Man About Town, St Louis Blues* 39 / *The Road To Singapore, Moon Over Burma* 40 / *The Road To Zanzibar* 41 / *The Fleet's In, Beyond The Blue Horizon, The Road To Morocco,* **Star Spangled Rhythm** 42 / *Dixie, Riding High* 43 / *Rainbow Island,* **And The Angels Sing** 44 / *Duffy's Tavern, Masquerade In Mexico, The Road To Utopia* 45 / *Variety Girl, The Road To Rio* 47 / *Lulu Belle* 48 / *Slightly French* 49 / *Here Comes The Groom* 51 / *The Road To Bali* 52. Retired to a ranch to raise a family, but emerged for a gag appearance in *The Road To Hong Kong* 62, and later starred on Broadway in 'Hello Dolly!'.

Landers, Lew. Bn 1901, as Lewis Friedlander. Director of budget films who turned out a few minor musicals for Republic and Columbia *The Girl From Havana, Sing Dance Plenty Hot* 40 / *The Redhead From Manhattan* 41 / *Cowboy Canteen, Doughboys In Ireland, Stars On Parade, Swing In The Saddle* 44.

Landis, Carole. Bn 1919, Fairchild, Wisconsin as Frances Ridate. Blonde actress-singer who played big roles in supporting musicals and vice versa *A Day At The Races*, **Hollywood Hotel, The Varsity Show** 37 | **The Golddiggers In Paris** 38 | *Road Show* 40 | *Cadet Girl*, **Moon Over Miami** 41 | *My Gal Sal, The Powers Girl, Orchestra Wives* 42 | *Wintertime* 43 | *Four Jills In A Jeep* 44. Died 1948.

Lane, Burton. Bn 2 February 1912, New York. Composer of many Broadway and Hollywood scores, under contract to Paramount 1936–41, in partnership with lyricists Harold Adamson (33–5) and Frank Loesser (38–41) (*q.v.*). Provided interpolated songs or complete scores for **Dancing Lady** 33 | *Palooka, Kid Millions, Bottoms Up, Strictly Dynamite* 34 | *Reckless, Here Comes The Band, Folies Bergère, The Smartest Girl In Town* 35 | *College Holiday* 36 | *Swing High Swing Low, Hideaway Girl, Artists And Models, Love On Toast, Double Or Nothing, Champagne Waltz* 37 | *The Cocoanut Grove, College Swing* 38 | *Some Like It Hot, Cafe Society, St Louis Blues* 39 | *Dancing On A Dime* 40 | **Babes On Broadway,** *Las Vegas Nights* 41 | *Presenting Lily Mars, Seven Sweethearts, Panama Hattie, Ship Ahoy,* **Dubarry Was A Lady,** *Meet The People,* **Thousands Cheer** 43 | *Hollywood Canteen, Rainbow Island* 44 | *This Time For Keeps* 46 | **Royal Wedding** 51 | **The Belle Of New York** 52 | **Give A Girl A Break** 53 | *Jupiter's Darling* 55 | **Finian's Rainbow** 68 | **On A Clear Day You Can See Forever** 69.

Lane, Lola. Bn 1909, Indianola, Iowa, as Dorothy Mulligan. Sister of Rosemary and Priscilla Lane (*q.v.*). Actress, former pianist for silent films. Leading lady in musicals *Fox Follies* 29 | *Good News, Let's Go Places* 30 | *Hollywood Hotel* 37 | *Mr Chump* 38 | *Why Girls Leave Home* 45.

Lane, Priscilla. Bn 1917, Indianola, Iowa, as Priscilla Mulligan. Sister of Lola and Rosemary Lane (*q.v.*). With Rosemary she was a singing act with Fred Waring's Pennsylvanians, appearing with the band in **The Varsity Show** 37. Later took mainly dramatic parts but was in musicals *A Cowboy From Brooklyn* 38 | **Blues In The Night** 41.

Lane, Rosemary. Bn 1914, Indianola, Iowa, as Rosemary Mulligan. Sister of Lola and Priscilla Lane (*q.v*). With Priscilla she was a singing act with Fred Waring's Pennsylvanians, appearing with the band in **The Varsity Show** 37. Later worked with both sisters in the *Four Daughters* series, but got more musical roles **Hollywood Hotel** 37 | **The Golddiggers In Paris** 38 | **The Boys From Syracuse** 40 | *Time Out For Rhythm* 41 | *All By Myself, Harvest Melody, Chatterbox* 43 | *Trocadero* 44.

Lanfield, Sidney. Bn 20 April 1899, Chicago. Director of leading Fox musicals of the thirties,

later doing comedies for Paramount and MGM. Musicals *Cheer Up And Smile* 30 | *Broadway Bad* 33 | *Moulin Rouge* 34 | **The King Of Burlesque** 35 | **Sing Baby Sing** 36 | *One In A Million*, **Wake Up And Live,** *Thin Ice*, **Love And Hisses** 37 | *Swanee River,* **Second Fiddle** 39 | **You'll Never Get Rich** 41 | *Let's Face It* 43 | *Bring On The Girls* 45 | *Skirts Ahoy* 52.

Lang, Walter. Bn 1897, Memphis, Tenn. Director of many of Fox's most memorable (and a few forgettable) musicals *Carnival, Hooray For Love* 35 | *The Bluebird* 39 | **Tin Pan Alley** 40 | **Moon Over Miami, A Weekend In Havana** 41 | *Song Of The Islands* 42 | *Coney Island* 43 | *Greenwich Village* 44 | **State Fair** 45 | **Mother Wore Tights** 47 | *When My Baby Smiles At Me* 48 | *You're My Everything* 49 | *On The Riviera* 51 | **With A Song In My Heart** 52 | **Call Me Madam** 53 | **There's No Business Like Show Business** 54 | **The King And I** 56 | **Can Can** 60 | *Snow White And The Three Stooges* 61.

Lange, Arthur. Bn 16 April 1889, Philadelphia. Composer-conductor, worked in both capacities for MGM (29–30), RKO (31–2), Fox (33–43), then as a freelance. Musical director of *Marianne*, **Hollywood Revue** 29 | *Chasing Rainbows, Good News, They Learned About Women* 30 | *The Best Of Enemies, Jimmy And Sally* 33 | *Stand Up And Cheer, Lottery Lover, Marie Galante* 34 | *Spring Tonic, The Little Colonel*, **Thanks A Million** 35 | **The Great Ziegfeld,** *Banjo On My Knee, Under Your Spell* 36 | *Sally Irene And Mary*, **On The Avenue** 37 | *Hold That Co-ed, Rebecca Of Sunnybrook Farm* 38 | *Let Freedom Ring, The Great Victor Herbert* 39 | *Lady Of Burlesque* 43 | *The Belle Of The Yukon* 44 | *It's A Pleasure* 45.

Langford, Frances. Bn 1913, Lakeland, Florida. Singer, on radio since 1931 and renowned for her work with Bob Hope. Did guest spots in major musicals and starred in minor ones for Columbia *et al*. **Every Night At Eight** (début), **The Broadway Melody Of 1936** 35 | *Collegiate, Hit Parade Of 1937, Palm Springs,* **Born To Dance** 36 | **Hollywood Hotel** 37 | *Dreaming Out Loud, Hit Parade Of 1941, Too Many Girls* 40 | *All-American Co-ed* 41 | **Yankee Doodle Dandy,** *Mississippi Gambler* 42 | *Never A Dull Moment, A Cowboy In Manhattan, Follow The Band,* **This Is The Army** 43 | *Career Girl, The Girl Rush* 44 | *Radio Stars On Parade, People Are Funny, Dixie Jamboree* 45 | *Bamboo Blonde, Beat The Band* 46 | *Melody Time* (soundtrack) 48 | *Make Mine Laughs* 49 | *Purple Heart Diary* 51 | **The Glenn Miller Story** 54.

Lanza, Mario. Bn 31 January 1921, South Philadelphia, as Alfred Arnold Cocozza. Tenor whose robust, florid singing of operatic arias and popular songs was featured in musicals of the early fifties, before problems of weight and a personal nature caused a gradual decline until his death on 7 October 1959. Films *That Midnight Kiss* (début) 49 / *The Toast Of New Orleans* 50 / *The Great Caruso* 51 / *Because You're Mine* 52 / *The Student Prince* (soundtrack singing for Edmund Purdom) 54 / *Serenade* 55 / *The Seven Hills Of Rome* 58 / *For The First Time* 59.

Lawford, Peter. Bn 7 September 1923, London. Actor, sometimes singer. Ex-child actor (début 1931) who went to Hollywood 1942, and *inter alia* did some useful work in MGM musicals *Two Sisters From Boston* 45 / *It Happened In Brooklyn* 46 / *Good News, On An Island With You* 47 / *Easter Parade* 48 / *Royal Wedding* 51 / *Pepe* 60.

Le Baron, William. Bn 16 February 1883, Elgin, Illinois. Producer who was writer/ass. producer on Broadway, magazine editor and director of Cosmopolitan Prodns 1914–24. Later executive producer with Paramount and Fox, promoting musical vehicles for Mae West, Mary Martin, Betty Grable, etc. *Rio Rita* 29 / *College Humour, I'm No Angel, She Done Him Wrong* 33 / *Bolero* 34 / *All The King's Horses, Coronado, Goin' To Town, Here Comes Cookie, Rose Of The Rancho, Rumba* 35 / *Give Us This Night, Klondike Annie, Poppy* 36 / *Rhythm On The River* 40 / *Kiss The Boys Goodbye, Las Vegas Nights, A Weekend In Havana* 41 / *Springtime In The Rockies, Song Of The Islands, Orchestra Wives, Footlight Serenade, Iceland* 42 / *Stormy Weather, Winter-time, The Gang's All Here* 43 / *Greenwich Village, Pin-Up Girl, Sweet And Lowdown* 44 / *Three Little Girls In Blue* 46 / *Carnegie Hall* 49. Died 9 February 1958.

Lee, Dixie. Bn 4 November 1911 as Wilma Wyatt. Blonde Fox starlet who married the unknown Bing Crosby and gave up a promising career that took her through musicals like *Fox Follies* 29 / *The Big Party, Cheer Up And Smile, Happy Days, Let's Go Places* 30 / *Love In Bloom, Redheads On Parade* 35. Died 1 November 1952.

Lee, Dorothy. Bn 23 May 1911, Los Angeles, as Marjorie Millsap. Singer-actress with Radio Pictures in early thirties; leading lady in many Wheeler and Woolsey comedies. Musicals *Syncopation* (début), *Rio Rita* 29 / *Dixiana, Cuckoos* 30 / *Girl Crazy* 32 / *Take A Chance* 33 / *Cockeyed Cavaliers, Hips Hips Hooray* 34 / *The Old Homestead, Rainmakers* 35 / *Too Many Blondes* 41.

Lee, Gypsy Rose. Bn 1914 as Louise Hovick. Actress-striptease performer. Long experience in vaudeville with her sister June Havoc (*q.v.*) and burlesque before going into films. Musicals *Sally Irene And Mary, Ali Baba Goes To Town, You Can't Have Everything* 37 / *My Lucky*

Star 38 / *Stage Door Canteen* 43 / *The Belle Of The Yukon* 44. Portrayed by Natalie Wood in her biopic *Gypsy* 62. Still active as actress and writer until her death from cancer on 26 April 1970.

Lee, Peggy. Bn 26 May 1920, Jamestown, N. Dakota, as Norma Egstrom. Singer-actress-composer-lyricist. One of the most highly respected singers in standard popular music. Although nominated for an Oscar in *Pete Kelly's Blues* 55, didn't follow up as an actress apart from TV guest roles. Other musicals *The Powers Girl* 42 / *Stage Door Canteen* 43 (both as Benny Goodman vocalist), *Mr Music* 50 / *The Jazz Singer* 52. Wrote songs for the latter and *Lady And The Tramp* (also sang on soundtrack) 55, and *Tom Thumb* 58.

Lee, Sammy. Bn 26 May, New York. Dance director who worked on Broadway as choreographer for Earl Carroll, Ziegfeld and the Astaires before going to Hollywood in early days of talkies. Dance direction on many musicals includes *It's A Great Life, Hollywood Revue* 29 / *It's Great To Be Alive, Jimmy and Sally, My Lips Betray, Dancing Lady, Adorable* 33 / *365 Nights In Hollywood* 34 / *The King of Burlesque* 35 / *Ali Baba Goes To Town, Heidi, The Life Of The Party, New Faces Of 1937* 37 / *Honolulu* 39 / *Hullabaloo* 40 / *Meet The People, Hit The Ice* 43 / *Two Girls And A Sailor, Carolina Blues* 44 / *Earl Carroll's Vanities* 45.

Lehman, Ernest. Bn *c.* 1916. Author-producer. Financial magazine editor, press agent and radio writer who started writing fiction, and went to Hollywood to adapt his stories for the screen. Among the heavy dramas has also done screen-plays for musicals *The King And I* 56 / *West Side Story* 61 / *The Sound of Music* 65. Produced *Hello Dolly!* 69.

Leigh, Janet. Bn 6 July 1927, Merced, Calif., as Jeanette Morrison. Actress. Discovered at eighteen by Norma Shearer and groomed for stardom, including experience in musicals *Words And Music* 48 / *Two Tickets To Broadway* 51 / *Walking My Baby Back Home* 53 / *Living It Up* 54 / *Pete Kelly's Blues, My Sister Eileen* 55, *Pepe* 60 / *Bye Bye Birdie* 63.

Leisen, Mitchell. Bn 6 October 1897, Menominee, Wash. Director. Former costume and set designer who brought a certain flair to Paramount films, including musicals *Murder At The Vanities* 34 / *The Big Broadcast Of 1937* 36 / *Swing High Swing Low, The Big Broadcast Of 1938* 37 / *Artists And Models Abroad* 38 / *Lady In The Dark* 43 / *Masquerade In Mexico* 45 / *Dream Girl* 48 / *Tonight We Sing* 53 / *The Girl Most Likely* 57.

Lemmon, Jack. Bn 8 February 1925, Boston.

Actor-pianist-composer. Superb interpreter of sophisticated comedy with matchless technique and timing. Worthy of a place in more musicals than the few he has done *My Sister Eileen,* **Three For The Show** 55 / *You Can't Run Away From It* 56 / **Some Like It Hot** 59 / **Pepe** 60.

Leonard, Robert Z. Bn 7 October 1889, Denver, Ill. Director. Actor on stage and in silents from 1910. Joined MGM as director and worked on mainly light subjects including many musicals *Marianne* 29 / **Dancing Lady** 33 / **The Great Ziegfeld** 36 / *Maytime, The Firefly* 37 / *The Girl Of The Golden West* 38 / *Broadway Serenade* (also prod) 39 / *New Moon* (also prod) 40 / **Ziegfeld Girl** (also prod) 41 / *A Weekend At The Waldorf* 45 / *Nancy Goes To Rio, In The Good Old Summertime* 49 / *The Duchess Of Idaho* 50 / *Everything I Have Is Yours* 52. Died 27 August 1968.

Lerner, Alan Jay. Bn 31 August 1918, New York. Lyricist-screenwriter-producer. Former radio scriptwriter whose collaboration with Frederick Loewe (*q.v.*) has produced some of the outstanding musicals of the fifties and sixties. Does his own screenplays, and is latterly producing. Wrote lyrics only for **The Belle Of New York** 52; screenplay only for *An American In Paris* 51; did both for **Royal Wedding** 51 / **Brigadoon** 55 / **Gigi** 58 / **My Fair Lady** 64 / **Camelot** 67, and also produced **On A Clear Day You Can See Forever, Paint Your Wagon** 69.

LeRoy, Mervyn. Bn 16 October 1900, San Francisco. Director-producer. Was in vaudeville, then came to films 1927 with Warner contract as director. His output has covered the whole spectrum of movie entertainment, but has found a special niche in musicals *Little Johnny Jones, Broadway Babies* 29 / *A Show Girl In Hollywood, Top Speed* 30 / *Big City Blues* 32 / **The Golddiggers**

Of 1933, *Hallelujah I'm A Bum* (also prod) 33 / *Happiness Ahead,* **Sweet Adeline** 34 / *Page Miss Glory* 35 / *Mr Dodd Takes The Air* (prod only), *The King And The Chorus Girl* (also prod) 37 / *Fools For Scandal* (also prod) 38 / **The Wizard Of Oz** (prod only), **At The Circus** (prod only) 39 / **Lovely To Look At** 52 / *Latin Lovers* 53 / **Rose Marie** (prod only) 54 / **Gypsy** (also prod) 62.

Leslie, Joan. Bn 26 January 1925, Detroit. Actress-dancer, first appeared as a child star under her real name of Joan Brodel. Danced with Astaire and co-starred in some prominent musicals but never achieved real stardom. Musicals **Yankee Doodle Dandy** 42 / **This Is The Army, The Sky's The Limit, Thank Your Lucky Stars** 43 / *Hollywood Canteen* 44 / **Where Do We Go From Here?, Rhapsody In Blue** 45 / *Two Guys From Milwaukee, Cinderella Jones* 46.

Levant, Oscar. Bn 27 December 1906, Pittsburgh. Composer-lyricist-pianist-conductor-author-actor whose mordant wit and deadpan delivery have saved many a film from boredom. Wrote scores for *Street Girl, Tanned Legs, Jazz Heaven* 29 / *Music Is Magic, In Person* 35 / *The Smartest Girl In Town* 36, usually with Sidney Clare lyrics. Appeared in musicals *The Dance Of Life* (début) 29 / *Rhythm On The River* 40 / *Kiss The Boys Goodbye* 41 / **Rhapsody In Blue** 45 / **Romance On The High Seas, You Were Meant For Me, The Barkleys Of Broadway** 48 / *An American In Paris* 51 / *The 'I Don't Care' Girl,* **The Band Wagon** 53.

Levin, Henry. Bn 1909. Director whose output has seldom risen above the routine. Musicals **Jolson Sings Again** 49 / *The Petty Girl* 50 / **The Farmer Takes A Wife** 52 / *Let's Be Happy* 56 / *April Love* 57 / *Where The Boys Are* 60 / *The Wonderful World Of The Brothers Grimm* 62.

Lewis, Jerry. Bn 16 March 1926, Newark, N.J., as Joseph Levitsch. Comedian-director-writer-producer. Fortuitous circumstance (the need to quell a noisy club audience) led to the ten-year partnership with Dean Martin (*q.v.*), which produced some, at first, good comedies. After the duo split in 1957 Lewis carried on, often writing, directing and producing. Musicals *My Friend Irma* 49 / *My Friend Irma Goes West, At War With The Army* 50 / *That's My Boy* 51 / *The Stooge, Sailor Beware, Jumping Jacks* 52 / *Scared Stiff* 53 / *Living It Up* 54 / *Hollywood Or Bust, Pardners, Artists And Models* 56 / *Cinderfella* 60 / *Ladies' Man* 61.

Lewis, Ted. Bn 6 June 1892, Circleville, Ohio, as Theodore Lewis Friedman. Clarinet-playing bandleader famous for catch-phrase 'Is everybody happy?' and his 'Me and my shadow' routine. Two films with the former name were built around him and his band, 29, 43; also appeared in musicals *The Show Of Shows* 29/*Here Comes The Band* 35/*Manhattan Merry-Go-Round* 37/*Hold That Ghost* 41/*Follow The Boys* 44.

Liberace. Bn 16 May 1919, Milwaukee, as Wladziu Valentino Liberace. Ostentatious pianist-showman famous on TV but seldom on screen. Musicals *Footlight Varieties* 51/*Sincerely Yours* 55/*Girl Crazy* 66.

Lightner, Winnie. Bn 17 September 1901, Greenport, Long Island, as Winifred Hanson. Singing star of early Warner musicals *The Golddiggers Of Broadway, The Show Of Shows* 29/*Hold Everything, The Life Of The Party, She Couldn't Say No* 30/*Manhattan Parade* 32/*Dancing Lady* 33. Died February 1971.

Lilley, Edward. Bn 1896. Director who filled Universal's quota of wartime 'quickies' and was never heard of again. Musicals *Babes On Swing Street, Honeymoon Lodge, Larceny With Music, Moonlight In Vermont, Never A Dull Moment, You're A Lucky Fellow Mr Smith* 43/*Sing A Jingle* (also prod), *My Gal Loves Music* (also prod), *Hi Good Lookin', Allergic To Love* 44/*Swing Out Sister* 45.

Lilley, Joseph J. Bn 16 August 1913, Providence, Rhode Island. Composer-conductor. Former radio writer who joined Paramount 1947, and usually drew assignments as musical director on Crosby, Hope, Martin and Lewis and Presley musicals *Variety Girl* 47/*Isn't It Romantic?* 48/*Red Hot And Blue* 49/*At War With The Army,* **Mr Music** 50/*Here Comes The Groom* 51/*Jumping Jacks, Just For You, The Road To Bali, Sailor Beware, The Stooge* 52/*Little Boy Lost, Scared Stiff, Those Redheads From Seattle* 53/ **White Christmas, The Country Girl,** *Living It Up,* **Red Garters** 54/*The Seven Little Foys* 55/ **Anything Goes** 56/*Beau James, Loving You* 57/ *King Creole* 58/*Li'l Abner* 59/*G.I. Blues* 60/ *Blue Hawaii,* **Gay Purr-ee,** *Girls Girls Girls* 62/ *Fun In Acapulco* 63/*Roustabout* 64/*Paradise Hawaiian Style* 66/*Easy Come Easy Go* 67/*Paint Your Wagon* 69. On some of these (especially later), he shared musical director duties, often on the choral side. Died 1 January 1971.

Lillie, Beatrice. Bn 1898, London. Veteran stage comedienne who made few, but generally very pointed, appearances in films. Musicals *The Show Of Shows* 29/*Doctor Rhythm* 38/ *Thoroughly Modern Millie* 67.

Livingston and Evans. Song-writers. Jay Livingston (bn 28 March 1915, McDonald, Pa.) met Ray Evans (bn 4 December 1915, New York) when college dance band musicians. Wrote special material; signed joint contract with Paramount 1945. Have written songs or complete scores for musicals *Footlight Glamour,*

On Stage Everybody, Swing Hostess 44/*Why Girls Leave Home, People Are Funny* 45/*Monsieur Beaucaire, The Stork Club* 46/*Dream Girl, Isn't It Romantic?, The Paleface* 48/*My Friend Irma* 49/ *My Friend Irma Goes West, Fancy Pants* 50/*Here Comes The Groom, Aaron Slick From Punkin Crick, That's My Boy* 51/*Here Come The Girls, Somebody Loves Me, Son of Paleface* 52/*The Stars Are Singing* 53/**Red Garters** 54/*All Hands On Deck* 61. Latterly writing title songs for films and TV.

Loesser, Frank. Bn 29 June 1910, New York. Composer-lyricist. For many years worked for Paramount as lyricist for other composers, especially with Burton Lane (38–41) (*q.v.* for credits). Eventually became a major Broadway and Hollywood show composer-lyricist. Musicals with other writers *Vogues Of 1938, Blossoms On Broadway* 37/*Swing That Cheer, Sing You Sinners, The Freshman Year, Thanks For The Memory* 38/*Hawaiian Nights, Man About Town* 39/*Buck Benny Rides Again, Moon Over Burma, A Night At Earl Carroll's* 40/*Glamour Boy, Sis Hopkins, Kiss The Boys Goodbye* 41/*Priorities On Parade, Happy Go Lucky, Beyond The Blue Horizon, Seven Days Leave, Sweater Girl, True To The Army* 42/**Thank Your Lucky Stars** 43/*Something For The Boys* 44. These were mostly interpolated songs; henceforth Loesser wrote complete scores single-handed *Christmas Holiday* 44/*The Perils Of Pauline, Variety Girl* 47/*Let's Dance, Neptune's Daughter, Red Hot And Blue* 49/ **Where's Charley?,** *Hans Christian Andersen* 52/**Guys And Dolls** 55/**How To Succeed In Business Without Really Trying** 66. Died 28 July 1969.

Loewe, Frederick. Bn 10 June 1904, Berlin, of Austrian parents. Composer who apart from *The Life Of The Party* 30, has worked in films exclusively with Alan Jay Lerner (*q.v.*). *Gigi* 58, was a screen original, other musicals were adaptations of their stage successes **Brigadoon** 55/**My Fair Lady** 64/**Camelot** 67/**Paint Your Wagon** 69. Semi-retired after illness.

Logan, Ella. Bn *c.* 1913, Glasgow. Singer-actress of well-known theatrical family. Went to America to appear with Fred Waring in 1932, but better known on radio and stage than in films. Musicals *Top Of The Town, 52nd Street* 37/ **The Goldwyn Follies** 38. Latterly in TV and clubs. Died 1 May 1969.

Logan, Joshua. Bn 1908. Broadway director-producer who, apart from dramatic films, has brought some of his stage musicals to the screen *South Pacific* 58/**Camelot** 67/**Paint Your Wagon** 69.

Lombardo, Guy. Bn 19 June 1902, London, Ontario. Bandleader who for over forty years has

produced 'The Sweetest Music This Side of Heaven'. Guest spots in musicals *Many Happy Returns* 34 / *Stage Door Canteen* 43 / *No Leave No Love, A Weekend At The Waldorf* 45.

London, Julie. Bn 26 September 1926, Santa Rosa, Calif. Former child actress who made a great name as a record singer yet only appeared in two musicals *On Stage Everybody* 44 / *The Girl Can't Help It* 57. Many dramatic roles.

Lord, Del. Bn 1895, Grimsley, Canada. Director who came up with Keystone, Sennett, etc., and turned out 'B' musicals for Columbia during the war years *Kansas City Kitty, She's A Sweetheart* 44 / *Let's Go Steady, I Love A Bandleader, The Blonde From Brooklyn* 45 / *Hit The Hay, Singin' In The Corn* 46 / *It's Great To Be Young* 47.

Loring, Eugene. Bn 1914 as Leroy Kerpestein. Dance director who came to prominence after the war, often in association with Fred Astaire. Musicals include **The Ziegfeld Follies** 44 / **Yolanda And The Thief** 45 / *Fiesta* 46 / *Mexican Hayride* 48 / *The Inspector General* 49 / **The Toast of New Orleans** 50 / **Deep In My Heart** 54 / **Funny Face, Meet Me In Las Vegas, Silk Stockings** 56 / **Pepe** 60. Appeared in *Something In The Wind* 47 / *Torch Song* 53.

Lubin, Arthur. Bn 25 July 1901, Los Angeles. Director, mainly on Universal minor musicals

and comedies *I'm Nobody's Sweetheart Now* 40 / *In The Navy, Keep 'em Flying, Ride 'em Cowboy, Hold That Ghost, Where Did You Get That Girl?, Buck Privates* 41 / *The Phantom Of The Opera* 43 / *Delightfully Dangerous* 44 / **New Orleans** 47 / *Hold On* 66.

Lubitsch, Ernst. Bn 28 January 1892, Berlin. Director of sparkling, sophisticated comedies and musicals, who learned his trade as a comedian and director in Germany, and brought many of his technical colleagues to Paramount. Musicals **The Love Parade** 29 / **Monte Carlo,** *The Playboy Of Paris,* **Paramount On Parade** 30 / *The Smiling Lieutenant* 31 / **One Hour With You** (part-dir) 32 / **The Merry Widow** 34 / *That Lady In Ermine* (also prod) 47 /. Died 1947.

Lynn, Diana. Bn 7 October 1926, Los Angeles, as Dolly Loehr. Actress-pianist. In many Paramount features as a teenager, including musicals *There's Magic In Music* (début) 41 / **Star Spangled Rhythm** 42 / **And The Angels Sing** 44 / *Out Of This World* 45 / *Variety Girl* 47 / *My Friend Irma* 49 / *My Friend Irma Goes West* 50 / *Meet Me At The Fair* 52. Has made records as pianist, and latterly done guest spots in TV series.

Lyon, Ben. Bn 6 February 1901, Atlanta, Georgia. Actor who left America in mid-thirties and settled in England with his wife Bebe Daniels (*q.v.*) to create a new radio and TV career. Musicals *Her Majesty Love, The Hot Heiress* 31 / *Dancing Feet* 36. As Fox's casting director he discovered and re-christened Marilyn Monroe.

Names beginning with Mc or Mac will be found together at the end of this section.

Magidson, Herb. Bn 7 January 1906, Braddock, Pa. Composer-lyricist. Wrote many film songs, mostly in collaboration with Con Conrad (34–6) and Allie Wrubel (37–51) (*q.v.* for credits). With other writers worked on musicals **The Show Of Shows** 29 / *George White's Scandals Of 1935* 35 / *Hats Off* 36 / *Priorities On Parade, Sleepytime Gal* 42 / *Hers To Hold* 43 / *Music In Manhattan, Rosie The Riveter* 44 / *Do You Love Me?* 46.

Malneck, Matty. Bn 10 December 1904, Newark, N.J. Composer-arranger-conductor. Former Paul Whiteman violinist-arranger who later appeared in films with his own orchestra *The East Side Of Heaven* 38 / *Man About Town* 39 / *Scatterbrain* 40 / *Trocadero* 44. Wrote songs for *Transatlantic Merry-Go-Round* 34 / *To Beat The Band* 35 / *St Louis Blues, Man About Town, Hawaiian Nights* 39 / *Let's Make Music* 40 / **Some Like It Hot** (also musical director) 59.

Mamoulian, Rouben. Bn 8 October 1898, Tiflis, Caucasus, Russia. Director from Broadway

Henry Mancini

who introduced his revolutionary stage techniques to the screen. Though he seldom excelled the brilliance of *Love Me Tonight* 32, his other musicals were always imaginative and individual *Applause* 29 | *The Gay Desperado* 36 | *High, Wide And Handsome* 37 | *Summer Holiday* 46 | *Silk Stockings* 56.

Mancini, Henry. Bn 16 April 1924, Cleveland, Ohio. Composer-conductor renowned for his theme and background music for romantic dramas of the sixties (also the innovator of 'crime jazz' in TV series), yet seldom involved with musicals. As Universal staff arranger (51–7) he scored *The Glenn Miller Story* 54 | *The Benny Goodman Story* 55 | *The Big Beat, Rock Pretty Baby* (also musical director) 57. Musical director for *High Time* 60 | *Darling Lili* (also wrote songs) 69.

Mann, Anthony. Bn 30 June 1906, Point Loma, Calif., as Emil Bundsmann. Director of Westerns and epics who started directorial life with 'B' musicals *Moonlight In Havana* 42 | *My Best Gal, Nobody's Darling* 43 | *Sing Your Way Home* 45 | *The Bamboo Blonde* 46 | *The Glenn Miller Story* 54 | *Serenade* 55. Died April 1967.

Manning, Irene. Bn *c*. 1914 as Inez Harvet. Ladylike blonde soprano who performed all too briefly for Warners in the forties. Musicals *Yankee Doodle Dandy* 42 | *The Desert Song* 43 | *Hollywood Canteen, Shine On Harvest Moon* 44. Later lived in England.

Marin, Edwin L. Bn 1899, Jersey City. Director whose output rarely rose above the mundane. Musicals *The Sweetheart Of Sigma Chi* 33 | *Listen Darling, Everybody Sing* 38 | *Hullabaloo* 40 | *Show Business* 44.

Marshall, George. Bn 1891, Chicago. Director who started as actor in silents, and later directed Westerns, big dramas and his fair share of musicals *She Learned About Sailors, 365 Nights In Hollywood* 34 | *Music Is Magic* 35 | *Can This Be Dixie?* 36 | *The Goldwyn Follies, Hold That Co-ed* 38 | *Pot Of Gold* 40 | *Star Spangled Rhythm* 42 | *Riding High, True To Life* 43 | *And The Angels Sing, The Incendiary Blonde* 44 | *Monsieur Beaucaire* 46 | *The Perils Of Pauline, Variety Girl* 47 | *My Friend Irma* 49 | *Fancy Pants* 50 | *Scared Stiff* 53 | *Red Garters* 54.

Martin, Dean. Bn 7 June 1917, Stubenville, Ohio, as Dino Crocetti. Singer-comedian turned actor. First known as half of Martin and Lewis: when his share of the partnership began decreasing he went solo, and contrary to show-business expectations the 'stooge' emerged as the real star. Musicals *My Friend Irma* 49 | *My Friend Irma Goes West, At War With The Army* 50 | *That's My Boy* 51 | *The Stooge, Sailor Beware, Jumping Jacks* 52 | *Scared Stiff* 53 | *Living It Up* 54 | *Hollywood Or Bust, Pardners, Artists And Models* 56 | *Ten Thousand Bedrooms* 57 | *Bells Are Ringing* 60 | *The Road To Hong Kong* (gag appearance) 62 | *Robin And The Seven Hoods* 64 | *Kiss Me Stupid* 65.

Martin, Hugh. Bn 11 August 1914, Birmingham, Alabama. Composer-lyricist. Former singer who met Ralph Blane (*q.v.*) in a Broadway show. They formed a singing group and collaborated on many stage and screen musicals *Best Foot Forward, Broadway Rhythm, Thousands Cheer* 43 | *Meet Me In St Louis, The Ziegfeld Follies* 44 | *Good News* 47 | *Athena* 54 | *The Girl Rush* 55 | *The Girl Most Likely* 57. Martin also did vocal arrangements for *The West Point Story* 50.

Martin, Mary. Bn 1 December 1914, Weatherford, Texas. Singing star whose speciality was combining arias with swing in Paramount musicals of the early forties *The*

Great Victor Herbert (début) 39 / *Love Thy Neighbour, Rhythm On The River* 40 / **The Birth Of The Blues,** *Kiss The Boys Goodbye* 41 / **Star Spangled Rhythm,** *Happy Go Lucky* 42 / *True To Life* 43 / **Night And Day** 45. Preferred the stage, where she achieved more permanent stardom.

Martin, Tony. Bn 25 December 1912, Oakland, Calif., as Al Norris. Singing actor who was a sax player before starring in many Fox musicals of the thirties (later with MGM) **Follow The Fleet** 35 / *Pigskin Parade, Poor Little Rich Girl,* **Sing Baby Sing,** *Banjo On My Knee* 36 / *Life Begins In College, Ali Baba Goes To Town, The Holy Terror,* **You Can't Have Everything,** *Sing And Be Happy, Sally Irene And Mary* 37 / *Thanks For Everything, Kentucky Moonshine* 38 / *Music In My Heart* 40 / **Ziegfeld Girl,** *The Big Store* 41; war service in AAF, Far East; awarded Bronze Star and citation; **Till The Clouds Roll By** 46 / *Casbah* 48 / *Two Tickets To Broadway* 51 / *Here Come The Girls* 52 / *Easy To Love* 53 / **Deep In My Heart** 54 / **Hit The Deck** 55 / *Let's Be Happy* 56. Later did a night-club act with his wife Cyd Charisse.

Martini, Nino. Bn 1904, Italy. Singing lead of a few Ruritanian musicals *Here's To Romance* 35 / *The Gay Desperado* 36 / *Music For Madame* 37.

Marx Brothers. Comedians. Groucho (Julius) bn 2 October 1895; Harpo (Arthur) bn 23 November 1893; Chico (Leonard) bn 22 August 1891; Zeppo (Herbert) bn 25 February 1901. Captured Broadway with their hilariously anarchic stunts, and went to Hollywood 1929 to film their stage shows. They never fully utilized the resources of the camera, earlier films being no more than photographed stage sets, but their zany comedy transcended this. The films were not musicals as such, but there were moments for Chico to play piano, Harpo to provide tranquil harp interludes, and Groucho to 'sing'. Later legitimate singing stars got into the action with production numbers which tended to hold things up. Principal films with music *Animal Crackers* 30 / *Horse Feathers* 32 / *Duck Soup* 33, then Zeppo left the act, *A Night At The Opera* 35 / *A Day At The Races* 37 / **At The Circus** 39 / *Go West* 40 / *The Big Store* 41 / *Love Happy* 48. Harpo alone in *Stage Door Canteen* 43. Groucho wrote the screenplay for *The King And The Chorus Girl* 37, appeared solo in *Copacabana* 47 / **Mr Music** 50 / *Double Dynamite* 51, later building a new career as M.C. on TV. Harpo died 28 September 1964; Chico in 1961.

Massey, Ilona. Bn 1912, Hungary as Ilona Hajmassy. Sultry-voiced blonde singer-actress whose accent and beauty spot appeared in musicals **Rosalie** 37 / *Balalaika* 39 / *Holiday In Mexico* 45 / *Northwest Outpost* 47 / *Love Happy* 48.

Maxwell, Marilyn. Bn 13 August 1920, Clarinda, Iowa. Blonde leading lady, formerly brunette singer Marvel Maxwell with Ted Weems' band. With Bob Hope on radio, made a number of films including musicals *Presenting Lily Mars* 42 / **Dubarry Was a Lady,** *Swing Fever,* **Thousands Cheer** 43 / *Lost In A Harem* 44 / **Summer Holiday,** *The Show-Off* 46. Now on TV.

Mayo, Archie. Bn 1896, New York. Director who began as scenarist-director of two-reel silents. Started in talkies with Warner, then Fox. Musicals *My Man* 28 / *Is Everybody Happy?* 29 / *Oh Sailor Behave* 30 / *Go Into Your Dance* 35 / *The Great American Broadcast* 41 / **Orchestra Wives** 42 / *Sweet And Lowdown* 44. Died 1968.

Mayo, Virginia. Bn 20 November 1920, St Louis as Virginia Jones. Blonde dancer-actress who first appeared in bits in *Sweet Rosie O'Grady* 43 / *Pin-Up Girl*, as one of the Goldwyn Girls in **Up In Arms** 44, rising to star in Danny Kaye's next films. Musicals *Seven Days Ashore* 44 / *Wonder Man* 45 / *The Kid From Brooklyn* 46 / *A Song Is Born* 48 / *Always Leave Them Laughing* 49 / **The West Point Story** 50 / *Starlift, She's Working Her Way Through College, Painting The Clouds With Sunshine* 51 / *She's Back On Broadway* 52.

Melchior, Lauritz. Bn 1890, Copenhagen. Wagnerian tenor, great in musical and physical stature, who turned up in benevolent roles in MGM musicals *The Thrill Of A Romance, Two Sisters From Boston* 45 / *This Time For Keeps* 46 / *Luxury Liner* 47 / *The Stars Are Singing* 53.

Melton, James. Leading tenor of the Metropolitan Opera whose good looks made him a temporary film star in the mid-thirties, but with no more permanent success than any other opera singer. Musicals **Stars Over Broadway**

Ethel Merman

35 | *Sing Me A Love Song* 36 | *Melody For Two*
37 | *The Ziegfeld Follies* 44. Died *c*. 1961.

Mercer, Johnny. Bn 18 November 1909.
Composer-lyricist. Former actor-singer-M.C.
with Paul Whiteman and Benny Goodman.
Noted for the unorthodoxy of his lyrics, he
worked most regularly with Richard Whiting
(37–8), Harry Warren (37–9) and Gene De Paul
(54–9) (*q.v.* for credits). With other composers
wrote songs for *College Coach* 33 | *Transatlantic
Merry-Go-Round* 34 | *To Beat The Band, Old Man
Rhythm* (also appeared in both) 35 | *Rhythm On
The Range* 36 | *Mr Chump* 38 | *Let's Make Music,
Second Chorus, You'll Find Out* 40 | **Blues In The
Night, The Birth Of The Blues,** *Navy Blues,
You're The One* 41 | **The Fleet's In, Star
Spangled Rhythm, You Were Never Lovelier**
42 | *True To Life, Riding High,* **The Sky's The
Limit** 43 | **Here Come The Waves,** *Out Of This
World* 45 | **Centennial Summer** 46 | *Always
Leave Them Laughing* 49 | *The Petty Girl* 50 | *Here
Comes The Groom* 51 | *Everything I Have Is Yours*
52 | *Dangerous When Wet* 53 | **Daddy Long
Legs,** *I'll Cry Tomorrow* 55 | **Merry Andrew**
58. Wrote lyrics for Henry Mancini's many
film themes and his songs for *Darling Lili* 69.

Merman, Ethel. Bn 16 January 1909, Astoria,
Long Island, as Ethel Agnes Zimmerman.
Brass-lunged singing star who concentrates on
stage but has made an impact whenever she does
appear in a screen musical *Follow The Leader*
30 | *Kid Millions,* **We're Not Dressing** 34 |
Strike Me Pink, **The Big Broadcast Of 1936** 35 |
Anything Goes 36 | *Happy Landing, Straight
Place And Show,* **Alexander's Ragtime Band**
38 | *Stage Door Canteen* 43 | **Call Me Madam** 53 |
There's No Business Like Show Business 54.

Merrill, Robert (Bob). Bn 17 May 1921,
Atlantic City. Composer-lyricist. Long experience
on staff of Columbia films and CBS TV and
records. Churned out record hits for Guy
Mitchell *et al* then turned to Broadway. Film
scores *The Wonderful World Of The Brothers
Grimm* 62 | **Funny Girl** 68.

Merry Macs. Top close harmony vocal group
of the thirties and forties, founded by the
McMichael brothers (Judd, Ted and the late
Joe), who did guest spots in Universal and
Paramount musicals *Love Thy Neighbour* 40 |
*Melody Lane, Moonlight In Hawaii, Ride 'em
Cowboy, San Antonio Rose* 41 | **Mr Music** 50.

Metro-Goldwyn-Mayer Pictures. Since the
advent of sound MGM and musicals have been
synonymous. The company was responsible for
one of the first big scale musicals, **The
Broadway Melody** 29, which set the pattern for
the extravaganzas produced by its rivals. In
later years, with Arthur Freed as the leading

producer of musicals (*q.v.*) MGM had access to
the best talent in the world, and stars like the
Marx Brothers, Judy Garland, Mickey Rooney,
Gene Kelly, Frank Sinatra, Esther Williams,
Mario Lanza, Eleanor Powell, Nelson Eddy,
Jeanette MacDonald, Howard Keel, Red
Skelton *et al*, such directors as Minnelli, Donen,
Thorpe, Berkeley, etc., musicians like Lennie
Hayton, Johnny Green, André Previn, Georgie
Stoll, Saul Chaplin, and dance directors Michael
Kidd, Donen, Bob Fosse, Hermes Pan, were
regularly employed under contracts. MGM
continued producing musicals after many of the
other companies had dropped out of the race,
and some of the best films of all came in the
fifties. But with the end of the musical era many
of the stars had left the screen, and the creative
talents turned to non-musical subjects. The
reign of 'King Leo' was over, but the 218

musical films listed in this book testify to the enduring quality of the work of his subjects.

Miller, Ann. Bn 12 April 1919, Houston, Texas, as Lucy Ann Collier. Vivacious dancer-singer who whirled through many musicals. Concentrating on tap rather than balletic style she never quite achieved the repute of others, but was never less than dependable. Musicals *The Life Of The Party* (début), *New Faces Of 1937*, *Radio City Revels* 37 / *Having Wonderful Time* 38 / *Melody Ranch, Hit Parade Of 1941, Too Many Girls* 40 / *Time Out For Rhythm, Go West Young Lady* 41 / *Priorities On Parade, True To The Army* 42 / *What's Buzzin' Cousin?, Reveille With Beverley* 43 / *Jam Session, Hey Rookie, Eadie Was A Lady, Carolina Blues* 44 / *The Thrill Of Brazil* 46 / **Easter Parade,** *The Kissing Bandit* 48 / **On The Town** 49 / *Two Tickets To Broadway, Texas Carnival* 51 / **Lovely To Look At** 52 / *Small Town Girl,* **Kiss Me Kate** 53 / **Deep In My Heart** 54 / **Hit The Deck** 55 / *The Opposite Sex* 56.

Miller, Glenn. Bn 1 March 1904, Clarinda, Iowa, as Alton Glenn Miller. Probably the most famous bandleader of all whose two films built around his band, **Sun Valley Serenade** 41, and **Orchestra Wives** 42, have become as legendary as the man himself, who vanished on a flight over the English Channel on 15 December 1944. Miller was portrayed by James Stewart, and his music re-created by former Miller musicians in *The Glenn Miller Story* 54.

Miller, Marilyn. Bn 1 September 1900, Evansville, Indiana as Marilyn Reynolds. Leading lady of Broadway who found fame in Shubert shows, and appeared briefly in musicals *Sally* 29 / *Sunny* 30 / *Her Majesty Love* 31. Died at the height of her fame on 7 April 1936, and was the subject of a biopic **Look For The Silver Lining** 49, with June Haver as the star.

Mills Brothers. Negro harmony group renowned in early days for their instrumental imitations. All born in Piqua, Ohio; John (father) on 11 February 1889, Herbert on 2 April 1912, Harry on 19 August 1913, Donald on 29 April 1915. Appeared in musicals *The Big Broadcast* 32 / *Operator 13, Strictly Dynamite,* **Twenty Million Sweethearts** 34 / *Broadway Gondolier* 35 / *Rhythm Parade* 42 / *Reveille With Beverley, Chatterbox* 43 / *When You're Smiling* 53 / *The Big Beat* 57. Since father's death, sons still active as a trio.

Minnelli, Vincente. Bn 28 February 1913, Chicago. Director whose previous experience as art director has often shown in the settings of his opulent MGM musicals. Like Busby Berkeley was sometimes accused of vulgarity and/or garishness, but his use of colour was always tremendously effective, and certainly his musicals are usually outstanding examples of the *genre* **Babes On Broadway** 41 / **Cabin In The Sky,** *Panama Hattie* 42 / **I Dood It** 43 / **The Ziegfeld Follies** (part-dir), **Meet Me In St Louis** 44 / **Yolanda And The Thief** 45 / **Till The Clouds Roll By** 46 / **The Pirate** 47 / **An American In Paris** 51 / **Lovely To Look At** 52 / **The Band Wagon** 53 / **Brigadoon, Kismet** 55 / **Gigi** 58 / **Bells Are Ringing** 60 / **On A Clear Day You Can See Forever** 69. Most later output non-musical.

Miranda, Carmen. Bn 9 February 1909, Marco Canavozes, Portugal as Maria do Carmo Miranda da Cuhnha. (Birth date is usually given as 1915 and has also appeared as 1904. The above is believed to be correct.) One of Brazil's top singers since about 1931, but came to prominence as 'The Brazilian Bombshell' in Fox's wartime musicals, in which her powers as a brilliant Latin singer were overshadowed by her fruity hats **Down Argentine Way** 40 / **That Night In Rio, A Weekend In Havana** 41 / **Springtime In The Rockies** 42 / **The Gang's All Here** 43 / *Four Jills In A Jeep, Greenwich Village, Something For The Boys* 44 / *Doll Face* 45 / *If I'm Lucky* 46 / *Copacabana* 47 / *A Date With Judy* 48 / *Nancy Goes To Rio* 49 / *Scared Stiff* 53. Died 5 August 1955.

Mirisch Corporation. Independent production company founded in 1957 by the Mirisch brothers, Harold and Marvin (former exhibitors) and Walter (producer). Specialized in 'quality' product, employing top talent like director Billy Wilder, and releasing through United Artists. Musicals **Some Like It Hot** 59 / **West Side Story** 61 / *Follow That Dream* 62 / *Kid Galahad* 64 / *Kiss Me Stupid* 65 / **How To Succeed In Business Without Really Trying** 66.

Vincente Minnelli / (off-set) An American in Paris

Mitchell, Guy. Bn 21 February 1927, Detroit as Al Cernik. Ebullient pop singer of the early fifties who bounced through two musicals *Those Redheads From Seattle* 53 / **Red Garters** 54. Latterly making a comeback on the club circuit and records.

Mitchell, Sidney D. Bn 15 June 1888, Baltimore. Lyricist who contributed now-familiar songs to many Fox musicals of the thirties *Broadway*, **Fox Follies** 29 / *Let's Go Places, The Life Of The Party* 30 / *Blonde Crazy* 31 / *Broadway Bad* 33 / *Down To Their Last Yacht, I Like It That Way* 34 / *Captain January* 35 / **Sing Baby Sing**, *Dancing Feet, Laughing Irish Eyes, Pigskin Parade, Sitting On The Moon* 36 / *One In A Million, Life Begins In College, Heidi, In Old Chicago, Thin Ice* 37 / *Kentucky Moonshine, Rebecca Of Sunnybrook Farm* 38 / *Johnny Doughboy* 42. Died 25 February 1942.

Monaco, James V. Bn 13 January 1885, Fornia, Italy. Composer who joined Paramount under contract in 1930. Wrote many of Bing Crosby's greatest songs with lyrics by Johnny Burke (38–41), then moved to Fox and wrote with Mack Gordon (42–5). Musicals **The Jazz Singer** 27 / *Let's Go Places* 30 / **The Big Broadcast** 32 / *Doctor Rhythm, The East Side Of Heaven, Sing You Sinners, The Star Maker* 38 / **The Road To Singapore**, *If I Had My Way, Rhythm On The River* 40 / *Six Lessons From Madame La Zonga*, **A Weekend In Havana** 41 / *Iceland* 42 / *Stage Door Canteen* 43 / *Sweet And Lowdown, Pin-Up Girl* 44 / *The Dolly Sisters* 45. Died 16 October 1945.

Monogram Pictures Corporation. A subsidiary company of Allied Artists that produced many quickies in the forties, mostly Westerns, crime films and distinctly low quality musicals, many produced by Sam Katzman (*q.v*) and catering for the bobbysox element.

Monroe, Marilyn. Bn 1 June 1926, Los Angeles, as Norma Jean Baker. Blonde comedienne-actress, sometime singer-dancer, whose too-short career gave us no chance to discover her true capabilities. Decorated several musicals *Ladies Of The Chorus* (début), *Love Happy* 48 / **Gentlemen Prefer Blondes** 53 / **There's No Business Like Show Business** 54 / **Some Like It Hot** 59 / *Let's Make Love* 60. Died 5 August 1962.

Montalban, Ricardo. Bn 25 November 1920, Mexico City. Actor who made his début in Mexican films at twenty-one and came to Hollywood 1946. Appeared in MGM dramas and musicals *Fiesta* (début) 46 / *On An Island With You* 47 / *The Kissing Bandit* 48 / *Neptune's Daughter* 49 / *Two Weeks With Love* 50 / *Latin Lovers* 53 / *The Singing Nun* 66 / **Sweet Charity** 68.

Moore, Constance. Bn 18 January 1919, Sioux City, Iowa. Singer-actress who starred in minor musicals of the war years *The Freshman Year, Swing That Cheer* 38 / *Laugh It Off, Charlie McCarthy Detective, Hawaiian Nights* 39 / *Argentine Nights, I'm Nobody's Sweetheart Now, Ma He's Making Eyes At Me* 40 / *Las Vegas Nights* 41 / *Delightfully Dangerous*, **Show Business**, *Atlantic City* 44 / *Earl Carroll's Vanities* 45 / *Earl Carroll's Sketchbook, Hit Parade Of 1947, Mexicana* 46.

Moore, Grace. Bn 5 December 1901, Jellico, Tenn. Metropolitan opera singer whose first films flopped, but made a successful comeback with **One Night Of Love** 34. Tendency to overweight caused problems and eventually returned to opera. Other musicals *A Lady's Morals, New Moon* 30 / *Love Me Forever* 35 / *The King Steps Out* 36 / *I'll Take Romance, When You're*

In Love 37. Killed in air crash at Copenhagen airport 1947. Her life story was filmed as *So This Is Love* 53, with Kathryn Grayson.

Moreno, Rita. Bn 11 December 1931, Humacao, Puerto Rico, as Rosita Dolores Anerio. Dancer-actress who came from TV spots to bits in 'B' pictures and later featured roles. Musicals *Pagan Love Song, **The Toast Of New Orleans** 50/**Singin' In The Rain** 52/Latin Lovers 53/ **The Vagabond King, The King And I** 56/ **West Side Story** 61.

Morgan, Dennis. Bn 20 December 1910, Prentice, Wisconsin, as Stanley Morner. Actor who studied for opera and sang on radio before screen début in 1936. Started with MGM, then under long contract to Warner. Musicals **The Great Ziegfeld** 36/The Desert Song, **Thank Your Lucky Stars** 43/Hollywood Canteen, Shine On Harvest Moon 44/ **The Time, The Place And The Girl,** Two Guys From Milwaukee 46/My Wild Irish Rose 47/One Sunday Afternoon, Two Guys From Texas 48/**It's A Great Feeling** 49/ Painting The Clouds With Sunshine 51.

Morgan, Helen. Bn 1900, Danville, Illinois. Singer whose Broadway appeal was transferred to the screen successfully. Probably the original 'torch singer'. Musicals *Applause, Roadhouse Nights,* **Glorifying The American Girl** 29/ *You Belong To Me, Marie Galante* 34/Go Into Your Dance, Frankie And Johnny, Sweet Music 35/ **Show Boat** 36. Died 8 October 1941 as a result of personal problems; these were partly revealed in her biopic **The Helen Morgan Story** 57, starring Ann Blyth as Helen Morgan.

Moross, Boris. Bn 1 January 1895, Alexandrovsk, Russia. Composer-conductor-producer. Was musical director of the Paramount theatre chain before joining the parent company's music department. Musical director of *Palm Springs,* **The Big Broadcast Of 1937** 36/ *Swing High Swing Low, This Way Please,* **The Big Broadcast Of 1938,** *The Thrill Of A Lifetime, Mountain Music, Turn Off The Moon,* **High, Wide And Handsome,** *Blossoms On Broadway, Vogues Of 1938* 37/*Tropic Holiday, Thanks For The Memory, Sing You Sinners, Romance In The Dark, Give Me A Sailor, Artists And Models Abroad* 38/*Cafe Society, Never Say Die* 39. Produced: *Second Chorus* 40/*Carnegie Hall* 49.

Munshin, Jules. Bn 1916. Comedian who shared top billing in four top musicals but never followed through **Take Me Out To The Ball Game, Easter Parade** 48/**On The Town** 49/ **Silk Stockings** 56/Ten Thousand Bedrooms 57. Died February 1970.

Murphy, George. Bn 4 July 1904, New Haven, Conn. Broadway dancer who was at one time a serious rival to Astaire, starring in many musicals *Kid Millions* (début) 34/*After The Dance* 35/ **The Broadway Melody Of 1938,** *Top Of The Town, You're A Sweetheart* 37/*Little Miss Broadway, Hold That Co-ed* 38/ **The Broadway Melody Of 1940,** *Little Nellie Kelly, Two Girls On Broadway* 40/*Rise And Shine* 41/**For Me And My Gal,** *The Mayor Of 44th Street, The Powers Girl* 42/**Broadway Rhythm, This Is The Army** 43/**Step Lively, Show Business** 44/*Big City* 48. Left show-business for politics.

Murphy, Ralph. Bn 1895, Rockville, Conn. Director-screenwriter. Did many adventure films, also several musicals, mainly inconsequential *Collegiate, Florida Special* 36/*Top Of The Town* 37/*You're The One, Las Vegas Nights, Glamour Boy* 41/*Salute For Three* 43/*Rainbow Island* 44/*Sunbonnet Sue, How Do You Do* 45/ *Mickey* 48. Died 1967.

Myrow, Josef. Bn 28 February 1910, Russia. Composer-lyricist. Former pianist who played with Philadelphia and Cleveland Symphony Orchestras, then radio and stage conductor. Started composing songs for films with Eddie De Lange (*If I'm Lucky* 46), then collaborated with Mack Gordon (*q.v.*) on scores for *Three Little Girls In Blue* 46/**Mother Wore Tights** 47/ *When My Baby Smiles At Me* 48/*The Beautiful Blonde From Bashful Bend* 49/*Wabash Avenue* 50/ *Call Me Mister* 51/*I Love Melvin* 53/*Bundle Of Joy* 56. Also worked with Ralph Blane on *The French Line* 53.

McCarey, Leo. Bn 3 October 1898, Los Angeles. Director-producer-scenarist. Started as gag writer-director with Mack Sennett, later established solid reputation as a craftsman. Output marked a policy of quality rather than quantity and most of his films are memorable. Musicals *Red Hot Rhythm* (dir) 29/*Let's Go Native* (dir) 30/ **The Kid From Spain** (dir-scr) 32/*Duck Soup* 33/*The Belle Of The Nineties* (dir) 34/*Love Affair* (prod-dir) 39/**Going My Way** (prod-dir-scr) 44/*The Bells Of St Mary's* (prod-dir-scr) 45. Died July 1969.

McDonald, Grace. Bn *c.* 1919, Boston. Dancer-actress. Sister of Ray McDonald (*q.v.*) whom she partnered in vaudeville in their teens. Stage success in 'Babes In Arms' and 'Very Warm For May' took her to Hollywood where she played lead in many 'B' musicals for Universal *et al Dancing On A Dime* 40/*Behind The Eight Ball, Give Out Sisters, Strictly In The Groove, What's Cookin'?* 42/*Always A Bridesmaid, Gals Incorporated, Get Going, It Ain't Hay, How's About It?* 43/*Follow The Boys, Hat Check Honey, Murder In The Blue Room, My Gal Loves Music, She's For Me* 44/*Honeymoon Ahead, See My Lawyer* 45.

MacDonald, Jeanette. Bn 18 June 1903,

Philadelphia. Soprano whose eternal brightness
bounced off Nelson Eddy's stolidity, but despite
the mis-match set the box-offices alight with
their annual operettas through the thirties.
Started as a chorus girl in 1920 and came to
films via Lubitsch, at which time the sparkle was
less forced. Musicals *The Love Parade* (début)
29/*Let's Go Native, Lottery Bride, Monte Carlo,
The Vagabond King* 30/*Love Me Tonight,
One Hour With You* 32/*The Cat And The
Fiddle, The Merry Widow* 34/*Naughty
Marietta, Rose Marie* 35/*San Francisco* 36/
The Firefly, Maytime 37/*The Girl Of The Golden
West, Sweethearts* 38/*Broadway Serenade* 39/
Bitter Sweet, New Moon 40/*I Married An Angel,
Cairo* 42/*Follow The Boys* 44/*Three Daring
Daughters* 48. Came back in the sixties doing
stage and cabaret singing till her death on
14 January 1965.

McDonald, Ray. Bn 1920, Boston. Dancer.
Brother of Grace McDonald (*q.v.*). Formed a
teenage vaudeville act with her, and entered
films soon after. Musicals *Babes On Broadway*
41/*Born To Sing, Presenting Lily Mars* 42/*Till
The Clouds Roll By* 46/*Good News* 47/*There's
A Girl In My Heart* 49/*All Ashore* 52. Married
Peggy Ryan (*q.v.*), left screen to start dance
schools. Died 1959.

McGuire, Marcy. Bn *c.* 1925. Perky, wide-
eyed teenage singer of several wartime musicals
Seven Days Leave 42/*Around The World, Follies
Girl, Higher And Higher* 43/*Seven Days
Ashore* 44/*Sing Your Way Home* 45/*Riverboat
Rhythm, It Happened In Brooklyn* 46/*Melody
Maker* 47/*Jumping Jacks* 52/*Mary Poppins* 64.

McHugh, Jimmy. Bn 10 July 1894, Boston.
Composer of some of the song hits of the
century, who worked on many films in colla-
boration with Dorothy Fields (30–5), and
Harold Adamson (37–8, 43–8), also with Frank
Loesser, Al Dubin, etc. Wrote songs or complete
scores for *Dance Fools Dance, Love In The Rough*
30/*Cuban Love Song, Flying High* 31/*Dancing
Lady* 33/*Nitwits, Hooray For Love, Every
Night At Eight* 35/*Dimples, The King of
Burlesque, Let's Sing Again* 36/*Banjo On My
Knee, You're A Sweetheart, Top Of The Town,
Hitting A New High* 37/*Merry-Go-Round Of 1938,
Mad About Music, That Certain Age* 38/*Down
Argentine Way, Buck Benny Rides Again, You'll
Find Out, You're The One* 40/*Seven Days Leave,
Happy Go Lucky, Hers To Hold* 42/*Higher And
Higher, Around The World* 43/*Something For The
Boys, Four Jills In A Jeep* 44/*Nob Hill, Bring On
The Girls, Doll Face, Radio Stars On Parade* 45/
Do You Love Me?, Hit Parade Of 1947 46/
Calendar Girl, If You Knew Susie 47/*A Date With
Judy* 48. Made guest appearance in *The Helen*

Morgan Story 57, in which many of his old
standards were used. Died 23 May 1969.

MacLaine, Shirley. Bn 24 April 1934,
Richmond, Virginia, as Shirley MacLaine
Beatty. Actress-comedienne-dancer. Former
Broadway chorus girl who has become out-
standing screen personality with zany sense of
humour. Musicals *Artists And Models* 56/*Can
Can* 60/*Sweet Charity* 68.

McLeod, Norman Z. Bn 30 September 1898,
Grayling, Michigan. Director, former cartoon
animator and scenarist. Did much of his work for
Paramount and proved a dependable director of
musicals *Along Came Youth* 31/*Horse Feathers*
32/*Many Happy Returns, Melody In Spring* 34/
Redheads On Parade, Coronado 35/*Pennies From
Heaven* 36/*Lady Be Good* 41/*Panama Hattie,
The Powers Girl* 42/*The Kid From Brooklyn* 46/
The Road To Rio 47/*The Paleface, Isn't It
Romantic?* 48/*Let's Dance* 49. Died 1964.

McLerie, Allyn. Bn 1 December 1926, Grand
Mere, Quebec. Dancer, who did useful work in a
few top musicals, but preferred the stage
Words And Music 48/*The Desert Song,
Where's Charley?* 52/*Calamity Jane* 53.

MacMurray, Fred. Bn 30 August 1908,
Kaukakee, Illinois. Actor, one of the screen's
evergreen leading men, adept at comedy or
drama. Came to Hollywood as saxophone player
with the California Collegians, broke into films
and stayed at the top. Musicals *To Beat The Band*
35/*Swing High Swing Low, Champagne Waltz* 37/
The Cocoanut Grove, Sing You Sinners 38/*Cafe
Society* 39/*Little Old New York* 40/*Star Spangled
Rhythm* 42/*And The Angels Sing* 44/*Where
Do We Go From Here?* 45/*The Happiest
Millionaire* 67.

McPhail, Douglas. Bn *c.* 1919. Robust young
baritone who did well in a few MGM musicals

Babes In Arms 39 / Little Nellie Kelly, The Broadway Melody Of 1940 40 / Born To Sing 42.

Macrae, Gordon. Bn 12 March 1921, East Orange, N.Y. Singer who came from dance bands and summer stock to a Warner contract. A rich and melodious baritone who could handle any kind of song, he failed to survive the end of the era of musicals Look For The Silver Lining 49 / Tea For Two, The Daughter of Rosie O'Grady, The West Point Story 50 / Starlift, On Moonlight Bay 51 / The Desert Song, By The Light Of The Silvery Moon, About Face 52 / Three Sailors And A Girl 53 / Oklahoma! 55 / Carousel, The Best Things In Life Are Free 56. Later went into cabaret and TV with his wife, Sheila.

Negulesco, Jean. Bn 29 February 1900, Craiova, Rumania. Director of over forty years' experience of all types of films, but only two musicals Daddy Long Legs 55 / The Pleasure Seekers 64.

Neilan, Marshall. Bn 1891. Successful director of silents who never really made the change to sound. Musicals Tanned Legs, The Vagabond Lover 29 / Sweethearts On Parade 30 / Swing It Professor 37. Died 1958.

Nelson, Gene. Bn 24 March 1920, Seattle, as Gene Berg. Dancer in many musicals, mostly for Warner, who began as a swimmer and skater but turned to dancing in the 'This Is The Army' show. Musicals I Wonder Who's Kissing Her Now (début) 47 / Tea For Two, The Daughter Of Rosie O'Grady, The West Point Story 50 / Lullaby Of Broadway, Painting The Clouds With Sunshine, She's Working Her Way Through College, Starlift 51 / She's Back On Broadway 52 / Three Sailors And A Girl 53 / So This Is Paris 54 / Oklahoma! 55. Directed Kissin' Cousins 64 / Harem Scarum, Your Cheatin' Heart 65.

Nelson, Ozzie. Bn 20 March 1906, Jersey City, as Oswald George Nelson. Bandleader since 1932, married Harriett Hilliard (q.v.) 1935, who then sang with his band on radio, screen and TV, and founded the Nelson Family (sons Rick and David). Musicals The Sweetheart Of The Campus 41 / Strictly In The Groove 42 / Honeymoon Lodge 43 / Hi Good Lookin', Take It Big 44 / People Are Funny 45 / Here Come The Nelsons 52.

Newman, Alfred. Bn 17 March 1901, New Haven, Conn. Composer-conductor. Ex-vaudeville and cinema pianist who also played with the N.Y. Philharmonic. Conducted vaudeville and Broadway shows, and went to Hollywood 1930. Musical director for United Artists in early thirties, also worked for RKO before joining Fox 1939. Credits as musical director too many to list, but principal musicals include Whoopee 30 / The Kid From Spain 32 / Kid Millions 34 / Strike Me Pink 35 / Born To Dance 36 / A Damsel In Distress, When You're In Love 37 / Alexander's Ragtime Band, The Goldwyn Follies 38 / The Star Maker 39 / The Broadway Melody Of 1940, Tin Pan Alley 40 / A Weekend In Havana, Moon Over Miami 41 / Orchestra Wives 42 / Coney Island 43 / Billy Rose's Diamond Horseshoe 45 / The Dolly Sisters 46 / My Blue Heaven 50 / With A Song In My Heart 52 / Call Me Madam 53 / Daddy Long Legs 55 / Carousel, The King And I 56 / South Pacific 58 / The Flower Drum Song 61 / State Fair 62 / Camelot 67. Died February 1970.

Newman, Emil. Fox musical director. Musicals Down Argentine Way 40 / Sun Valley Serenade 41 / Iceland 42 / Hello, Frisco, Hello 43 / Something For The Boys, Sweet And Lowdown, Pin-Up Girl, Four Jills In A Jeep 44 / Doll Face, Where Do We Go From Here? 45 / Wake Up And Dream, If I'm Lucky, Do You Love Me? 46 / Carnival In Costa Rica 47 / A Song Is Born 48 / Just For You 52.

Newman, Lionel. Bn 4 January 1916, New Haven. Conn. Composer-conductor – the third Newman to get musical director credits on Fox musicals. Joined the company from 'Earl Carroll's Vanities' as pianist and rose to Head Of Music Department. Musical director of Give My Regards To Broadway 48 / I'll Get By 50 / Meet Me After The Show, Golden Girl 51 / The Farmer Takes A Wife 52 / Gentlemen Prefer Blondes 53 / There's No Business Like Show Business 54 / Love Me Tender, The Best Things In Life Are Free 56 / April Love, The Girl Can't Help It, Sing Boy Sing 57 / Mardi Gras 58 / Say One For Me 59 / Let's Make Love, Flaming Star 60 / Swingin' Along 62 / The Pleasure Seekers 64 / Doctor Dolittle 67 / Hello Dolly! 69.

Nicholas Brothers. Negro dancing act who

did their (often incredibly agile) speciality in musicals *The Big Broadcast Of 1936* 35 / *Down Argentine Way, Tin Pan Alley* 40 / *The Great American Broadcast, Sun Valley Serenade* 41 / *Orchestra Wives* 42 / *Stormy Weather* 43 / *Carolina Blues* 44 / *The Pirate* 47.

Niesen, Gertrude. Bn at sea between London and New York, brought up in Brooklyn. Broadway singer who was a radio favourite in the forties and appeared in musicals *Top Of The Town, Start Cheering* 37 / *A Night At Earl Carroll's* 40 / *Rookies On Parade* 41 / *This Is The Army, Thumbs Up* 43.

Noble, Ray. Bn 1907, Brighton, England. Bandleader-composer of international reputation in the thirties. Went to USA 1934 and formed band later taken over by Glenn Miller. Famous for radio work with Edgar Bergen. Appeared in musicals (sometimes wrote songs also) *The Big Broadcast Of 1936* 35 / *A Damsel In Distress* 37 / *Here We Go Again* 42 / *Lake Placid Serenade, Out Of This World* 45. Now in retirement in the Channel Islands.

North, Sheree. Bn 17 January 1930, Los Angeles, as Dawn Bethel. Blonde dancer-singer-actress signed by Fox, allegedly as a threat to keep Marilyn Monroe in order. Musicals *Living It Up* 54 / *The Best Things In Life Are Free* 56 / *Mardi Gras* 58. Now does relatively small acting bits.

Nugent, Elliot. Bn 20 September 1899, Dover, Ohio. Director. Former child vaudevillian who acted in many MGM films before turning playwright. Directed and produced Broadway shows then turned to screen directing. Musicals (as actor) *So This Is College* 29 (as director) *She Loves Me Not, Strictly Dynamite* 34 / *Love In Bloom* 35 / *Give Me A Sailor* 38 / *Never Say Die* 39 / *Up In Arms* 44 / *Welcome Stranger* 47 / *Just For You* 52.

Oakie, Jack. Bn 12 November 1903, Sedalia, Missouri, as Lewis D. Offield. Chubby comedian who was practically a fixture in Paramount and other musicals of the thirties *Hit The Deck, Street Girl, Sweetie* 29 / *Paramount On Parade, Close Harmony, Let's Go Native* 30 / *Dude Ranch, June Moon* 31 / *College Humour, Sitting Pretty, Too Much Harmony* 33 / *Shoot The Works, Murder At The Vanities, College Rhythm* 34 / *The Big Broadcast Of 1936, The King of Burlesque* 35 / *That Girl From Paris, Colleen, Collegiate, Florida Special* 36 / *Champagne Waltz, Hitting A New High, Radio City Revels* 37 / *Thanks For Everything* 38 / *Tin Pan Alley, Young People* 40 / *Navy Blues, Rise And Shine, The Great American Broadcast* 41 / *Footlight Serenade, Iceland, Song Of The Islands* 42 / *Hello, Frisco, Hello, Something To Shout About, Wintertime* 43 / *The Merry*

Monahans, On Stage Everybody, Sweet And Lowdown, Bowery To Broadway 44 / *That's The Spirit* 45 / *When My Baby Smiles At Me* 48.

O'Brien, Virginia. Bn 1923, Los Angeles. Singer-comedienne, whose deadpan delivery of ballads and comedy songs was a passable gimmick in MGM's wartime musicals, but didn't make for lasting fame *Hullabaloo* (début) 40 / *Lady Be Good, The Big Store* 41 / *Panama Hattie, Ship Ahoy* 42 / *Dubarry Was A Lady, Thousands Cheer, Meet The People* 43 / *Two Girls And A Sailor, The Ziegfeld Follies* 44 / *The Harvey Girls* 45 / *The Show-Off, Till The Clouds Roll By* 46.

O'Connor, Donald. Bn 30 August 1925, Chicago. One of Hollywood's greatest all-round talents as singer, dancer and comedian. Survived the transitions from child star to teenage star to adult star; even survived a film series with a talking mule. Musicals *Melody For Two* (début) 37 / *Sing You Sinners* 38 / *On Your Toes* 39 / *Get Hep To Love, Give Out Sisters, Private Buckaroo, When Johnny Comes Marching Home* 42 / *Top Man, Mr Big* 43 / *Patrick The Great, The Merry Monahans, Follow The Boys, A Chip Off The Old Block, Bowery To Broadway, This Is The Life* 44 / *Something In The Wind* 47 / *Feudin' Fussin' And A-Fightin', Are You With It?* 48 / *Yes Sir That's My Baby* 49 / *Double Crossbones, The Milkman* 50 / *Singin' In The Rain* 52 / *I Love Melvin, Call Me Madam, Walking My Baby Back Home* 53 / *There's No Business Like Show Business* 54 / *Anything Goes* 56.

O'Curran, Charles. Dance director of musicals *Swingtime Johnny* 43 / *Seven Days Ashore, Music In Manhattan, Moonlight And Cactus* 44 / *Bamboo Blonde* 46 / *If You Knew Susie* 47 / *Somebody Loves Me* 52 / *Bells Are Ringing* 60 / *Girls Girls Girls* 62 / *Fun In Acapulco* 63.

often in singing partnership with Jane Frazee (*q.v.*). Musicals *Cain And Mabel* 36 | *The Lady Objects* 38 | *Dancing On A Dime* 40 | *Almost Married, Don't Get Personal, Melody Lane, San Antonio Rose* 41 | *Pardon My Sarong, Hellzapoppin', Get Hep To Love, What's Cookin'?* 42 | *A Cowboy In Manhattan, Crazy House, Get Going, Hi Buddy, Hi' Ya Chum, How's About It?, Mr Big* 43 | *Follow The Boys* 44 | **Can't Help Singing,** *Shady Lady* 45. Took up straight acting but parts got smaller; you may have missed him in **Bye Bye Birdie** 63.

Pan, Hermes. Bn 1910, Nashville, Tenn. Doyen of Hollywood dance directors, associated with Fred Astaire throughout his career. Pan's colourful routines were a feature of Fox musicals of the forties; contemporary work shows he has moved with the times. Musicals include **Flying Down To Rio** 33 | **The Gay Divorcee, Roberta** 34 | **Top Hat, Follow The Fleet,** *Old Man Rhythm, In Person, I Dream Too Much* 35 | **Swing Time** 36 | **A Damsel In Distress, Shall We Dance?,** *Radio City Revels* 37 | **Carefree** 38 | *Second Chorus* 40 | *Rise And Shine, That Night In Rio, A Weekend In Havana, Sun Valley Serenade, Moon Over Miami** 41 | **Springtime In The Rockies,** *Song Of The Islands, My Gal Sal*,* **Footlight Serenade** 42 | *Sweet Rosie O'Grady*, Coney Island* 43 | *Pin-Up Girl*, Irish Eyes Are Smiling* 44 | **Billy Rose's Diamond Horseshoe** 45 | **Blue Skies** 46 | *The Shocking Miss Pilgrim, I Wonder Who's Kissing Her Now* 47 | **The Barkleys Of Broadway,** *That Lady In Ermine* 48 | *Let's Dance* 49 | **Three Little Words** 50 | *Excuse My Dust, Texas Carnival* 51 | **Lovely To Look At** 52 | **Kiss Me Kate** 53 | *The Student Prince* 54 | *Jupiter's Darling,* **Hit The Deck** 55 | **Meet Me In Las Vegas, Silk Stockings** 56 | **Pal Joey** 57 | **Porgy And Bess** 59 | **Can Can** 60 | *The Flower Drum Song* 61 | **My Fair Lady** 64 | **Finian's Rainbow** 68 | *Darling Lili* 69. Also appeared in films marked *.

Panama, Norman. Bn 1914. Director-producer-screenwriter, in partnership with Melvin Frank (*q.v.*) for twenty-five years. In the musical field, wrote together screenplays for *Happy Go Lucky,* **Star Spangled Rhythm** 42 | **Thank Your Lucky Stars** 43 | **And The Angels Sing** 44 | *The Road To Utopia* 45 | *Monsieur Beaucaire* 46 | **White Christmas** 54. Wrote, directed and produced in tandem *Knock On Wood* 54 | *The Court Jester* 56 | *Li'l Abner* 59 | **The Road To Hong Kong** 62.

Paramount Picture Corporation. Founded in 1912 by Adolph Zukor (as Famous Players, later to merge with Jesse Lasky's company) who retained an active interest as executive producer till 1935, Paramount has, according to the index

O'Driscoll, Martha. Bn 4 March 1922, Tulsa, Oklahoma. Singer-actress who appeared in a number of second-string films without leaving anything for posterity to remember. Musicals *Collegiate* (début) 36 | *Champagne Waltz* 37 | *Mad About Music* 38 | *Youth On Parade* 42 | *Crazy House* 43 | *Allergic To Love, Ghost Catchers, Hi Beautiful, Weekend Pass, Follow The Boys* 44 | *Here Come The Co-eds, Shady Lady* 45 | *Carnegie Hall* 49.

Oreste. Bn 26 July 1926, Malta, as Oreste Kirkop. Opera singer with probably the shortest cinematic career in history. It began and ended with **The Vagabond King** 56.

Paige, Janis. Bn 16 September 1922, Tacoma, Wash., as Donna Mae Jaden. Leading lady who came from the Tacoma Opera Company to Broadway and Hollywood success as singer, but more often straight actress. Musicals *Hollywood Canteen* (début), *Bathing Beauty* 44 | *Two Guys From Milwaukee,* **The Time, The Place And The Girl** 46 | *Love And Learn* 47 | **Romance On The High Seas,** *One Sunday Afternoon* 48 | *Two Gals And A Guy* 51 | **Silk Stockings** 56 | *Follow The Boys* 63. Lately concentrated on stage work.

Paige, Robert. Bn 2 December 1910, Indianapolis, as John Arthur Paige. Singing actor who made the Universal studios his home and 'B' musicals his *milieu* during the war years,

Larry Parks/Jolson Sings Again

in this book, produced more musicals (243) than even MGM, but the quality of the product was not so consistently high, although in the early thirties Lubitsch's glamorous productions with Chevalier, MacDonald, Colbert *et al* did set a high standard. Later in the decade Bing Crosby's series of musicals established the company as a good box-office attraction, emphasized by the 'Road' series. Bob Hope's comedies and semi-musicals, together with those of Betty Hutton, Buddy Rogers, Martin and Lewis, Mae West, Elvis Presley, Dorothy Lamour, Marx Brothers, Danny Kaye and many others, have always ensured a steady output of good entertainment. In the non-musical field, Paramount's biggest money-spinners were probably Cecil B. de Mille's epics. After many years devoid of musicals, the company, now owned by Gulf and Western Industries, and with its own music publishing and recording interests, has come back to the scene with *Paint Your Wagon* and *On A Clear Day You Can See Forever.*

Parks, Larry. Bn 3 December 1914, Olathe, Kansas. Actor who spent years in second features before being sprung on the world as Al Jolson in *The Jolson Story* 46, and *Jolson Sings Again* 49. Other musicals *You Were Never Lovelier* 42 / *Is Everybody Happy?*, *Reveille With Beverley* 43 / *She's A Sweetheart*, *Stars On Parade, Hey Rookie* 44 / *Down To Earth* 47. His career never recovered from the Jolson casting and a smear during the McCarthy trials, and he developed a stage act with his wife Betty Garrett (*q.v.*).

Pasternak, Joe. Bn 19 September 1901, Szilgy Somlyo, Hungary. Producer responsible for discovering Deanna Durbin while with Universal. Moved to MGM 1942 and shared the company's musical production with Arthur Freed, Pasternak's being the less opulent films of mass, or family, appeal *Three Smart Girls, 100 Men And A Girl* 36 / *Mad About Music, That Certain Age* 38 / *Three Smart Girls Grow Up, First Love* 39 / *The Underpup, It's A Date, A Little Bit Of Heaven, Spring Parade* 40 / *Nice Girl* 41 / *It Started With Eve, Presenting Lily Mars, Seven Sweethearts* 42 / *Thousands Cheer* 43 / *Anchors Aweigh, Music For Millions, Two Girls And A Sailor* 44 / *No Leave No Love, The Thrill Of A Romance, Two Sisters From Boston, Holiday In Mexico* 45 / *This Time For Keeps* 46 / *The Unfinished Dance, Three Daring Daughters, On An Island With You, Luxury Liner* 47 / *Big City, A Date With Judy, The Kissing Bandit* 48 / *Nancy Goes To Rio, **That Midnight Kiss,** In The Good Old Summertime* 49 / *Summer Stock, The Duchess Of Idaho, **The Toast Of New***

Orleans 50 / *The Great Caruso, Rich Young And Pretty, The Strip* 51 / *Skirts Ahoy, Because You're Mine, **The Merry Widow*** 52 / *Easy To Love, Latin Lovers, Small Town Girl* 53 / *The Student Prince, The Flame And The Flesh, Athena* 54 / ***Hit The Deck, Love Me Or Leave Me*** 55 / *The Opposite Sex, **Meet Me In Las Vegas*** 56 / *Ten Thousand Bedrooms, This Could Be The Night* 57 / *Where The Boys Are* 60 / ***Billy Rose's Jumbo*** 62 / *Looking For Love* 64 / *Girl Happy* 65 / *Spinout* 66.

Payne, John. Bn 28 May 1912, Roanoke. Va. Actor-singer. Millionaire's son who did it the hard way, coming up through stock companies. Début 1936, and was to be found in many Fox wartime musicals *Hats Off* 36 / *Love On Toast, College Swing, The Garden Of The Moon* 38 / ***Tin Pan Alley*** 40 / *The Great American Broadcast, **A Weekend In Havana, Sun Valley Serenade*** 41 / *Footlight Serenade, Iceland, **Springtime In The Rockies*** 42 / ***Hello, Frisco, Hello*** 43 / *Wake Up And Dream, The Dolly Sisters* 46. Latterly on TV (own series and guest spots).

Pearce, Bernard (Babe). Dance director of musicals *Holiday Inn* 42 / *The Kid From Brooklyn* 46 / *The Road To Rio* 47 / *Casbah* 48.

Pennington, Ann. Bn 1895. Star of 'George White's Scandals', etc. who was in a few early musicals *Tanned Legs, Is Everybody Happy?*, ***The Golddiggers Of Broadway*** 29 / *Happy Days* 30.

Perlberg, William. Bn 22 October 1896, New York. Producer with Fox and Paramount whose name was usually coupled with director George Seaton (*q.v.*). Was mainly concerned with quality rather than routine product, including musicals *The King Steps Out* 36 / *The Lady Objects* 38 / ***Hello, Frisco, Hello,** Sweet Rosie O'Grady, Coney Island* 43 / ***Billy Rose's Diamond Horseshoe, Where Do We Go From Here?*** 45 / *The Shocking Miss Pilgrim* 47 / *I'll Get By, Wabash Avenue* 50 / *Aaron Slick From Punkin Crick* 51 / *Somebody Loves Me* 52 / *Little Boy Lost* 53 / ***The Country Girl*** 54. Died October 1968.

Pinza, Ezio. Bn 18 May 1892, Rome. Operatic *basso* who made surprise Broadway appearance in 'South Pacific' after twenty-two years at the Met., but was unsuccessful in 'popular' appearances in screen musicals *Carnegie Hall* 49 / *Mr Imperium* 51 / *Tonight We Sing* 53. Died 1957.

Platt, Marc. Bn 2 December 1913, Pasadena, Calif. Vastly under-rated dancer who never achieved his rightful place in screen history, his first musicals not being up to his talent *Tonight And Every Night* 44 / *Tars And Spars* 46 / *When A Girl's Beautiful, Down To Earth* 47 / ***Seven Brides For Seven Brothers*** 54 / ***Oklahoma!*** 55. Later stage choreographer.

Metropolitan Opera *I Dream Too Much* 35 /
That Girl From Paris 36 / *Hitting A New High*
37 / *Carnegie Hall* 49. Has since sung with her
husband Andre Kostelanetz on his world tours.

Porter, Cole. Bn 9 June 1892, Peru, Indiana.
The supreme sophisticate among Broadway and
Hollywood composers, and one of the 'big
five' (Gershwin, Kern, Rodgers and Berlin).
His urbane, witty lyrics were a perfect match for
the rich harmonies of his music, and although
constant pain caused his standard to fall in the
last decade or so of his life his output was still
far above that of lesser writers. Musicals varied
between adaptations of his stage shows and
original screen scores *The Battle Of Paris* 31 /
The Gay Divorcee 34 / ***Anything Goes*** 36, 56 /
Born To Dance 36 / ***Rosalie*** 37 / ***The Broadway
Melody Of 1940*** 40 / ***You'll Never Get Rich***
41 / *Panama Hattie* 42 / ***Dubarry Was A Lady,***
Let's Face It, ***Something To Shout About*** 43 /
The Pirate 47 / *Mexican Hayride* 48 / ***Kiss Me
Kate*** 53 / ***High Society, Silk Stockings*** 56 / ***Les
Girls*** 57 / ***Can Can*** 60. Cary Grant played
Porter in a most unlikely biopic ***Night And
Day*** 45. Died 15 October 1964.

Porter, Jean. Bn 8 December 1924, Cisco,
Texas. Song-and-dance girl, former child
performer, who was seen in a few musicals in
supporting roles ***Babes On Broadway*** (début)
42 / *Bathing Beauty* 44 / *Easy To Wed* 45 / *Betty Co-ed*
46 / *Two Blondes And A Redhead, Little Miss
Broadway, Sweet Genevieve* 47.

Potter, Henry C. Bn 1906, New York.
Director of superior comedies and a few
musicals *Romance In The Dark* 38 / ***The Story
Of Vernon And Irene Castle*** 39 / *Second Chorus*
40 / *Hellzapoppin'* 42 / ***Three For The Show*** 55.
A former stage director, has since returned to
Broadway.

Powell, Dick. Bn 14 November 1904,
Mountain View, Ark. Singer-actor-director-
producer. Difficult to imagine any Warner
musicals of the Busby Berkeley era without him
(there were just a few). Moved to Paramount
for a couple of years, then shed his college boy
image for that of the tough private eye, and
never sang another song. Musicals *Blessed Event*
(début) 32 / ***Footlight Parade, Forty-Second
Street, The Golddiggers Of 1933*** 33 / *College
Coach,* ***Flirtation Walk, Dames,*** *Happiness
Ahead,* ***Wonder Bar, Twenty Million Sweet-
hearts*** 34 / *Broadway Gondolier,* ***The Golddiggers
Of 1935*** / *Page Miss Glory,* ***Thanks A Million,***
Shipmates Forever 35 / *Colleen,* ***Stage Struck,***
Hearts Divided 36 / ***The Golddiggers Of 1937,
Hollywood Hotel,*** *The Singing Marine,* ***The
Varsity Show, On The Avenue*** 37 / *A Cowboy
From Brooklyn, Hard To Get, Going Places* 38 /

Pollack, Lew. Bn 16 June 1895, New York.
Composer-lyricist. Wrote theme music for silent
films, then many songs for Shirley Temple and
other Fox musicals *Blue Skies* 29 / *Our Little Girl,
Captain January,* ***The King of Burlesque***
35 / *Pigskin Parade, Song And Dance Man,* ***Sing
Baby Sing*** 36 / *Heidi, Thin Ice, Life Begins In
College, One In A Million, In Old Chicago* 37 /
*Hold That Co-ed, Kentucky Moonshine, Rebecca Of
Sunnybrook Farm, Straight Place And Show* 38 /
Tahiti Honey, Hi Buddy, Jitterbugs 43 / *Lady Let's
Dance, Music In Manhattan, Seven Days Ashore,
Sweethearts Of The USA, The Girl Rush* 44 /
Vacation In Reno, The Bamboo Blonde 46 / *Make
Mine Laughs* 49. Died 18 January 1946.

Pons, Lily. Bn 12 April 1898 (has also been
given as 1904), Cannes, France. Soprano known
as 'The Pocket Prima Donna' who made
several interesting films before returning to the

Naughty But Nice 39 | *In The Navy* 41 | *Happy Go Lucky* 42 | *Meet The People*, **Star Spangled Rhythm,** *Riding High, True To Life* 43. Produced and directed *You Can't Run Away From It* 56. Later TV executive producer. Died of cancer June 1963.

Powell, Eleanor. Bn 21 November 1912, Springfield, Ohio. Dancer, discovered by Gus Edwards, who made stage début in the 'Scandals'. Brought an immaculate tap style to all too few musicals **The Broadway Melody Of 1936,** *George White's Scandals Of 1935* 35 | **Born To Dance** 36 | **The Broadway Melody Of 1938,** *Rosalie* 37 | *Honolulu* 39 | **The Broadway Melody Of 1940** 40 | **Lady Be Good** 41 | *Ship Ahoy* 42 | **I Dood It, Thousands Cheer** 43 | *Sensations Of 1945* 44 | *The Duchess Of Idaho* 50.

Powell, Jane. Bn 1 April 1929, Saskatchewan, Canada as Suzanne Bruce. Singer-actress, a former child star on radio, who weathered the transition to adult star better than most. Musicals *The Song Of The Open Road* (début), *Delightfully Dangerous* 44 | *Holiday In Mexico* 45 | *Three Daring Daughters, Luxury Liner* 47 | *A Date With Judy* 48 | *Nancy Goes To Rio* 49 | *Two Weeks With Love* 50 | **Royal Wedding,** *Rich Young And Pretty* 51 | *Small Town Girl, Three Sailors And A Girl* 53 | **Deep In My Heart,** *Athena,* **Seven Brides For Seven Brothers** 54 | **Hit The Deck** 55 | *The Girl Most Likely* 57. Now in cabaret and TV.

Power, Tyrone. Bn 5 May 1913, Cincinnati. Actor, best known for dramatic roles, but who also served his time in Fox musicals **Flirtation Walk** 34 | *In Old Chicago, Thin Ice* 37 | **Alexander's Ragtime Band** 38 | **Rose Of Washington Square, Second Fiddle** 39 |. Played the title role in *The Eddie Duchin Story* 56. Died from a heart attack on location, 1958.

Preisser, June. Bn *c.* 1922, New Orleans. Blonde singing and dancing starlet who started out well but wound up in Sam Katzman quickies. Musicals **Babes In Arms** (début), *Dancing Co-ed* 39 | **Strike Up The Band** 40 | *Sweater Girl* 42 | *Babes On Swing Street* 43 | *Murder In The Blue Room* 44 | *I'll Tell The World, Let's Go Steady* 45 | *Junior Prom, High School Hero, Freddie Steps Out* 46 | *Sarge Goes To College, Two Blondes And A Redhead* 47 | *Campus Sleuth* 48.

Preminger, Otto. Bn 5 December 1906, Vienna. Director of allegedly dictatorial methods which nevertheless produced results. Did few, but generally memorable, musicals *Under Your Spell* 36 | **Centennial Summer** (also prod) 46 | *That Lady In Ermine* 48 | **Carmen Jones** (also prod) 54 | **Porgy And Bess** 59.

Presley, Elvis. Bn 8 January 1935, Tupelo, Missouri. Rock and roll singer who started it

all. Early films had some dramatic pretensions, but later starred in a series of computerized musicals *Love Me Tender* 56 | *Loving You, Jailhouse Rock* 57 | *King Creole* 58 | *Flaming Star, G.I. Blues* 60 | *Wild In The Country* 61 | *Blue Hawaii, Follow That Dream, Girls Girls Girls, Kid Galahad* 62 | *Fun In Acapulco, It Happened At The World's Fair* 63 | *Viva Las Vegas, Kissin' Cousins, Roustabout* 64 | *Tickle Me, Girl Happy, Harem Scarum* 65 | *Frankie And Johnny, Paradise Hawaiian Style, Spinout* 66 | *Easy Come Easy Go, Double Trouble* 67 | *Speedway, Stay Away Joe, Clambake* 68.

Presnell, Harve. Bn *c.* 1937. Singer-actor from stage and Robert Shaw Chorale who has been presented as a leading man in the Howard Keel tradition, not too successfully to date. Musicals **The Unsinkable Molly Brown** (début) 64 | *Girl Crazy* 66 | **Paint Your Wagon** 69.

Preston, Robert. Bn 8 June 1917, Newton Highlands, Mass., as Robert Preston Messervey. Actor, long renowned for parts in gangster, Western and epic films, who suddenly gained musical fame on stage and screen in **The Music Man** 62. Had previously appeared in non-musical roles in musicals *Moon Over Burma* 40 | *Variety Girl* 47 | *Big City* 48.

Previn, André. Bn 6 April 1929, Berlin. Pianist-composer-conductor-arranger. Child prodigy who joined MGM at fifteen arranging Jose Iturbi's 'jazz' piano features. Was conducting film scores at twenty-one and winning Oscars at twenty-nine, all the while making hit jazz records. Played piano for Sinatra on soundtrack of **It Happened In Brooklyn** 46, and appeared in *Fiesta* 46. Musical director of **Three Little Words** 50 | **Give A Girl A Break,** *Small Town Girl,* **Kiss Me Kate** 53 | *Invitation To*

The Dance 54/*It's Always Fair Weather* (also wrote songs), *Kismet* 55/*Silk Stockings* 56/ *Gigi* 58/*Porgy And Bess* 59/*Pepe* (also wrote songs and played on soundtrack), *Bells Are Ringing* 60/*My Fair Lady* 64/*Kiss Me Stupid* 65/*Thoroughly Modern Millie* 67. Wrote additional songs for *Paint Your Wagon* 69. Now resident in England as conductor of London Symphony Orchestra.

Previn, Charles. Bn 11 January 1888, Brooklyn. Composer-conductor. Former pianist-conductor in all fields from vaudeville to opera. As Head of Music Department at Universal received musical director credit on most of that company's films in the thirties and forties. Later conductor at Radio City Music Hall.

Prima, Louis. Bn 7 December 1911, New Orleans. Jazz trumpeter-bandleader who made guest appearances in musicals *Rhythm On The Range* 36/*You Can't Have Everything, Manhattan Merry-Go-Round, Start Cheering* 37/ *Rose Of Washington Square* 39/*Hey Boy Hey Girl* 59/*The Jungle Book* (sang on soundtrack) 67.

Prinz, Eddie. Dance director of *Hellzapoppin', Moonlight In Havana, Behind The Eight Ball* 42/*Tea For Two, The West Point Story* 50.

Prinz, Le Roy. Bn 1895. Dance director of polished and imaginative routines, who worked regularly for Paramount till 1942 then moved to Warners for whom he did some of his best work *Too Much Harmony* 33/*Bolero* 34/*The Big Broadcast Of 1936, All The King's Horses, Stolen Harmony* 35/*Every Day's A Holiday, The Big Broadcast Of 1938, This Way Please, Waikiki Wedding, Turn Off The Moon, High, Wide And Handsome, The Thrill Of a Lifetime* 37/*Artists And Models Abroad, Give Me A Sailor* 38/*Too Many Girls, The Road To Singapore, Buck Benny Rides Again* 40/*The Road to Zanzibar, Time Out For Rhythm* 41/*Yankee Doodle Dandy* 42/ *The Desert Song, Thank Your Lucky Stars, This Is The Army* 43/*Hollywood Canteen, Shine On Harvest Moon* 44/*Rhapsody In Blue, Night And Day* 45/*The Time, The Place And The Girl* 46/*My Wild Irish Rose* 47/*Two Guys From Texas, April Showers* 48/*My Dream Is Yours, Look For The Silver Lining, Always Leave Them Laughing, It's A Great Feeling* 49/*The Daughter Of Rosie O'Grady, Tea For Two, The West Point Story* 50/*On Moonlight Bay, Painting The Clouds With Sunshine, Lullaby Of Broadway, I'll See You In My Dreams, Starlift, She's Working Her Way Through College* 51/*April In Paris, About Face, The Desert Song, She's Back on Broadway, The Jazz Singer* 52/*The Eddie Cantor Story, Three Sailors And A Girl* 53/*Lucky Me* 54/*The*

Helen Morgan Story 57/*South Pacific* 58. Also directed *All-American Co-ed, Fiesta* 41, and produced the latter.

Prowse, Juliet. Bn 25 September 1937, Bombay, India. Long-limbed dancer who did little in films before returning to the night-club and stage world. Musicals *G.I. Blues, Can Can* 60/*The Right Approach* 61.

Pryor, Roger. Former dance bandleader-singer who turned actor and starred in (mostly second string) films, including musicals *Moonlight And Pretzels* 33/*The Belle Of The Nineties, Gift Of Gab, I Like It That Way, Romance In The Rain, Wake Up And Dream* 34/ *The Girl Friend, To Beat The Band* 35/*Sitting On The Moon* 36/*Thumbs Up* 43.

Quine, Richard. Bn 12 November 1920, Detroit. Actor, later director. Musicals (as actor) *Babes On Broadway* 41/*For Me And My Gal* 42/*Words And Music* 48 (as director) *The Sunny Side Of The Street, Purple Heart Diary* 51/ *All Ashore, Rainbow Round My Shoulder* 52/ *Cruisin' Down The River* 53/*So This Is Paris* 54/ *My Sister Eileen* 55. Latterly directing major non-musical films.

Rainger, Ralph. Bn 7 October 1901, New York. Composer. Contract writer with Paramount 1930-9, then with Fox, always in collaboration with Leo Robin (*q.v.*), with whom he wrote many of Bing Crosby's best-known songs. Musicals *Queen High, Be Yourself* 30/*Big City Blues, The Big Broadcast* 32/*Torch Singer, She Done Him Wrong, Bedtime Story, The Way To Love, International House* 33/*Here Is My Heart, Little Miss Marker, She Loves Me Not, Shoot The Works* 34/*The Big Broadcast Of 1936, Millions In The Air, Rose Of The Rancho, Rumba* 35/*The Big Broadcast Of 1937, College Holiday Palm Springs, Poppy, Three Cheers For Love* 36/ *The Big Broadcast Of 1938, Swing High Swing Low, Waikiki Wedding, Hideaway Girl, Artists And Models, Blossoms On Broadway* 37/*Thanks For The Memory, Tropic Holiday, Artists And Models Abroad, Give Me A Sailor, Her Jungle Love* 38/ *Never Say Die, Gulliver's Travels, Paris Honeymoon* 39/*Moon Over Miami, Rise And Shine, Cadet Girl, Tall Dark And Handsome* 41/*Footlight Serenade, My Gal Sal* 42/*Coney Island, Riding High* 43. Killed in air crash 23 October 1942.

Rall, Tommy. Bn 27 December 1929, Kansas City. Dancer who made an impression with his talent and vitality in *Kiss Me Kate* 53/ *Invitation To The Dance, Seven Brides For Seven Brothers* 54/*My Sister Eileen* 55, then returned to the stage, making brief appearances in *Merry Andrew* 58, and (uncredited) *Funny Girl* 68.

Ralston, Vera Hruba. Bn 12 July 1919,

Prague. Czech skating champion who went to USA with ice show, and remained on outbreak of war. Starred in several skating films, including musicals *Ice-Capades* 41 / *Ice-Capades Revue* 42 / *Lake Placid Serenade* 45 / *Murder In The Music Hall* 46. Later in Westerns. Married Republic chief Herbert Yates and retired.

Ramirez, Carlos. Bn Bogota, Colombia. South American tenor who came to appear at Radio City and stayed to do guest spots in MGM musicals *Anchors Aweigh, Bathing Beauty,* **Two Girls And A Sailor** 44 / *Easy To Wed, Holiday In Mexico,* **Night And Day, Where Do We Go From Here?** 45 / *Latin Lovers* 53.

Rasch, Albertina. Choreographer and ballet mistress, active in early musicals **Hollywood Revue** 29 / *Broadway To Hollywood, Going Hollywood* 33 / *The Cat And The Fiddle* 34 / **The Broadway Melody Of 1936,** *After The Dance* 35 / *The King Steps Out* 36 / **Rosalie,** *The Firefly* 37, etc.

Ratoff, Gregory. Bn 20 April 1897, Samara, Russia, as Eugene Leontovitch. Explosive actor, ex-Moscow Arts Theatre and N.Y. Yiddish Players, who later turned director for Fox. Musicals (as actor) *I'm No Angel, Sitting Pretty, Broadway Thru' A Keyhole* 33 / **George White's Scandals,** *Let's Fall In Love* 34 / **The King Of Burlesque,** *Under Your Spell* 35 / **Sing Baby Sing** 36 / *Sally Irene and Mary* 37 / (as screen-writer) **You Can't Have Everything** 37 / (as director) **Rose Of Washington Square** 39 / *Footlight Serenade* 42 / *The Heat's On* (also prod) / **Something To Shout About** (also prod) 43 / *Irish Eyes Are Smiling* 44 / **Where Do We Go From Here?** 45 / *Do You Love Me?* 46 / *Carnival In Costa Rica* 47. Died 1961.

Ray, Leah. Singer, formerly with Phil Harris band, who had feature roles in musicals *Bedtime Story* 33 / *The Holy Terror, One In A Million, Sing And Be Happy, Thin Ice,* **Wake Up And Live** 37 / *Happy Landing, Walking Down Broadway* 38.

Raye, Don. Bn 16 March 1909, Washington, D.C., as Donald McRae Wilhoite Jnr. Composer-lyricist who went to Hollywood 1940, wrote two film scores with Hughie Prince *Argentine Nights* 40 / *Buck Privates* 41, then teamed up with Gene De Paul (*q.v.* for credits) at Universal, with whom he collaborated till 1951.

Raye, Martha. Bn 27 August 1916, Butte, Montana, as Margie Reed. Comedienne of the expressive mouth and lusty larynx, who cornered the market in roles of man-hungry females. Musicals *Rhythm On The Range* (début), *College Holiday,* **The Big Broadcast Of 1937** 36 / *Waikiki Wedding, Mountain Music, Hideaway Girl,*

Double Or Nothing, Artists And Models, **The Big Broadcast Of 1938** 37 / *Tropic Holiday, Give Me A Sailor, College Swing* 38 / *Never Say Die* 39 / **The Boys From Syracuse** 40 / *Navy Blues, Keep 'em Flying* 41 / *Hellzapoppin'* 42 / *Pin-Up Girl, Four Jills In A Jeep* 44 / **Billy Rose's Jumbo** 62.

Raymond, Gene. Bn 13 August 1908, New York, as Raymond Guion. Actor. Favourite leading man of the thirties, especially in musicals **Flying Down To Rio** 33 / *I Am Suzanne, Transatlantic Merry-Go-Round* 34 / *Hooray For Love* 35 / *The Smartest Girl In Town, That Girl From Paris, Walking On Air* 36 / *The Life Of The Party* 37 / *She's Got Everything* 38. Grew old gracefully in character parts, including **Hit The Deck** 55 / *I'd Rather Be Rich* 64. Now often seen on TV.

Regan, Phil. Bn 28 May 1908, Brooklyn. Singer-actor. Started singing in the N.Y. Police Glee Club, entered films 1934. Appeared in a few major productions, but generally played leads in 'B' films, including musicals *Student Tour,* **Sweet Adeline, Dames** 34 / *Broadway Hostess, Go Into Your Dance, We're In The Money,* **In Caliente** 35 / *Happy Go Lucky, Hit Parade Of 1937, Laughing Irish Eyes* 36 / *Manhattan Merry-Go-Round* 37 / *Outside Of Paradise* 38 / *Las Vegas Nights* 41 / *Sweet Rosie O'Grady* 43 / *Sunbonnet Sue* 45 / *The Sweetheart Of Sigma Chi, Swing Parade* 46 / **Three Little Words** 50.

Reisner, Charles F. Bn March 1887, Minneapolis. Director, formerly in vaudeville, who started in films as comedy writer. Naturally concentrated on comedies, with a few musicals included in his output **Hollywood Revue** 29 / *Chasing Rainbows, Love In The Rough* 30 / *Stepping Out, Flying High* 31 / *Student Tour* 34 / *Manhattan Merry-Go-Round* 37 / *The Big Store* 41 / *Meet The People* 43 / *Lost In A Harem* 44. Died 1962.

Republic Pictures Corporation. Probably the movie industry's major source of 'B' pictures, none of which survived beyond its year of copyright (except maybe some of the Roy Rogers Westerns which still crop up after a quarter of a century). The company was founded in 1935 by Herbert J. Yates, who continued to run it personally till its conversion to television features twenty-two years later. Republic contributed 56 musicals to this book, but there were many more in old and modern Western style (*Barnyard Follies* kind of thing) that did not warrant inclusion.

Revel, Harry. Bn 21 December 1905, London. Composer. From 1933–9 the team of Gordon and Revel provided numerous hit songs in their scores for Paramount and Fox musicals (see

also Mack Gordon). Revel later collaborated with Mort Greene (41–2) and Paul Francis Webster (43–6). Wrote songs or complete scores for *Sitting Pretty, Broadway Thru' A Keyhole* 33 | *She Loves Me Not,* **We're Not Dressing,** *College Rhythm, Shoot The Works,* **The Gay Divorcee** 34 | *Paris In The Spring, Two For Tonight, Love In Bloom, Stolen Harmony,* **The Big Broadcast Of 1936** 35 | *Poor Little Rich Girl, Palm Springs, Collegiate, Florida Special, Stowaway* 36 | **Love And Hisses,** *Sally Irene And Mary,* **Wake Up And Live,** *In Old Chicago,* **You Can't Have Everything,** *Ali Baba Goes To Town* 37 | *Rebecca Of Sunnybrook Farm, Hold That Co-ed, Thanks For Everything, Josette, My Lucky Star, Love Finds Andy Hardy* 38 | **Rose Of Washington Square,** *Tail Spin* 39 | *Two Girls On Broadway, Moon Over Burma* 40 | *Four Jacks And A Jill* 41 | *Sing Your Worries Away, Joan Of Ozark, The Mayor Of 44th Street, Beyond The Blue Horizon, Here We Go Again, Moonlight Masquerade* 42 | *It Ain't Hay, Hit The Ice* 43 | *Ghost Catchers, Minstrel Man* 44 | *I'll Tell The World* 45 | *The Stork Club* 46. After leaving films, recorded experimental music. Died 3 November 1958.

Reynolds, Debbie. Bn 1 April 1932, El Paso, Texas, as Mary Frances Reynolds. Singer-dancer-actress, one time french horn and bassoon player in Burbank Youth Symphony Orchestra. Delightfully fresh and extremely talented young performer who later became more stylized. Musicals *The Daughter Of Rosie O'Grady* (début), *Two Weeks With Love,* **Three Little Words** 50 | *Mr Imperium* 51 | *Skirts Ahoy,* **Singin' In The Rain** 52 | *The Affairs Of Dobie Gillis,* **Give A Girl A Break,** *I Love Melvin* 53 | *Athena* 54 | **Hit The Deck** 55 | *Bundle Of Joy,* **Meet Me in Las Vegas** 56 | *Say One For Me* 59 | **Pepe** 60 | *The*

Unsinkable Molly Brown 64 | *The Singing Nun* 66.

Reynolds, Marjorie. Bn 1921 as Marjorie Goodspeed. Blonde leading lady of Paramount films of the forties, including musicals **Holiday Inn, Star Spangled Rhythm** 42 | *Dixie* 43 | *Bring On The Girls, Meet Me On Broadway, Duffy's Tavern* 45 | *Monsieur Beaucaire* 46 | **That Midnight Kiss** 49.

Rhodes, Betty Jane. Bn 21 April 1921, Rockford, Illinois. Radio and night-club singer who came to films at fifteen, and sang in Paramount musicals *The Life Of The Party* 37 | *Oh Johnny* 40 | **The Fleet's In,** *Priorities On Parade, Sweater Girl,* **Star Spangled Rhythm** 42 | *Salute For Three* 43 | *Rainbow Island, You Can't Ration Love* 44.

Richman, Harry. Bn 10 August 1894, Cincinnati. Stage entertainer of the Ziegfeld–George White era, song-writer, singer, comedian, pianist, also the first aviator to make a two-way flight of the Atlantic. Made two musicals *Puttin' On The Ritz* 30 | *The Music Goes Round* 36, but found no niche in films and returned to the night-club world.

Riddle, Nelson. Bn 1 June 1921, Hackensack, N.J. Composer-conductor, famed for Sinatra recording accompaniments. Started in films as arranger, including musicals *All Ashore* 52 | **The Pajama Game,** *The Joker Is Wild* 57 and other Sinatra films. Musical director of **Pal Joey** 57 | **Merry Andrew,** *St Louis Blues* 58 | *Hey Boy Hey Girl, Li'l Abner* 59 | **Can Can** 60 | **Robin And The Seven Hoods** 64 | **How To Succeed In Business Without Really Trying** 66 | **Paint Your Wagon, On A Clear Day You Can See Forever** 69.

Ritz Brothers. Comedians. Al, bn 1901; Jim, bn 1903; Harry, bn 1906. Tried to emulate the

Debbie Reynolds | Say One For Me; Nelson Riddle

Marx Brothers' success, but although clever enough lacked the essentially anarchic qualities of their precursors. On their own level had some good musical routines, and assisted many Fox musicals of the thirties *Sing Baby Sing* 36/ *One In A Million, Life Begins In College,* **You Can't Have Everything, On The Avenue** 37/ *Straight Place And Show, Kentucky Moonshine,* **The Goldwyn Follies** 38/*Pack Up Your Troubles, The Three Musketeers* 39/*Argentine Nights* 40/*Behind The Eight Ball* 42/*Hi' Ya Chum, Never A Dull Moment* 43. A recent comeback in night-clubs was cut short by Al's death in 1965.

RKO Radio Pictures. Although better known as distributors of the work of big independents like Disney and Goldwyn, RKO nevertheless contributed much to the screen as producers. In the musical field, the early Wheeler and Woolsey musical comedies were followed by the commercially and artistically successful Fred Astaire-Ginger Rogers series of gay adventures, the introduction of Lily Pons and Frank Sinatra, the less imposing but financially successful films of Bobby Breen, Kay Kyser, Wally Brown and Alan Carney, etc. Other stars promoted by RKO were Cary Grant, Lucille Ball, Leon Errol *et al.* The company started in silent days through a merger of RCA and the Keith-Orpheum theatre chain (hence the initials), and was eventually bought out in the fifties by Lucille Ball and Desi Arnaz (as the Desilu TV company), both of whom had, interestingly enough, started as feature players there.

Roach, Hal. Bn 19 January 1892, Elmira, N.Y. Producer, started in Universal Westerns as an actor. Joined forces with Harold Lloyd to make one film, the profits from which set up the Hal Roach Studios. Best known for his many Laurel and Hardy shorts, but also produced major comedies and a few musicals *Stepping Out* 31/ *Babes In Toyland* 34/*The Bohemian Girl* 35/ *Nobody's Baby, Pick A Star* 37/*Road Show* 40/ *All-American Co-ed* 41. Distributed through MGM, later with United Artists.

Robbins, Gale. Bn 7 May 1924, Mitchell, Indiana. Singer-actress, ex-radio and dance bands, who played feature roles in musicals **The Barkleys Of Broadway** 48/*Oh You Beautiful Doll* 49/**Three Little Words** 50/**The Belle Of New York** 52/**Calamity Jane** 53.

Robbins, Jerome. Bn 1918. Stage choreographer, director and former dancer who transferred two of his greatest stage hits to the screen *The King And I* 56/**West Side Story** 61.

Roberti, Lyda. Bn 1910, Warsaw, Poland. Singer of thirties musicals who spent her youth

touring Europe with circus parents, and was later discovered in a Chicago revue and taken to Hollywood. Musicals **The Kid From Spain** 32/ *Torch Singer* 33/*College Rhythm* 34/*George White's Scandals Of 1935,* **The Big Broadcast Of 1936** 35/*Nobody's Baby, Pick A Star* 37. Died 12 March 1938.

Roberts, Allan. Bn 12 March 1905, Brooklyn. Composer-lyricist, former burlesque writer who joined Mike Todd 1936 for four years, then awarded a Columbia contract writing film songs with Doris Fisher (*q.v.* for credits). Also collaborated with Lester Lee on musicals *When A Girl's Beautiful* 47/*Ladies Of The Chorus, Lulu Belle* 48/*Slightly French, Holiday In Havana* 49/*Purple Heart Diary* 51. Later pianist and writer of special material for Sid Caesar, Red Buttons, Eddie Cantor, etc., Died 14 January 1966.

Robeson, Paul. Bn 9 April 1898, Princeton, N.J. Magnificent Negro bass singer who made many 'African' films in England. In Hollywood repeated his stage successes in *Emperor Jones* 33/ **Show Boat** 36. Later concentrated on the concert stage and politics.

Robin, Leo. Bn 6 April 1900, Pittsburgh. Lyricist best known for collaboration with Ralph Rainger (32–42) (*q.v.* for credits). Also worked regularly with Richard Whiting (29–33, 35–6) (*q.v.* for credits). After Rainger's death was in great demand by other top writers including Harry Warren, Jerome Kern, Harold Arlen, Jule Styne *et al*, and contributed songs to *Wintertime,* **The Gang's All Here** 43/ *Greenwich Village* 44/*Wonder Man* 45/ **Centennial Summer, The Time, The Place And The Girl** 46/*Something In The Wind* 47/ *That Lady In Ermine, Casbah* 48/*Two Tickets To Broadway, Meet Me After The Show* 51/*Just For You* 52/*Small Town Girl,* **Gentlemen Prefer Blondes** 53/*My Sister Eileen,* **Hit The Deck** 55. Prior to the Rainger collaboration, Robin had written for early musicals *Syncopation, Hit The Deck* 29/**The Vagabond King** 30/*You Belong To Me* 34/*It's A Great Life* 36/*Champagne Waltz* 37.

Robinson, Bill 'Bojangles'. Bn 1878. Legendary Negro dancer, twice commemorated in song titles, whose screen appearances were mainly limited to servile roles in Shirley Temple musicals, although these did at least give him a chance to show his paces *Hooray For Love, The Littlest Rebel, The Little Colonel,* **The Big Broadcast Of 1936** 35/*Dimples* (choreographer only) 36/*Just Around The Corner, Rebecca Of Sunnybrook Farm* 38/**Stormy Weather** 43. All Harlem turned out to mourn when he died 25 November 1949.

Rodgers, Richard. Bn 28 June 1902, New York. Composer of some of Broadway's greatest shows of the century, who worked with only two lyricists, Lorenz Hart until 1942 (*q.v.*) and Oscar Hammerstein II from 1943 till the latter's death in 1960. Has since collaborated with Stephen Sondheim, but not in films. Film scores by Rodgers and Hart *Follow Through, Heads Up, Leathernecking, Melody Man, Spring Is Here* 30 / *The Hot Heiress* 31 / **Love Me Tonight,** *The Phantom President* 32 / *Hallelujah I'm A Bum,* **Dancing Lady** 33 / **Mississippi** 35 / *The Dancing Pirate* 36 / *Fools For Scandal* 38 / **Babes In Arms, On Your Toes** 39 / *Too Many Girls,* **The Boys From Syracuse** 40 / *They Met In Argentina* 41 / *I Married An Angel* 42 / **Pal Joey** 57 / **Billy Rose's Jumbo** 62. Some of their songs were used in **Gentlemen Marry Brunettes** 55, and all their biggest numbers were featured in the all-star biopic **Words And Music** 48, with Tom Drake as Rodgers and Mickey Rooney as Hart. Scores by Rodgers and Hammerstein **State Fair** 45, 62 / **Oklahoma!** 55 / **Carousel, The King And I** 56 / **South Pacific** 58 / **The Flower Drum Song** 61 / **The Sound Of Music** 65.

Rogell, Albert S. Bn 21 August 1901, Oklahoma. Director who started as cameraman, progressed to direction of 'B' dramas, Westerns and a few musicals *Start Cheering* 37 / *Laugh It Off, Hawaiian Nights* 39 / *Argentine Nights, I Can't Give You Anything But Love* 40 / *Priorities On Parade, Sleepytime Gal, True To The Army, Youth On Parade* 42 / *Hit Parade Of 1943* 43 / *Earl Carroll's Sketchbook* 46.

Rogers, Charles 'Buddy'. Bn 13 August 1904, Olathe, Kansas. Singer-actor who alternated leading a successful dance band with starring in Paramount musicals *Illusion* 29 / *Follow Through, Heads Up, Close Harmony, Safety In Numbers,* **Paramount On Parade** 30 / *Along Came Youth* 31 / *The Best Of Enemies,* **Take A Chance** 33 / *Old Man Rhythm* 35 / *This Way Please* 37 / *Sing For Your Supper* 41.

Rogers, Ginger. Bn 16 July 1911, Independence, Missouri, as Virginia Katherine McMath. Actress-singer-dancer, one of Hollywood's major personalities. Starred on Broadway in 'Girl Crazy' at nineteen and joined the trek West the same year. Did several films before achieving that perfect blend with Fred Astaire, following which she virtually bowed out of musicals and became one of the best all-round actresses on the scene. Musicals *Young Man Of Manhattan* (début), *Follow The Leader, Queen High* 30 / *Broadway Bad, Sitting Pretty,* **Forty-Second Street, The Golddiggers Of 1933, Flying Down To Rio** 33 / **Twenty Million Sweethearts, The Gay Divorcee, Roberta** 34 / *In Person,* **Top Hat, Follow The Fleet** 35 / **Swing Time** 36 / **Shall We Dance?** 37 / *Having Wonderful Time,* **Carefree** 38 / **The Story Of Vernon And Irene Castle** 39 / **Lady In The Dark** 43 / *A Weekend At The Waldorf* 45 / **The Barkleys Of Broadway** 48. Forty years after her Broadway début she was still holding the stage, in the London production of 'Mame'.

Ginger Rogers / *Lady in the Dark; Cesar Romero*

Romberg, Sigmund. Bn 29 July 1887, Nagy Kaniza, Hungary. Composer of perennial operettas, many of which were filmed *The Desert Song* 29, 43, 52 | *The New Moon* 30, 40 | *Viennese Nights* 30 | *Children Of Dreams* 31 | *The Night Is Young* 34 | *Maytime* 37 | *The Girl Of The Golden West* 38 | *Let Freedom Ring* 39 | *Up In Central Park* 48 | *The Student Prince* 54. Died 9 November 1951, and three years later MGM assembled an all-star cast for his biopic *Deep In My Heart* 54, with Jose Ferrer as Romberg.

Rome, Harold. Bn 27 May 1908, Hartford, Conn. Composer-lyricist better known for stage productions. Former architect who started writing summer camp shows and developed further with Army revues. Wrote songs for musicals *Babes On Broadway* 41 | *Thousands Cheer* 43 | *Junior Prom* 46 | *Call Me Mister* 51, the only one of his shows to be filmed. ('Fanny' was filmed without the songs, which were used purely as background themes.)

Romero, Alex. Dance director of *The Affairs Of Dobie Gillis* 53 | *Love Me Or Leave Me* 55 | *The Wonderful World Of The Brothers Grimm* 62 | *Double Trouble* 67, etc.

Romero, Cesar. Bn 15 February 1907, New York. Versatile actor, specializing equally well in comedy, villainy or music. Former Broadway dancer who rarely danced on screen despite presence in many musicals (mostly for Fox) *Happy Landing, My Lucky Star* 38 | *A Weekend In Havana, The Great American Broadcast, Romance Of The Rio Grande, Tall Dark And Handsome* 41 | *Springtime In The Rockies, Orchestra Wives* 42 | *Wintertime, Coney Island* 43 | *Carnival In Costa Rica* 47 | *That Lady In Ermine* 48 | *The Beautiful Blonde From Bashful Bend* 49 | *Happy Go Lovely* 51 | *Pepe* 60. Found new

fame with the young generation as The Joker in TV's 'Batman'.

Rooney, Mickey. Bn 23 September 1920, Brooklyn, as Joe Yule, Jnr. Multi-talented actor-singer-dancer-drummer-pianist-composer-lyricist who has done all these things since he entered silent films at six. As a child acted as Mickey McGuire ('Our Gang' and his own series), and first reached a wide audience with MGM as Andy Hardy. His partnership with Judy Garland was a natural, resulting in some fine musicals. Musical credits *Broadway To Hollywood* 33 | *I Like It That Way* 34 | *Thoroughbreds Don't Cry* 37 | *Love Finds Andy Hardy* 38 | *Babes In Arms* 39 | *Andy Hardy Meets Debutante, Strike Up The Band* 40 | *Babes On Broadway* 41 | *Thousands Cheer, Girl Crazy* 43 | *Summer Holiday* 46 | *Words And Music* 48 | *The Strip* 51 | *All Ashore* 52.

Rose, Billy. Bn 6 September 1899, New York. Best known as a showman, club owner and producer (e.g. *Billy Rose's Diamond Horseshoe* and *Billy Rose's Jumbo*), but was also a prolific song-writer who contributed to early musicals *The Singing Fool* 28 | *The King Of Jazz, Be Yourself, Take A Chance* 30 | *Millions In The Air* 35. Died 10 February 1966.

Rose, David. Bn 15 June 1910, London. Composer-conductor. Former arranger for bands, radio and TV, first known for light music themes like 'Holiday For Strings'. Wrote songs for musicals *Never A Dull Moment* 43 | *Wonder Man* 45 | *The Unfinished Dance* 47, then joined MGM as musical director for *Rich Young And Pretty, Texas Carnival* 51 | *Everything I Have Is Yours* 52 | *Jupiter's Darling* 55. Latterly on TV.

Rose, Jack. Bn 1911. Producer-screenwriter, always partnered by Melville Shavelson (*q.v.*).

Together they wrote *It's A Great Feeling* 49/ *The Daughter Of Rosie O'Grady* 50/*I'll See You In My Dreams*, On Moonlight Bay 51/ *April In Paris* 52, for Warners, *Riding High* 50, and *Living It Up* 54, for Paramount, then signed a deal with Paramount to write their own films with Rose producing and Shavelson directing. Musicals thus produced *The Seven Little Foys* 55/*Beau James* 57/*The Five Pennies* 59. Rose also co-wrote *The Road To Rio* 47/ *The Paleface* 48.

Rosen, Milton. Bn 2 August 1906, New York. Composer-conductor. For many years Assistant Head of Music at Universal. Wrote many background scores, also songs for musicals with lyrics by Everett Carter (*q.v.*) *I'm Nobody's Sweetheart Now* 40/*Sing Another Chorus, Six Lessons From Madame La Zonga, Too Many Blondes, Where Did You Get That Girl?* 41/*Juke*

Box Jenny, Mississippi Gambler 42/Babes On Swing Street, A Cowboy In Manhattan, Gals Incorporated, Follow The Band, Get Going, Hi Buddy, Hi' Ya Sailor, Swingtime Johnny 43/Hat Check Honey, Moon Over Las Vegas, My Gal Loves Music, South Of Dixie, Slightly Terrific, Twilight On The Prairie, Weekend Pass 44/I'll Tell The World, Honeymoon Ahead, See My Lawyer, The Senorita From The West, Shady Lady 45. Background score for *Swing Out Sister* 45. Musical director of *On Stage Everybody* 44/*Cuban Pete* 46/ *Slightly Scandalous* 47/*The Milkman* 50.

Ross, Herbert. Bn Brooklyn. Choreographer-director, who came to Hollywood after experience on Broadway as dancer and choreographer, and a spell with the Ballet Theatre. Dance direction on *Carmen Jones* 54/*Doctor Dolittle* 67/*Funny Girl* 68, and directed *Goodbye Mr Chips* 69.

Ross, Lanny. Bn 19 January 1906, Seattle, Wash. Singer, ex-radio, one of the tenors who proliferated in pre-war Hollywood but vanished with the years. Musicals *Melody In Spring, College Rhythm* 34/*The Lady Objects* 38/ *Gulliver's Travels* (soundtrack) 39/*Stage Door Canteen* 43.

Ross, Shirley. Bn 1909, Omaha, as Bernice Gaunt. Singer-actress who came from Gus Arnheim's band and appeared in many Paramount films, including musicals *Hollywood Party*, **The Merry Widow** 34/**The Big Broadcast Of 1937**, San Francisco 36/*Waikiki Wedding, Hideaway Girl*, **The Big Broadcast Of 1938**, Blossoms On Broadway 37/*Thanks For The Memory* 38/*Cafe Society, Paris Honeymoon, Some Like It Hot* 39/*A Song For Miss Julie* 45.

Roth, Lilian. Bn 11 December 1910, Boston. Singer whose disturbed life was the subject of a biopic *I'll Cry Tomorrow* 55, in which Susan

Jane Russell/Gentlemen Prefer Blondes; Rosalind Russell/Gypsy

Hayward played the lead. After life as a child star, Lilian Roth joined Paramount 1929 and starred in early musicals *Illusion* (début), **The Love Parade** 29 / *Animal Crackers*, **Paramount On Parade,** *Honey*, **The Vagabond King** 30 / **Take A Chance** 33. The biopic revealing her cure from alcoholism enabled her to make a comeback in cabaret.

Rowland, Roy. Bn 31 December 1910, New York. Director who served his time in MGM short subjects, later did features, including musicals *Hollywood Party* 34 / *Two Weeks With Love* 50 / *Excuse My Dust* 51 / **Hit The Deck** 55 / **Meet Me In Las Vegas** 56 / *The Seven Hills Of Rome* 58.

Ruggles, Wesley. Bn 1889, Los Angeles. Director, with experience in stage musicals and stock, who joined the Keystone Cops before going behind the camera. Did many films for Paramount, including musicals *Street Girl* 29 / *Honey* 30 / *College Humour*, *I'm No Angel* 33 / *Bolero*, *Shoot The Works* 34 / *Sing You Sinners* (also prod) 38. Later concentrated on romances and comedies.

Russell, Andy. Bn 16 September 1920, Los Angeles, as Andrew Rabago. Singer who had a brief vogue in the mid-forties and guested in musicals *The Stork Club*, *Breakfast In Hollywood*, **Make Mine Music** (soundtrack) 46 / *Copacabana* 47

Russell, Jane. Bn 21 September 1921, Bonidji, Minnesota. Actress discovered by Howard Hughes who eventually overcame the hilarious badness of 'The Outlaw' and developed as a light comedy performer, also capable of holding her own in musicals *The Paleface* 48 / *Double Dynamite* 51 / *The Las Vegas Story, Son of Paleface* 52 / **Gentlemen Prefer Blondes,** *The French Line* 53 / **Gentlemen Marry Brunettes** 55.

Russell, Rosalind. Bn 4 June 1911, Waterbury, Conn. Actress with a wonderful flair for sophisticated comedy; specialized in revealing the romantic heart beneath the business executive's well-tailored suit. Was featured in a few musicals: *The Night Is Young* 34 / *Reckless* 35 / *The Girl Rush* 55, but her only completely musical role was in **Gypsy** 62.

Ryan, Peggy. Bn 28 August 1924, Long Beach, Calif., as Margaret Irene Ryan. Dancer-comedienne from vaudeville (as a child performer) invariably found in Universal's teenage musicals of the forties *Top Of The Town* (début) 37 / *When Johnny Comes Marching Home, Get Hep To Love, Give Out Sisters, Private Buckaroo* 42 / *Babes On Swing Street, Mr Big* 43 / *On Stage Everybody, Patrick The Great, The Merry Monahans, Follow The Boys, A Chip Off The Old Block, This Is The Life, Bowery to Broadway, Top*

Man 44 / *Here Come The Co-eds, That's The Spirit* 45 / *There's A Girl In My Heart* 49 / *All Ashore* 52. Retired to marry her dancing partner Ray McDonald (*q.v.*) and start dancing schools.

St Clair, Malcolm. Bn 1899, Los Angeles. Director who did most of his work in silents and early talkies, although still active to the forties. Musicals *Montana Moon, Dangerous Nan McGrew* 30 / *Jitterbugs, Swing Out The Blues* 43. Died 1952.

Salter, Hans J. Bn 14 January 1896, Vienna. Composer-conductor. Assistant conductor Vienna Volksoper and Berlin State Opera, head of UFA music department. Came to USA 1937 and joined Universal. Wrote many background scores, and songs for *That Night With You, That's The Spirit* 45. Musical director of *I'm Nobody's Sweetheart Now, Margie* 40 / *Hold That Ghost, Where Did You Get That Girl?* 41 / *A Cowboy In Manhattan, Get Going, His Butler's Sister, Hi' Ya Chum, Hi' Ya Sailor, Never A Dull Moment* 43 / *Twilight On The Prairie, Allergic To Love, Christmas Holiday, Hat Check Honey, Hi Good Lookin', The Merry Monahans, Pardon My Rhythm* 44 / **Can't Help Singing,** *Easy To Look At, I'll Tell The World, See My Lawyer* 45 / *Follow That Dream* 62. May have actually been musical director on other Universal features nominally credited to music head Charles Previn. Later providing music for TV series.

Sandrich, Mark. Bn 1901, New York. Director from silent days. Worked consistently with Pandro S. Berman at RKO especially on the Astaire-Rogers series, later joining Paramount as producer-director. Musicals *Melody Cruise* 33 / **The Gay Divorcee,** *Cockeyed Cavaliers, Hips Hips Hooray* 34 / **Follow the Fleet, Top Hat,** 35 / **Shall We Dance?** 37 / *Carefree* 38 / *Man About Town* 39; following films produced also *Love Thy Neighbour, Buck Benny Rides Again* 40 / **Holiday Inn** 42 / **Here Come The Waves** 45. Died 1945.

Sands, Tommy. Born 27 August 1937, Chicago. Pop singer of the rock and roll era who made a few musicals for more advanced tastes *Sing Boy Sing* 57 / *Mardi Gras* 58 / *Babes In Toyland* 61.

San Juan, Olga. Bn 1927. Petite singer-dancer; never a star but a useful second lead in musicals *Rainbow Island* 44 / *Out Of This World, Duffy's Tavern* 45 / **Blue Skies,** *Cross My Heart* 46 / *Variety Girl* 47 / *The Countess Of Monte Cristo,* **One Touch of Venus,** *Are You With It?* 48 / *The Beautiful Blonde From Bashful Bend* 49. Retired after marriage to Edmond O'Brien.

Santell, Alfred. Bn 1896, San Francisco. Director. Former architect and short story writer who joined Sennett as gag-writer-cum-actor, later becoming director with Warner. Musicals *Having Wonderful Time, The Cocoanut Grove* 38 / *Beyond The Blue Horizon* 42 / *Mexicana* (also prod) 45.

Santley, Joseph. Bn 1890, Salt Lake City.

219

Prolific director of 'B' films including musicals
Swing High 30 | *Dancing Feet, Laughing Irish Eyes,
Walking On Air* 36 | *Swing Sister Swing, She's Got
Everything* 38 | *Dancing On A Dime, Melody And
Moonlight, Melody Ranch, Music In My Heart* 40 |
*Puddin' Head, Rookies On Parade, Sis Hopkins,
Ice-Capades* 41 | *Joan of Ozark, Yokel Boy* 42 | *Thumbs
Up* (also prod), *Chatterbox, Here Comes Elmer,
Sleepy Lagoon* 43 | *Rosie The Riveter, Brazil, Three
Little Sisters* 44 | *Earl Carroll's Vanities* 45 | *Make
Believe Ballroom* 49 | *When You're Smiling* 53.

Sawyer, Geneva. Dance director formerly
active with Fox. Musicals include *Love and
Hisses* 37 | *Little Miss Broadway, Hold That Co-ed*
38 | *The Bluebird* 39 | **Down Argentine Way,**
Young People 40 | *Jitterbugs* 43.

Scharf, Walter. Bn 1 August 1910, New York.
Composer-conductor. Former accompanist to
Kate Smith and Rudy Vallee. Head of Republic
music department 1942–6, with Universal
1948–50, later with RKO and Paramount. Wrote
many background scores, arranged for many
films, and received musical director credit on
musicals *Ice-Capades Revue, Johnny Doughboy* 42 |
*Hit Parade Of 1943, Nobody's Darling, Sleepy
Lagoon, Thumbs Up, Chatterbox* 43 | *Brazil,
Casanova In Burlesque, Atlantic City* 44 | *Earl
Carroll's Vanities, Mexicana, Lake Placid Serenade*
45 | *Murder In The Music Hall* 46 | *Are You With

It?, Casbah, The Countess Of Monte Cristo, Mexican
Hayride* 48 | *Yes Sir That's My Baby* 49 | *Two
Tickets to Broadway* 51 | **Hans Christian
Andersen** 52 | *Hollywood Or Bust, Artists And
Models, Bundle Of Joy* 56 | *The Joker Is Wild* 57 |
Cinderfella 60 | *Ladies' Man* 61 | *Tickle Me* 65 |
Funny Girl 68.

Schertzinger, Victor. Bn 8 April 1890,
Mahanoy City, Penn. Director-composer-
lyricist-conductor. Usually wrote the songs for
his own films, sometimes the background scores,
including the first original film music in 1916 (for
Thomas Ince's 'Civilization'). Wrote songs only
for **The Love Parade** 29 | *Glamour Boy* 41.
Directed only *Safety In Numbers,* **Paramount
On Parade** 30 | *The Mikado* 39 | **The Birth of the
Blues,** *The Road to Zanzibar* 41. Did both for
Heads Up 30 | *The Cocktail Hour* 33 | **One Night
Of Love** 34 | *Let's Live Tonight, Love Me Forever*
35 | *The Music Goes Round* 36 | *Something To Sing
About* 37 | **The Road To Singapore,** *Rhythm On
The River* 40 | *Kiss The Boys Goodbye* 41 | **The Fleet's
In** 42. Died 26 October 1941.

Schwartz, Arthur. Bn 25 November 1900,
Brooklyn. Composer-producer. Ex-lawyer who
started writing songs in law school. Collaborated
with Howard Dietz (*q.v.*) on many Broadway
shows (28–36), and has concentrated on stage
work with other lyricists until the present day. In
films, wrote scores for musicals *Queen High,
Follow The Leader, Lottery Bride* 30 | *That Girl From
Paris, Under Your Spell* 36 | *Navy Blues* 41 | *Cairo*
42 | **Thank Your Lucky Stars** 43 | **The Time,
The Place and The Girl** 46 | *Dancing In The
Dark* 49 | *Excuse My Dust* 51 | **The Band Wagon,**
Dangerous When Wet 53. Produced **Thank Your
Lucky Stars** 43 | **Cover Girl** 44 | **Night And
Day** 45.

Scott, Hazel. Bn 11 June 1920, Trinidad.
Night-club pianist who made guest appearances
in forties musicals **Something To Shout About,
I Dood It,** **The Heat's On, Broadway Rhythm,
Thousands Cheer** 43 | **Rhapsody In Blue** 45.

Seaton, George. Bn 1911. Director-scenarist,
formerly on stage. Wrote many films prior to
directing, including musicals *A Day At The
Races* 37 | **That Night In Rio** 41 | *Coney Island* 43.
In association with producer William Perlberg
(*q.v.*) wrote and directed musicals **Billy Rose's
Diamond Horseshoe** 45 | *The Shocking Miss
Pilgrim* 47 | *Aaron Slick From Punkin Crick* 51 |
Somebody Loves Me 52 | *Little Boy Lost* 53 | **The
Country Girl** 54.

Segal, Vivienne. Bn 1897, Philadelphia.
Operetta and Ziegfeld Follies star of the twenties
who starred in early musicals, and was still active
on stage and screen in the fifties. Musicals *The
Golden Dawn* (début), *The Bride Of The Regiment,*

Artie Shaw

Song of The West, Viennese Nights 30 / *The Cat And The Fiddle* 34 / *Golden Girl* 51.

Seiler, Lewis. Bn 1891, New York. Director who started in silents with Westerns after experience as scenarist and assistant director. Did several forgettable musicals *Paddy O'Day* 35 / *First Baby* 36 / *Turn Off The Moon* 37 / *Something For The Boys* 44 / *Doll Face* 45 / *If I'm Lucky* 46. Died 1964.

Seiter, William A. Bn 1892, New York. Director, ex-silent two-reelers, who dealt mainly in comedies and musicals *Footlights And Fools, Smiling Irish Eyes* 29 / *Sunny* 30 / *Girl Crazy* 32 / *Hello Everybody* 33 / **Roberta** 34 / *In Person* 35 / *Stowaway, Dimples* 36 / *Life Begins In College, The Life Of The Party, Sally Irene And Mary* 37 / *Thanks For Everything* 38 / *It's A Date* 40 / *Nice Girl* 41 / **You Were Never Lovelier,** *Broadway* 42 / *The Belle Of The Yukon, Four Jills In A Jeep* 44 / *It's A Pleasure, That Night With You* 45 / *I'll Be Yours* 47 / **One Touch Of Venus,** *Up In Central Park* 48. Died 1964.

Shavelson, Melville. Bn 1917. Director-scenarist, for many years in partnership with producer-writer Jack Rose (*q.v.* for credits). Prior to this collaboration had written independently, including musicals *Wonder Man* 45 / *The Kid From Brooklyn* 46 / *Always Leave Them Laughing* 49.

Shaw, Artie. Bn 23 May 1910, New York, as Arthur Arshawsky. Clarinet-playing band-leader of the swing era who appeared in only two musicals *Dancing Co-ed* 39 / *Second Chorus* 40.

Shaw, Winifred 'Wini'. Bn 1910, San Francisco as Winifred Lei Momi. Singing lead of Warner musicals *Million Dollar Ransom, Wake Up and Dream, Sweet Adeline, Gift of Gab* 34 / *In Caliente,* *Broadway Hostess,* **The Golddiggers Of 1935** 35 / **The Singing Kid** 36 / **Ready, Willing And Able,** *Melody for Two* 37.

Sheridan, Ann. Bn 21 February 1915, Dallas, Texas, as Clara Lou Sheridan. Actress whose career proceeded uneventfully for years till the 'Oomph Girl' publicity campaign paid off for her and Warners. Sometimes sang a bit, danced a bit and acted in musicals *Bolero* (debut), *Shoot the Works, College Rhythm, Murder At The Vanities* 34 / **Mississippi,** *Rumba* 35 / **Sing Me A Love Song** 36 / *Broadway Musketeers, A Cowboy From Brooklyn* 38 / **Naughty But Nice** 39 / *Navy Blues* 41 / *Juke Girl* 42 / **Thank Your Lucky Stars** 43 / *Shine On Harvest Moon* 44 / *The Opposite Sex* 56. Was already dying of cancer when commencing TV series in 1966, and died January 1967.

Sherman, Richard M. and Robert B. Richard bn 12 June 1928, Robert 19 December 1925, both in New York as sons of veteran song-writer Al Sherman. Have been writing interpolated songs for Disney films since 1959, and complete scores

for musicals *Summer Magic, In Search Of The Castaways* 63 / **Mary Poppins** 64 / *The Jungle Book,* **The Happiest Millionaire** 67 / *The One And Only Genuine Original Family Band,* **Chitty Chitty Bang Bang** 68 / *The Aristocats* 70.

Shore, Dinah. Bn 1 March 1917, Winchester, Tenn., as Frances Rose Shore. Radio and recording singer of the early forties who leapt straight into screen stardom, but was eventually wasted by Hollywood. Musicals **Thank Your Lucky Stars** (début) 43 / *The Belle Of The Yukon,* **Up In Arms,** *Follow The Boys* 44 / **Make Mine Music** (soundtrack), *Fun And Fancy Free* (soundtrack), **Till The Clouds Roll By** 46 / *So Dear To My Heart* (soundtrack) 49 / *Aaron Slick From Punkin Crick* 51. Left the screen and built up a big TV reputation.

Sidney, George. Bn 4 October 1911, New York. Director for many years with MGM, for whom he handled top musicals **Thousands Cheer** 43 / **The Ziegfeld Follies,** *Bathing Beauty,* **Anchors Aweigh** 44 / **The Harvey Girls,** *Holiday In Mexico* 45 / *Good News* 47 / **Annie Get Your Gun** 49 / **Show Boat** 51 / **Kiss Me Kate** 53 / *Jupiter's Darling* 55 ; moved to Columbia *The Eddie Duchin Story* 56 / *Pal Joey* 57 / *Pepe* (also prod) 60 / *Bye Bye Birdie* 63 / *Viva Las Vegas* (also prod) 64 / *Half A Sixpence* 67.

Siegel, Sol C. Bn 1903, New York. Producer, contracted at various times to Paramount, Fox and MGM, specializing in light entertainment, including musicals *Melody Ranch* 40 / *Glamour Boy* 41 / *Priorities On Parade, Sweater Girl, True To The Army* 42 / *Rainbow Island* 44 / **Blue Skies** 46 / *Welcome Stranger, The Perils Of Pauline* 47 / *My Blue Heaven* 50 / *On The Riviera* 51 / **Call Me Madam, Gentlemen Prefer Blondes** 53 / **There's No Business Like Show Business** 54 / **High Society** 56 / **Les Girls** 57 / **Merry Andrew** 58.

Silvers, Louis. Bn 6 September 1889, New York. Composer-conductor. Former vaudeville pianist and producer who entered silent films with D. W. Griffith. Musical director with Columbia 1934–5, Fox from 1936. Conducted musicals *The Jazz Singer* 27 / *Smiling Irish Eyes* 29 / *Dancing Lady* 33 / *One Night Of Love* 34 / *Captain January, Love Me Forever* 35 / *Poor Little Rich Girl, Sing Baby Sing, Dimples, Stowaway* 36 / *Heidi, In Old Chicago, Life Begins In College, Ali Baba Goes To Town, Wake Up And Live, One In A Million, Thin Ice, Love And Hisses* 37 / *Straight Place And Show, Happy Landing, Kentucky Moonshine, Just Around The Corner, Little Miss Broadway, My Lucky Star, Thanks For Everything* 38 / *Second Fiddle, Tail Spin, Swanee River, Rose Of Washington Square* 39 / *The Powers Girl* 42. Died 26 March 1954.

Silvers, Phil. Bn 11 May 1911, Brooklyn. Burly bespectacled comedian who provided comic relief without starring for twenty years, until he became a world-wide hit on TV as Sergeant Bilko. Musicals *Hit Parade Of 1941* 40 / *Lady Be Good* 41 / *My Gal Sal, Footlight Serenade* 42 / *Coney Island* 43 / *Cover Girl, Four Jills In A Jeep, Something For The Boys* 44 / *Where Do We Go From Here?, Billy Rose's Diamond Horseshoe* 45 / *If I'm Lucky* 46 / *Summer Stock* 50 / *Lucky Me* 54 / *A Funny Thing Happened On The Way To The Forum* 67.

Silvers, Sid. Bn 1 January 1907, Brooklyn. Comedian-screenwriter. Many years in vaudeville then started writing for stage. Appeared in films in thirties but turned more to writing. Acted in musicals *The Show Of Shows* 29 / *Dancing Sweeties* 30 / *My Weakness* 33 / *Bottoms Up, Transatlantic Merry-Go-Round* 34 / *The Broadway Melody Of 1936* 35 / *Born To Dance* 36 / *52nd Street, The Broadway Melody Of 1938* 37. Co-

wrote screenplays for *The Broadway Melody Of 1936* 35 / *Born To Dance* 36 / *The Broadway Melody Of 1938* 37 / *For Me And My Gal, The Fleet's In* 42.

Simms, Ginny. Bn *c.* 1913, Fresno, Calif. Singer with dance bands since 1931. Was with Kay Kyser's band when they made their first films, and went solo, appearing mainly in 'B' features *That's Right You're Wrong* 39 / *You'll Find Out* 40 / *Playmates* 41 / *Seven Days Leave, Here We Go Again* 42 / *Broadway Rhythm, Hit The Ice* 43 / *Shady Lady, Night And Day, No Leave No Love* 45 / *Disc Jockey* 51.

Simon, S. Sylvan. Bn 1910. Director of mainly routine films. Musicals *Dancing Co-ed* 39 / *Two Girls On Broadway* 40 / *Rio Rita* 42 / *Song Of The Open Road* 44 / *The Thrill Of Brazil* 46.

Sinatra, Frank. Bn 12 December 1915, Hoboken, N.J. Singer-actor-producer-director-lyricist-executive. The original 'swoon crooner' of the forties has developed into a powerful figure in the entertainment industry. Appears to have lost interest in musicals, but has contributed more than his fair share since he first appeared on screen as vocalist (anonymous) with the Tommy Dorsey band in *Las Vegas Nights* 41, and *Ship Ahoy* 42. Did a one-song guest spot in *Reveille With Beverley* 43, then signed by RKO, later moving to MGM. Musicals *Higher And Higher* 43 / *Step Lively, Anchors Aweigh* 44 / *It Happened In Brooklyn, Till The Clouds Roll By* 46 / *The Kissing Bandit, Take Me Out To The Ball Game* 48 / *On The Town* 49 / *Double Dynamite* 51 / *Meet Danny Wilson* 52 / *Young At Heart* 54 / *Guys And Dolls* 55 / *High Society* 56 / *The Joker Is Wild, Pal Joey* 57 / *Can Can, Pepe* 60 / *The Road to Hong Kong* (gag appearance) 62 / *Robin And The Seven Hoods* (also prod) 64. Announced retirement, 1971.

Sinatra, Nancy. Bn May 1940, New Jersey. Singer, daughter of Frank, not yet a film name but has appeared in a few features including musicals *The Swingin' Set* 66 / *Speedway* 68.

Skelton, Red. Bn 18 July 1910, Vincennes, Indiana, as Richard Skelton. Rubber-faced comic who made numerous MGM films during the forties, an amiable clown whose grotesqueries may not have appealed to every taste. Musicals *Having Wonderful Time* (début) 38 / *Lady Be Good* 41 / *Panama Hattie, Ship Ahoy* 42 / *Dubarry Was A Lady, Thousands Cheer, I Dood It* 43 / *Bathing Beauty, The Ziegfeld Follies* 44 / *The Show-Off* 46 / *Neptune's Daughter* 49 / *Three Little Words, Duchess of Idaho,* 50 / *Texas Carnival, Excuse My Dust* 51 / *Lovely To Look At* 52. Later films diminished in importance, but found success on TV.

Slack, Freddie. Bn 7 August 1910, La Crosse,

Wis. Pianist-bandleader in the boogie-woogie idiom who guested in musicals *Reveille With Beverley, Babes On Swing Street,* **The Sky's The Limit** 43 / *Follow The Boys, Hat Check Honey, Seven Days Ashore* 44 / *High School Hero* 46. Died 10 August 1965.

Smith, Ethel. Bn 22 November 1910, Pittsburgh. Girl organist whose looks and technical skill made her a decorative addition to musicals *Bathing Beauty* 44 / *Easy To Wed, George White's Scandals of 1945* 45 / *Cuban Pete* 46 / *Melody Time* (soundtrack) 48.

Smith, 'Whispering' Jack. The very first crooner, who made his name in the twenties as 'The Whispering Baritone', and appeared in early musicals *The Big Party, Cheer Up And Smile, Happy Days* 30.

Smith, Kate. Bn 1 May 1906, Greenville, Va. Big (in every sense) radio singer of the thirties who tried her luck in films but wasn't built for screen stardom. Musicals **The Big Broadcast** 32 / *Hello Everybody* 33 / **This Is The Army** 43. Still active on radio, records and TV.

Sondheim, Stephen. Bn 22 March 1930, New York. Composer-lyricist-librettist. A protégé of Oscar Hammerstein from his schooldays, became a TV scriptwriter ('Topper', etc.), then on to Broadway. Wrote lyrics for **West Side Story** 61 / **Gypsy** 62; music and lyrics for *A Funny Thing Happened On The Way To The Forum* 67.

Sothern, Ann. Bn 22 January 1909, Valley City, N. Dakota, as Harriet Lake. Actress, sometime singer, specializing in comedy roles. Musicals **The Show Of Shows** 29 / *Broadway Thru' A Keyhole* 33 / *Let's Fall In Love, Kid Millions, Melody In Spring* 34 / *Folies Bergère, The Girl Friend* 35 / *The Smartest Girl In Town, Walking On Air, Hooray For Love* 36 / *She's Got Everything* 38 / **Lady Be Good** 41 / *Panama Hattie,* **Rio Rita** 42 / **Thousands Cheer** 43 / *April Showers,* **Words And Music** 48 / *Nancy Goes To Rio* 49. Has had own TV series, and is also music publisher, cattle owner and producer.

Stanton, Robert. Bn *c.* 1920. Singer-actor, brother of Dick Haymes. Originally appeared as Bob Haymes then changed to Stanton. Sang with bands of Carl Hoff, Bob Chester and Freddie Martin before starring in Columbia 'B' musicals *Swing Out The Blues, Is Everybody Happy?* 43 / *Beautiful But Broke, Hey Rookie* 44 / *The Blonde From Brooklyn* 45 / *Sing While You Dance* 46 / *It's Great To Be Young* 47 / *So Dear To My Heart* (soundtrack) 49. Now song-writer.

Steele, Tommy. Bn 17 December 1936, Bermondsey, London, as Thomas Hicks. Former rock and roll singer who appeared in British pop music films. Now appeals to a wider audience in adult musicals **The Happiest Millionaire,** *Half A Sixpence* 67 / **Finian's Rainbow** 68.

Steiner, Max. Bn 10 May 1888, Vienna. Composer-conductor. Wrote and conducted his first operetta at fourteen, conducted musical shows in England, France and USA before joining RKO as Head of Music (29–36). Musical director of **Rio Rita** 29 / *Dixiana* 30 / **Flying Down To Rio,** *Melody Cruise* 33 / *Hips Hips Hooray,* **The Gay Divorcee,** *Down To Their Last Yacht,* **Roberta,** *Strictly Dynamite* 34 / **Top Hat,** *I Dream Too Much,* **Follow The Fleet** 35. Then went to Warner, turning out many magnificent scores for dramatic films (best remembered for 'Gone With The Wind' done on loan-out to Selznick). Did many background scores and arrangements for musicals, and may actually have conducted, but musical director credit always went to Leo F. Forbstein as Head Of Music.

Stept, Sam. Bn 18 September 1897, Odessa, Russia. Composer, former accompanist to Mae West and bandleader. Wrote songs for musicals *Lucky In Love, Mother's Boy, Syncopation* 29 / *A Show Girl In Hollywood, Big Boy* 30 / *Shady Lady* 33 / *Baby Take A Bow* 34 / *Dancing Feet, Happy Go Lucky, Hit Parade Of 1937, Laughing Irish Eyes, Sitting On The Moon* 36 / *23½ Hours Leave* 37 / *Having Wonderful Time* 38 / *Hullabaloo* 40 / *When Johnny Comes Marching Home, Johnny Doughboy, Yokel Boy* 42. Died 1 December 1964.

Stevens, George. Bn 1905, Oakland, Calif. Director who came to Hollywood at eighteen and worked his way through the industry. Directed early musicals for RKO: *Nitwits* 35 / **Swing Time** 36 / **A Damsel In Distress** 37, but has done nothing in the *genre* since, concentrating on polished comedies and, later, more dramatic epics.

Stevens, Rise. Bn 1913, New York. Metropolitan opera star who followed many of her fellow divas in and out of Hollywood, leaving only the memory of *The Chocolate Soldier* 41 / **Going My Way** 44 / *Carnegie Hall* 49. Still active, starred in a 1966 revival of 'The King and I'.

Stewart, Freddie. Singer, formerly with Tommy Dorsey's band, who starred for Sam Katzman in a number of mid-forties teenage romps, and has never been heard from since *Freddie Steps Out, High School Hero, Junior Prom* 46 / *Sarge Goes To College* 47 / *Campus Sleuth* 48.

Stockwell, Larry. Another of the popular tenors of the thirties of whom little is now known. Best remembered (anonymously) as the voice of the Prince in *Snow White And The Seven Dwarfs* 37. Also appeared in *Here Comes The Band* 35.

Stokowski, Leopold. Bn 18 April 1887,

London, as Leopold Antonio Stanislaw Boleslawowicz Stokowski of Polish father and Irish mother. One of the world's great conductors, who tried to popularize classical music in films *The Big Broadcast Of 1937, 100 Men And A Girl* 36 / *Fantasia* 40 / *Carnegie Hall* 49.

Stoll, Georgie. Bn 1905. Composer-conductor. Musical director of 66 MGM musicals, all of which cannot be listed here. Credits include *Listen Darling* 38 / *Ice Follies Of 1939* 39 / *Go West, Little Nellie Kelly* 40 / *The Big Store* 41 / *Ship Ahoy, Panama Hattie* 42 / *Music For Millions* 44 / *Holiday In Mexico, The Thrill Of A Romance* 45 / *This Time For Keeps* 46 / *Luxury Liner* 47 / *A Date With Judy, The Kissing Bandit* 48 / *Neptune's Daughter* 49 / *Two Weeks With Love, The Duchess Of Idaho* 50 / *Glory Alley, Skirts Ahoy* 52 / *I Love Melvin* 53 / *Athena, The Student Prince* 54 / *Ten Thousand Bedrooms* 57 / *The Seven Hills Of Rome* 58 / *For The First Time* 59 / *Where The Boys Are* 60 / *Viva Las Vegas* 64 / *Girl Happy* 65.

Stoloff, Morris. Bn 1 August 1898, Philadelphia. Composer-conductor. Ex-violinist, member Los Angeles Philharmonic, founded Stoloff String Quartet. Came to films as concertmaster of Paramount studio orchestra and has been Head of Music at Columbia since 1936, receiving musical director credit on the majority of the company's films since then.

Stone, Andrew. Bn 1902. Producer-director who has operated independently, releasing through the majors. Musicals *The Great Victor Herbert* 39 / *There's Magic In Music* 41 / *Stormy Weather* (dir only) 43 / *Sensations Of 1945* 44. Latterly working in Britain, returning to musicals with *Song Of Norway* 70.

Stothart, Herbert. Bn 11 September 1885, Milwaukee. Composer-conductor. Ex-music

professor at Wisconsin University who became successful Broadway conductor. Joined MGM 1929 as General Music Director, wrote songs for and conducted *A Lady's Morals* 29 / *Cuban Love Song* 31. Musical director of more conservative types of musical and operettas *Devil May Care* 29 / *New Moon, Rogue Song, Montana Moon* 30 / *Going Hollywood* 33 / *The Merry Widow,* The Cat And The Fiddle, The Night Is Young 34 / *Naughty Marietta,* Here Comes The Band, *Rose Marie,* A Night At The Opera 35 / *San Francisco* 36 / *Maytime, Rosalie,* The Firefly 37 / *Sweethearts, The Girl Of The Golden West* 38 / *Broadway Serenade, Balalaika, The Wizard Of Oz* 39 / *Bitter Sweet, New Moon* 40 / *The Chocolate Soldier, Ziegfeld Girl* 41 / *Cairo, Rio Rita,* I Married An Angel 42 / *Thousands Cheer* 43 / *Three Daring Daughters, The Unfinished Dance* 47. Died 1 February 1949.

Strayer, Frank R. Bn 21 September 1891, Altoona, Pa. Director. Joined Metro as assistant director after World War I service, and became fully-fledged director with Fox in 1930. Output mainly routine (e.g. the 'Blondie' series) including second string musicals *Let's Go Places* 30 / *Dance Girl Dance* 33 / *Blondie Goes Latin, Go West Young Lady* 41 / *Footlight Glamour* 44 / *The Senorita From The West* 45. Died 1964.

Streisand, Barbra. Bn 24 April 1942, Brooklyn. Singer-comedienne who took Broadway by storm in the sixties and repeated her success in screen version of *Funny Girl* 68. *Hello Dolly!, On A Clear Day You Can See Forever* 69.

Stromberg, Hunt. Bn 12 July 1894, Louisville Kentucky. Producer, for many years executive with MGM, latterly independent. In the musical field was mainly concerned in production of the Nelson Eddy-Jeanette MacDonald series *Naughty Marietta, Rose Marie* 35 / *The Great Ziegfeld* 36 / *The Firefly, Maytime* 37 / *Sweethearts* 38 / *I Married An Angel* 42 / *Lady Of Burlesque* 43. Died 23 August 1968.

Strouse and Adams. Song-writers. Charles Strouse (composer) bn 7 June 1929, New York; Lee Adams (lyricist) bn 14 August 1924, Mansfield, Ohio. Adams was writing summer camp shows, Strouse composing for TV and newsreels, when they came together in the fifties to write for Broadway. Their only show filmed to date is *Bye Bye Birdie* 63, but they also wrote *The Night They Raided Minsky's* 68, specially for the screen.

Styne, Jule. Bn 31 December 1905, London. Composer. Played piano with Chicago Symphony Orchestra at nine, later led his own band. Went to Hollywood as vocal coach with Fox, for whom he started writing songs with Sidney Clare *Hold That Co-ed, Kentucky Moonshine,*

Pack Up Your Troubles 38. With other lyricists wrote interpolated songs for *Sing Dance Plenty Hot, The Girl From Havana, Melody And Moonlight, Ice-Capades* 40 / *Melody Ranch, Rookies On Parade, Puddin' Head, Angels With Broken Wings, Hit Parade Of 1941, Sis Hopkins* 41 / *Beyond The Blue Horizon, Sweater Girl, Ice-Capades Revue, Priorities On Parade, Sleepytime Gal, The Powers Girl, Salute For Three, Hit Parade Of 1943* 42; formed a regular partnership with Sammy Cahn (*q.v.*) and wrote songs or complete scores for *Johnny Doughboy, Youth On Parade, Here Comes Elmer, Larceny With Music, Let's Face It, Thumbs Up* 43 / **Step Lively,** *Knickerbocker Holiday, Follow The Boys, Carolina Blues,* **Anchors Aweigh,** *Tonight And Every Night* 44 / *Cinderella Jones, Earl Carroll's Sketchbook, The Kid From Brooklyn, The Stork Club,* **It Happened In Brooklyn,** *The Sweetheart Of Sigma Chi, Tars And Spars* 46 / *Glamour Girl, Ladies' Man* 47 / **Romance On The High Seas,** *Two Guys From Texas* 48 / **It's A Great Feeling** 49 / *Double Dynamite,* **The West Point Story** 50; collaborated with Leo Robin for *Meet Me After The Show, Two Tickets To Broadway* 51 / **Gentlemen Prefer Blondes** 53 / *My Sister Eileen* 55; with other lyricists for *Living It Up* 55 / **Bells Are Ringing** 60 / **Gypsy** 62 / *The Pleasure Seekers* (Cahn) 64 / **Funny Girl** 68.

Sutherland, Edward A. Bn 5 January 1895, London. Director. Learned the comedy business in vaudeville, with the Keystone Cops, and as a director with Chaplin. Directed many comedies, also musicals *The Dance Of Life* 29 / *Close Harmony,* **Paramount On Parade** 30 / *June Moon,* **Palmy Days** 31 / *International House, Too Much Harmony* 33 / **Mississippi** 35 / *Poppy* 36 / *Every Day's A Holiday, Champagne Waltz* 37 / **The Boys From Syracuse, One Night In The Tropics** 40 / *Sing Your Worries Away* 42 / *Dixie* 43 / *Follow The Boys* 44.

Swarthout, Gladys. Bn 25 December 1904, Deepwater, Missouri. Opera star who had more success than most with ventures into popular music. Her screen career with Paramount covered a film a year for four years *The Rose Of The Rancho* 35 / *Give Us This Night* 36 / *Champagne Waltz* 37 / *Romance In The Dark* 38. Later branched out into radio and TV. Died 8 July 1969.

Tamblyn, Russ. Bn 30 December 1934. Actor-dancer who started as a child (Rusty Tamblyn). Though an expert dancer, has diversified his screen activities and now operates almost exclusively as an actor. Remembered for musicals **Seven Brides For Seven Brothers** 54 / **Hit The Deck** 55 / *Tom Thumb* 58 / **West Side Story** 61 / *The Wonderful World Of The Brothers Grimm* 62 / *Follow The Boys* 63.

Tashlin, Frank. Bn 1913. Producer-director-screenwriter. A successful cartoonist who started writing for Bob Hope and Martin and Lewis films, some of which he directed. Musicals (as writer only) *Monsieur Beaucaire* 46 / *Variety Girl* 47 / *Love Happy, The Paleface,* **One Touch Of Venus** 48 / (as director-writer) *Son Of Paleface* 52 / *Hollywood Or Bust, Artists And Models* 56 / *The Girl Can't Help It* (also prod) 57 / *Say One For Me* 59 / *Cinderfella* 60.

Taurog, Norman. Bn 23 February 1899, Chicago. Director. Child actor in silents at fourteen, later became prop man, then stage and screen actor, and director of Larry Semon comedies. Directed musicals *Lucky Boy* 29 / *Follow The Leader* 30 / *The Phantom President* 32 / *The Way To Love, Bedtime Story* 33 / **We're Not Dressing,** *College Rhythm* 34 / *Strike Me Pink,* **The Big Broadcast Of 1936** 35 / *Rhythm On The Range* 36 / **You Can't Have Everything** 37 / *Mad About Music* 38; moved to MGM **The Broadway Melody Of 1940,** *Little Nellie Kelly* 40 / *Presenting Lily Mars* 42 / *Girl Crazy* 43 / *Big City, Words And Music* 48 / *That Midnight Kiss* 49 / *The Toast Of New Orleans* 50 / *Rich Young And Pretty* 51; back to Paramount *Jumping Jacks, The Stooge* 52 / *The Stars Are Singing* 53 / *Living It Up* 54 / *Pardners* 56; freelanced *Bundle Of Joy* 56 / *G.I. Blues* 60 / *All Hands On Deck* 61 / *Blue Hawaii, Girls Girls Girls* 62 / *Palm Springs Weekend, It Happened At The World's Fair* 63 / *Tickle Me* 65 / *Spinout* 66 / *Double Trouble* 67 / *Speedway* 68.

Teagarden, Jack. Bn 20 August 1905, Vernon, Texas. Jazz trombonist who did guest appearances in musicals **The Birth Of The Blues** 41 / *So's Your Uncle* 43 / *Hi Good Lookin', Twilight On The Prairie* 44 / *The Strip* 51 / *Glory Alley* 52 / *Jazz On A Summer's Day* 60. Died 15 January 1964.

Temple, Shirley. Bn 23 April 1928, Santa Monica, Calif. Curly-headed moppet of considerable song-and-dance talent who was discovered by Charles Lamont for Educational Films shorts, and by Jay Gorney for Fox feature films. A box-office winner from 1934 on in musicals *Stand Up And Cheer, Little Miss Marker, Baby Take A Bow* 34 / *Bright Eyes, Our Little Girl, The Little Colonel, Curly Top, The Littlest Rebel, Captain January* 35 / *Poor Little Rich Girl, Dimples, Stowaway* 36 / *Heidi* 37 / *Rebecca Of Sunnybrook Farm, Just Around The Corner, Little Miss Broadway* 38 / *Young People, The Bluebird* 40. Made the transition to teenage actress but retired for marriage, only returning for special TV storyteller appearances. Now in politics.

Thomas, Danny. Bn 1914, Ohio, as Amos Jacobs. Comedian, long admired in clubs and vaudeville, who only came to films (briefly) late

in his career. Musicals *The Unfinished Dance*
(début) 47/*Big City* 48/*I'll See You In My
Dreams, Call Me Mister* 51/*The Jazz Singer* 52.
Now a TV tycoon with his own production
company (Dick Van Dyke Show, etc.).

Thompson, Kay. Bn 9 November 1913,
St Louis. Comedienne-author-composer-
choreographer-arranger-vocal coach-singer-
dancer-conductor-pianist. A useful asset to a film
studio, who has functioned in all capacities at
one time or another. Appeared as pianist with
the St Louis Symphony; was Fred Waring's
choral director; wrote best-selling 'Eloise' books
and portrayed the character on radio; did
successful night-club act, etc. Wrote songs for
The Thrill Of A Romance, No Leave No Love 45;
arranger on **Broadway Rhythm, I Dood It** 43/
The Ziegfeld Follies 44 (also wrote songs)
The Harvey Girls 45; appeared in *Manhattan
Merry-Go-Round* 37/*The Kid From Brooklyn* 46/
Funny Face 56.

Thorpe, Richard. Bn 24 February 1896,
Hutchinson, Kansas, as Rollo Thorpe. Director.
Followed the familiar pattern of vaudeville,
musical comedy, acting in silents, then directing,
first as freelance then with MGM from 1935.
Musicals **Two Girls And A Sailor** 44/*The Thrill
Of A Romance* 45/*Fiesta, This Time For Keeps*
46/*On An Island With You* 47/*A Date With Judy*
48/**Three Little Words** 50/*The Great Caruso*
51/*The Student Prince, Athena* 54/*Jailhouse Rock,
Ten Thousand Bedrooms* 57/*Follow The Boys, Fun
In Acapulco* 63.

Tibbett, Laurence. Bn 16 November 1896,
Bakersfield, Calif. Operatic baritone, adept as an
actor in his own field, who never mastered
screen acting. Musicals *The Rogue Song* (début),
New Moon 30/**Cuban Love Song** 31/*Under Your
Spell* 36. Starred every week on radio's 'Hit
Parade' in 1945. Died 15 July 1960.

Tierney, Harry. Bn 21 May 1890, Perth
Amboy, N.J. Composer of twenties' Broadway
shows who joined RKO to adapt his own
musicals and write for Wheeler and Woolsey
Rio Rita 29, 42/*Dixiana* 30/*Irene* 40. When
screen musicals slumped in 1932 he returned to
New York but wrote little else. Died 22 March
1965.

Todd, Thelma. Bn 1908, Lawrence, Mass.
Blonde comedienne in two-reelers, and some-
times romantic lead in Marx Brothers films and
musicals *Follow Through* 30/*The Hot Heiress* 31/
Horse Feathers 32/*Sitting Pretty* 33/*Bottoms Up,
Cockeyed Cavaliers, Hips Hips Hooray, Palooka* 34/
*After The Dance, Two For Tonight, The Bohemian
Girl* 35. Died 16 December 1935.

Tomlin, Pinky. Bn 9 September 1907, Eureka
Springs, Arkansas. Singer-comedian-composer-

bandleader. Well-known recording artist in
thirties who appeared in minor musicals *King
Solomon Of Broadway, Paddy O'Day* 35/*With Love
And Kisses* 36/*Swing It Professor, Swing While
You're Able* 37. Retired and is now president of
an oil properties firm.

Tormé, Mel. Bn 13 September 1925, Chicago,
as Melvin Howard Tormé. Singer who was on
radio as a child, came to films as a teenager and
developed into one of the finest singers and
most literate composers in popular music.
Appeared in musicals **Higher And Higher** 43/
Pardon My Rhythm 44/*Let's Go Steady* 45/*Good
News* 47/**Words And Music** 48/*So Dear To My
Heart* (soundtrack) 49/*The Duchess Of Idaho* 50/
Girls' Town 59. Now developing as a dramatic
actor on TV.

Trotti, Lamar. Bn 1900. Screenwriter turned
producer, for many years with Fox. In the
musical field wrote *First Baby, Can This Be Dixie*
36/*In Old Chicago* 37/*Hold That Co-ed*, **Alexander's Ragtime Band** 38/*When My Baby
Smiles At Me* 48/*My Blue Heaven* 50/**There's
No Business Like Show Business** 54;
Wrote and produced **Mother Wore Tights**
47/*You're My Everything* 49/*Stars And Stripes
Forever*, **With A Song In My Heart** 52.
Died 1952.

Tucker, Sophie. Bn 1884, Poland, as Sophia
Abuza. Great stage and cabaret entertainer with
a unique narrative song delivery of perfect
timing and projection. Musicals *Honky Tonk* 29/
Thoroughbreds Don't Cry, **The Broadway Melody
Of 1938** 37/*Follow The Boys, Sensations Of 1945*
44/*The Joker Is Wild* 57. Was still an active
performer till her death in February 1966.

Tuttle, Frank. Bn 1893, New York. Director
of many Paramount films who started with the
company as continuity writer. Did Bing
Crosby films and other musicals *Sweetie* 29/
Paramount On Parade, *Love Among The
Millionaires* 30/*Dude Ranch* 31/**The Big
Broadcast** 32/**Roman Scandals** 33/*Here Is My
Heart* 34/*All The King's Horses, Two For Tonight*
35/*College Holiday* 36/*Waikiki Wedding* 37/
Doctor Rhythm 38/*Paris Honeymoon, Charlie
McCarthy Detective* 39/*Suspense* 46. Died 1963.

Twentieth Century-Fox Film Corporation.
(Referred to herein as 'Fox'). One of the leaders
in musical production, being early in the field
with **Fox Follies** 29, named after production
head/founder William Fox. On the 1935 merger
of the company with 20th Century Pictures that
company's Joseph M. Schenck took over as top
man with Darryl F. Zanuck as boss of
production. At this time Shirley Temple's films
were hugely successful, and the mid-thirties saw
some fine Fox musicals with Tony Martin,

Joan Davis, The Ritz Brothers and Alice Faye, who also carried over into what was perhaps Fox's peak period. Their luscious Technicolor extravaganzas, always featuring some permutation of Faye, Betty Grable, John Payne, Carmen Miranda, Cesar Romero and Don Ameche, were wonderful escapist fare during the war years, and introduced some songs which have proved memorable. The fifties saw Zanuck replaced by Spyros Skouras, the introduction of Marilyn Monroe and CinemaScope and yet more musicals, this time with Dan Dailey, Ethel Merman, Donald O'Connor, Gordon Macrae *et al.*, and the collected works of Rodgers and Hammerstein, with *The Sound of Music* as one of the greatest box-office hits.

United Artists Corporation. One of the most powerful companies at the time of writing, with some big hits to show after a recent reorganization, but for most of its existence U-A was purely a distribution company with little production of its own. The Artists who United were Charles Chaplin, Mary Pickford, Douglas Fairbanks and D. W. Griffith, who got together to handle their own films, distributing others along the way, including, eventually, much of the output of Goldwyn and Selznick. Still distributors to a certain extent (e.g. Mirisch Corporation films), but with their own record label and music company U-A are now obviously calling the tune.

Universal-International / Universal Pictures. One of the earliest film production companies, founded by Carl Laemmle in 1912, with many changes in its fortunes and the quality of its output, especially in the musical field. There were occasional pre-war highspots, but for many years the studios were busy churning out 'B' films with performers like Jane Frazee, Robert Paige, The Andrews Sisters, Donald O'Connor, Peggy Ryan, Gloria Jean, Susanna Foster and numerous other young singers hoping to win some of the honour and glory of the fading Deanna Durbin. None of them did, and by the beginning of the fifties Universal-International, as it was by then known, was creaking along on increasingly corny Abbott and Costello features. The musical output picked up later with biographically inaccurate but musically precise biographies of Glenn Miller and Benny Goodman, and when MCA (Music Corporation Of America) took over in 1962 the comany was in better shape, since when Ross Hunter's box-office bonanzas (including *Thoroughly Modern Millie*) and a forward-looking board have made Universal City one of the most active factories in the world.

Vallee, Rudy. Bn 28 July 1901, Island Pond,

Vermont, as Hubert Prior Vallee. Actor-singer, though it is many years since the latter applied. In the twenties Vallee was a flappers' heart-throb as the singing leader of The Connecticut Yankees dance band, and starred in numerous films. After war service returned to screen in burlesques of his former self. Musicals *The Vagabond Lover* (début), **Glorifying The American Girl** 29 / *International House* 33 / **George White's Scandals** 34 / *Sweet Music* 35 / **The Golddiggers In Paris** 38 / **Second Fiddle** 39 / *Time Out For Rhythm, Too Many Blondes* 41 / *Happy Go Lucky* 42 / *People Are Funny* 45 / *The Beautiful Blonde From Bashful Bend* 49 / **Gentlemen Marry Brunettes** 55 / **The Helen Morgan Story** 57 / **How To Succeed In Business Without Really Trying** 66 / *The Night They Raided Minsky's* (soundtrack) 68.

Van Dyke, Dick. Bn 13 December 1925, Danville, Ill. Long-service comedian who reached fame fairly late in life, via his TV show which ran for years. A unique and talented peformer who still seeks as good a role on screen as on TV. Musicals **Bye Bye Birdie** 63 / **Mary Poppins** 64 / **Chitty Chitty Bang Bang** 68.

Van Dyke, W. S. Bn 1890, San Diego. Director of many MGM films, who was usually associated with the Eddy-MacDonald musicals after his initial film in the *genre* **Cuban Love Song** 31 / **Rose Marie, Naughty Marietta** 35 /

San Francisco 36 / **Rosalie** 37 / *Sweethearts* 38 /
Bitter Sweet 40 / *Cairo, I Married An Angel* 42.
Died 1944.

Van Heusen, Jimmy. Bn 26 January 1913,
Syracuse, N.J. Composer who gave Crosby and
Sinatra some of their best film songs, the former's
written with lyricist Johnny Burke from 1940 to
1953, the latter's in collaboration with Sammy
Cahn (1956 to date). Wrote scores for *Love Thy
Neighbour* 40 / *The Road To Zanzibar, Playmates* 41 /
My Favourite Spy, The Road To Morocco 42 / **Lady
In The Dark**, *Dixie* 43 / **And The Angels Sing**,
The Belle Of The Yukon, **Going My Way** 44 / *The
Bells Of St Mary's, Duffy's Tavern, The Road To
Utopia* 45 / *Cross My Heart* 46 / *The Road to Rio,
Variety Girl, Welcome Stranger* 47 / *A Connecticut
Yankee At King Arthur's Court* 48 / *Top O'
The Morning* 49 / **Mr Music**, *Riding High* 50 / *The
Road To Bali* 52 / *Little Boy Lost* 53 / **Anything
Goes**, *Pardners* 56 / *The Joker Is Wild* 57 / *Say One
For Me* 59 / *High Time, Let's Make Love* 60 / **The
Road To Hong Kong** 62 / *The Pleasure Seekers,
Robin And The Seven Hoods 64 /
Thoroughly Modern Millie 67 / **Star!** 68.

Velez, Lupe. Bn 18 July 1908, San Louise
Potosi, Mexico, as Guadelupe Velez de Villa
Lobos. Actress, appropriately named 'Mexican
Spitfire' after her film character. Started opposite
Laurel and Hardy in shorts, then on to musicals
Cuban Love Song 31 / *Mr Broadway* 33 / *Hollywood
Party, Palooka, Strictly Dynamite* 34 / *The Girl
From Mexico* 39 / *Redhead From Manhattan,
Playmates, Six Lessons From Madame La Zonga* 41.
Died 14 December 1944.

Vera-Ellen. Bn 16 February 1921, Cincinnati,
as Vera Ellen Westmeyr Rohe. Petite dancer,
discovered by Goldwyn, who did some superb
work in major musicals *Wonder Man* (début) 45 /
The Kid From Brooklyn, Three Little Girls in Blue

46 / *Carnival In Costa Rica* 47 / **Words And Music,**
Love Happy 48 / **On The Town** 49 / **Three Little
Words** 50 / *Happy Go Lovely* 51 / **The Belle Of
New York** 52 / **Call Me Madam** 53 / *White
Christmas* 54 / *Let's Be Happy* 56.

Vidor, Charles. Bn 27 April 1900, Budapest,
Hungary. Director who did some of his most
memorable work in musicals **Cover Girl** 44 /
A Song To Remember 45 / **Hans Christian
Andersen** 52 / **Love Me Or Leave Me** 55 / *The
Joker Is Wild* 57. Died 1959 while making
Song Without End.

Vidor, King. Bn 8 February 1895, Galveston,
Texas, as King Wallis Vidor. Long-service
(since 1915) director who worked his way up
from the bottom. Only musical *Hallelujah* 29.

Von Sternberg, Josef. Bn 1894, Vienna.
Director who discovered Dietrich. Had
enormous ego, hated actors (see his autobiography). Even he did a musical *The King
Steps Out* 36. Died December 1969.

Walker, Hal. Bn 1896. Director of many
Paramount films by Hope and Crosby, Martin
and Lewis, etc. Musicals *The Road To Utopia,
Duffy's Tavern, Out Of This World* 45 / *The Stork
Club* 46 / *My Friend Irma Goes West, At War With
The Army* 50 / *That's My Boy* 51 / *The Road To
Bali, Sailor Beware* 52. Died 1956.

Walker, Nancy. Bn 10 May 1922, Philadelphia.
Stocky comedienne, better known on stage, who
made a few musicals *Best Foot Forward* (début),
Girl Crazy, Broadway Rhythm 43 / *Lucky Me* 54.

Wallace, Oliver. Bn 6 August 1887, London.
Composer-conductor. The first cinema organist
in history (Seattle, 1910), later worked exclu-
sively with Disney, writing songs and acting as
musical director for *Dumbo* 41 / *Fun And Fancy
Free* 46 / *Ichabod And Mr Toad* 49 / *Cinderella* 50 /
Alice In Wonderland 51 / *Peter Pan* 53 / *Lady And
The Tramp* 55. Died 16 September 1963.

Wallace, Richard. Bn 1894, Sacramento City.
Director who started as assistant cutter with
Sennett, worked his way through all studio
routine and directed features from 1927.
Musicals *Innocents Of Paris, Shopworn Angel* 29 /
Blossoms On Broadway 37 / *The Underpup* 39 /
Because Of Him 45. Died 1951.

Waller, Thomas 'Fats'. Bn 21 May 1904, New
York. Negro singer-pianist, who made jazz
palatable to a wide public. A great entertainer
seen only in **The King of Burlesque,** *Hooray
For Love* 35, **Stormy Weather** 43. Died
15 December 1943.

Wallis, Hal B. Bn 14 September 1898,
Chicago. Producer, heavily committed with
production of musicals for Warner, later with
Paramount. Responsible for introduction of
many present-day stars; in our field gave first

opportunities to Martin and Lewis, Presley, etc.
Musicals *Green Pastures* 36 / **The Golddiggers Of
1937, Hollywood Hotel, Ready, Willing And
Able,** *The Singing Marine* 37 / *Swing Your Lady,
A Cowboy From Brooklyn, Going Places, Hard To
Get,* **The Golddiggers In Paris** 38 / **Naughty
But Nice, On Your Toes** 39 / *The Strawberry
Blonde, Navy Blues,* **Blues In The Night** 41 / *Juke
Girl,* **Yankee Doodle Dandy** 42 / **This Is The
Army** 43 / **Night And Day** 45 / *My Friend Irma*
49 / *My Friend Irma Goes West* 50 / *That's My Boy*
51 / *The Stooge, Sailor Beware, Jumping Jacks* 52 /
Scared Stiff 53 / *Artists And Models, Hollywood Or
Bust* 56 / *Loving You* 57 / *King Creole* 58 / *G.I. Blues*
60 / *Blue Hawaii, Girls Girls Girls* 62 / *Roustabout*
64 / *Paradise Hawaiian Style* 66 / *Easy Come
Easy Go* 67.

Walsh, Raoul. Bn 11 March 1889, New York.
Director who learned his trade as actor and
scenarist with D. W. Griffith (*Birth Of A Nation,*
etc.). Lost an eye in 1928 and concentrated on
directing, including musicals *Going Hollywood*
33 / **Every Night At Eight** 35 / *Klondike Annie*
36 / *Artists And Models, Hitting A New High* 37 /
College Swing 38 / *St Louis Blues* 39 / *The Strawberry
Blonde* 41 / *The Man I Love* 46 / *One Sunday
Afternoon* 48 / *Glory Alley* 52.

Walston, Ray. Bn 1917. Fast-talking comedy
actor renowned on stage and screen as 'Luther
Billis' in *South Pacific* 58. Other musicals
Damn Yankees 58 / *Say One For Me* 59 / *Kiss Me
Stupid* 65 / **Paint Your Wagon** 69.

Walters, Charles. Bn 17 November 1911,
Pasadena, Calif. Director who started as a stage
dancer and joined MGM as choreographer and
director of dance sequences in musicals *Presenting
Lily Mars* (début), *Seven Days Leave* 42 / **Dubarry
Was A Lady, Girl Crazy,** *Best Foot Forward,
Meet The People,* **Broadway Rhythm** 43 / **Meet
Me In St Louis, The Ziegfeld Follies** 44 / **The
Harvey Girls** 45 / **Summer Holiday** 46; full
direction of *Good News* 47 / **The Barkleys Of
Broadway, Easter Parade** 48 / **Summer Stock**
50 / *Texas Carnival* 51 / **The Belle Of New York**
52 / *Easy To Love, Lili, Torch Song, Dangerous
When Wet* 53 / **High Society** 56 / **Billy Rose's
Jumbo** 62 / **The Unsinkable Molly Brown** 64.

Wanger, Walter. Bn 16 October 1894, San
Francisco. Producer of many major films over
the past thirty-five years, including musicals
Every Night At Eight 35 / *Palm Springs* 36 /
Vogues Of 1938 / 52nd Street 37. Died 1968.

Waring, Fred. Bn 9 June 1900, Tyrone, Penn.
Orchestra leader whose Pennsylvanians have
been part of American popular music history
since the late twenties. Musicals *Syncopation* 29 /
The Varsity Show 37 / *Melody Time* (soundtrack) 48.

Warner Brothers Pictures Inc. The company
that started the story of the Hollywood musical
with **The Jazz Singer** is still in the forefront
with **Camelot** and **Finian's Rainbow.** It
started in 1923 with a joint effort by Albert,
Harry, Sam and Jack Warner (the latter still
active as producer, *viz.* **My Fair Lady**) who
pioneered the Vitaphone sound system a few
years later and gave films a voice. Warners were
ahead of the field in the thirties with their long
series of spectacular musicals directed by Lloyd
Bacon, Ray Enright *et al.* with the incredible
dance and song routines created by Busby
Berkeley. Though this was Warners' peak period
there has never been a time when musicals were
not a prime part of the company's output, be
they the Porter and Gershwin biopics of the
mid-forties, the cosy comedies with music of
Dennis Morgan, Jack Carson, Doris Day, Gene
Nelson, Gordon Macrae and Frank Sinatra, or
the prestige musicals like **Yankee Doodle
Dandy,** which did its share of Oscar-winning.
Latterly Warners have joined forces with Seven
Arts Films and Reprise Records.

Warren, Gloria. Bn *c.* 1926. One of the many
teenage sopranos spawned during the Deanna
Durbin era who never quite made it. Musicals
Always In My Heart 42 / *Cinderella Swings It* 44.

Warren, Harry. Bn 24 December 1893,
Brooklyn. Parents Italian immigrants, real name
Guaragna. Composer of more good film songs
than one can count, most of them from the
Berkeley era when he collaborated with Al Dubin
(*q.v.*) (32–8) and the Fox wartime musicals with
Mack Gordon (*q.v.*) (40–5). Also used Johnny
Mercer (*q.v.*) as lyricist (films marked *).
Complete credits *Spring Is Here* 30 / *Crooner* 32 /
**Roman Scandals, The Golddiggers Of 1933,
Footlight Parade, Forty-Second Street** 33 /

Mae West/Belle of the Nineties

Wonder Bar, Dames, Twenty Million Sweethearts, Moulin Rouge 34 | Page Miss Glory, *In Caliente,* Go Into Your Dance, Broadway Gondolier, *The Golddiggers Of 1935,* Shipmates Forever, Sweet Music, **Stars Over Broadway** 35 | Colleen, Hearts Divided, **Sing Me A Love Song,** Cain And Mabel 36 | Melody For Two, **The Golddiggers Of 1937,** Mr Dodd Takes The Air, The Singing Marine* 37 | Hard To Get, **The Golddiggers In Paris,** The Garden Of The Moon*, A Cowboy From Brooklyn*, Going Places* 38 | **Naughty But Nice*,** Honolulu 39 | Young People, **Tin Pan Alley, Down Argentine Way** 40 | The Great American Broadcast, **Sun Valley Serenade, That Night In Rio, A Weekend In Havana** 41 | Iceland, Song Of The Islands, **Orchestra Wives, Springtime In The Rockies** 42 | **Hello, Frisco, Hello,** Sweet Rosie O'Grady, **The Gang's All Here** 43 | The **Ziegfeld Follies** 44 | Billy Rose's Diamond **Horseshoe, The Harvey Girls*, Yolanda And The Thief** 45 | Summer Holiday 46 | The **Barkleys Of Broadway** 48 | My Dream Is Yours* 49 | Pagan Love Song, **Summer Stock** 50 | Painting The Clouds With Sunshine, Lullaby Of Broadway, Texas Carnival 51 | **The Belle Of New York*,** Just For You, Skirts Ahoy 52 | Artists And Models 56 | Cinderfella 60 | Ladies' Man 61.

Waters, Ethel. Bn 31 October 1900, Chester, Penn. Negro vaudeville and blues singer whose influence has been felt in many other coloured singers. Sang in early musicals and developed as a legitimate actress in her 40s. Musicals *On With The Show* (début) 29 | Rufus Jones For President 31 | Gift Of Gab 34 | **Cabin In The Sky,** Cairo 42 | Stage Door Canteen 43.

Webster, Paul Francis. Bn 20 December 1907, New York. Lyricist adept at converting title themes into pop hits. Worked with composers Harry Revel (43–5) and Sammy Fain (53–8) (*q.v.* both for credits). Other musicals *Our Little Girl* 35 | Rainbow On The River 36 | Make A Wish 37 | Vogues Of 1938, Breaking The Ice 38 | Fisherman's Wharf 39 | Presenting Lily Mars, Seven Sweethearts 42 | **Thousands Cheer** 43 | How Do You Do 45 | The Great Caruso 51 | **The Merry Widow** 52 | The Student Prince, **Rose Marie** 54 | Let's Be Happy 56.

Weill, Kurt. Bn 2 March 1900, Dessau, Germany. Composer whose operas caused a sensation in Germany of the twenties, but who had to leave for America with the rise of the Nazis. Gradually turned from his ragtimey operatic style (e.g. *Die Dreigroschenoper*) to the more sophisticated music of Broadway and Hollywood. Wrote original scores for *You And Me,* **The Goldwyn Follies** 38 | **Where Do We Go From Here?** 45; stage works filmed **Lady In The Dark** 43 | Knickerbocker Holiday 44 | One

Touch Of Venus 48 | The Threepenny Opera 64. Died 3 April 1950.

West, Mae. Bn 17 August 1892, Brooklyn. Actress-author-producer whose 'Come up and see me sometime' was an early movie catchphrase. Constantly emphasized yet debunked sex both in her own screenplays and performances. Musicals *She Done Him Wrong, I'm No Angel* 33 | The Belle Of The Nineties 34 | Goin' To Town 35 | Go West Young Man, Klondike Annie 36 | Every Day's A Holiday 37 | The Heat's On 43. Returned to stage and TV, made rock and roll records at seventy-five and screen comeback at seventy-seven.

Wheeler and Woolsey. Comedians. Bert Wheeler, bn 31 August 1900, Patterson, N.J., and Bob Woolsey bn 14 August 1889, Oakland, Calif. Broadway comics who came to Hollywood to repeat their stage success in **Rio Rita** 29, and stayed to make many comedies for RKO, including musicals *Cuckoos, Dixiana* 30 | Girl Crazy 32 | Cockeyed Cavaliers, Hips Hips Hooray 34 | Nitwits, Rainmakers 35. Woolsey died 31 October 1938; Wheeler turned to straight acting, appeared in *Las Vegas Nights* 41. Later in small TV parts. Died 18 January 1968.

Whelan, Tim. Bn 1893. Director who started career in England and returned to Hollywood on outbreak of war. Did some musicals for MGM and RKO *Seven Days Leave* (also prod) 42 | Swing Fever, **Higher And Higher** 43 | **Step Lively** 44. Died 1957.

White, Onna. Choreographer of *The Music Man* 62 | **Bye Bye Birdie** 63 | Oliver! 68.

Whiteman, Paul. Bn 28 March 1891, Denver. Portly bandleader mistakenly named 'The King Of Jazz' (though it made a good peg on which to hang one of the earliest musicals). Bing, the Dorseys and many others started in his orchestra. Musicals **The King Of Jazz** 30 | Thanks A

Million 35 | *Strike Up The Band* 40 | *Lady
Let's Dance, Atlantic City* 44 | *Rhapsody In Blue*
45 | *The Fabulous Dorseys* 47. Died 29 December
1967.

Whiting, Richard. Bn 12 November 1891,
Peoria, Ill. Composer, former publisher and
Broadway writer who joined Paramount when
talkies came in and wrote many individual songs
and complete scores, collaborating with Leo
Robin (29–33, 35–6), Sidney Clare (34–6),
Johnny Mercer (37–8), for *Sweetie, Why Bring
That Up?, Dance Of Life, Innocents Of Paris* 29 |
*Paramount On Parade, The Playboy Of Paris,
Close Harmony, Monte Carlo, Dangerous Nan
McGrew, Safety In Numbers, Let's Go Native* 30 |
Along Came Youth, Dude Ranch 31 | *One Hour
With You* 32 | *Adorable, My Weakness,* **Take A
Chance** 33 | *Bottoms Up, She Learned About
Sailors, 365 Nights In Hollywood, Transatlantic
Merry-Go-Round* 34 | *The Littlest Rebel, Bright
Eyes, Coronado, Here Comes Cookie,* **The Big
Broadcast Of 1936** 35 | *Rhythm On The Range,*
Anything Goes, *First Baby,* **Sing Baby Sing** 36 |
**Hollywood Hotel, The Varsity Show, Ready,
Willing And Able** 37 | *A Cowboy From Brooklyn*
38. Died 10 February 1938.

Whorf, Richard. Bn 1906. Former stage and
screen actor (mostly in second leads) who joined
MGM as director. Musicals (as actor) *Blues In*

The Night 41 | *Juke Girl,* **Yankee Doodle
Dandy** 42 | *Christmas Holiday* 44 | (as director) *It
Happened In Brooklyn, Till The Clouds
Roll By* 46 | *Luxury Liner* 47. Died 1966.

Wiere Brothers. Musical comedians who did
guest spots in *Vogues Of 1938* 37 | *The Great
American Broadcast* 41 | *The Road To Rio* 47 |
Double Trouble 67.

Wilder, Billy. Bn 1906, Austria. Director-
screenwriter. Worked on Vienna newspaper,
then in Berlin. Sent stories to Hollywood, but
when none were accepted came over in person
to break down the doors. Often working with
producer-writer Charles Brackett came to
specialize in satire and *film noir,* but also managed
a few musicals *The Emperor Waltz* 48 | *Some
Like It Hot* 59 | *Kiss Me Stupid* 65. Also produced
last two.

Williams, Andy. Bn 3 December 1930, Wall
Lake, Iowa. Recording and TV singer whose
first film work was at fourteen when he dubbed
the singing for Lauren Bacall in *To Have And
Have Not.* Appeared with The Williams Brothers
in *My Best Gal* 44 | *Something In The Wind* 47.
Only solo film to date *I'd Rather Be Rich* 64.

Williams, Esther. Bn 8 August 1922, Los
Angeles. Champion swimmer around whom
MGM built a series of spectacular musicals:
Bathing Beauty, **The Ziegfeld Follies** 44 | *Easy To*

Natalie Wood with Mervyn LeRoy | *(rehearsing)*
Gypsy

Ready, Willing And Able 37 | *Tail Spin* 39 |
My Favourite Spy, Footlight Serenade 42 | *Hollywood Canteen* 44, **Night And Day** 45 |*It's A Great Feeling* 49 | *Here Comes The Groom, Starlift* 51 | **Hans Christian Andersen,** *Just For You* 52.

Wymore, Patrice. Bn 17 December 1926, Miltonville, Kansas. Broadway singer-dancer who found a spot in Warner musicals, often in bitchy roles **Tea For Two** 50 | *Starlift, She's Working Her Way Through College,* **I'll See You In My Dreams** 51 | *She's Back On Broadway* 52.

Wynn, Ed. Bn 9 November 1886, Philadelphia, as Isaiah Edwin Leopold. Comedian of the Ziegfeld era, on radio in thirties and TV in fifties. Walt Disney helped him to a screen comeback in the sixties. Musicals *Follow The Leader* 30 | *Stage Door Canteen* 43 | *Alice in Wonderland* (soundtrack) 51 | *Cinderfella* 60 | *Babes In Toyland* 61 | **Mary Poppins** 64. Father of Keenan Wynn. Died 19 June 1966.

Yacht Club Boys. Favourite night-club entertainers of the thirties who did guest spots in musicals **Thanks A Million** 35 | **The Singing Kid,** *Pigskin Parade,* **Stage Struck** 36 | *The Thrill Of A Lifetime* 37 | *Artists And Models Abroad, The Cocoanut Grove* 38.

Yarborough, Jean. Bn 22 August 1902, Marianna, Ark. Director who started as gagman-director with Roach and Sennett. Kept busy at Universal during the war churning out 'B' movies, including musicals *Follow The Band, Get Going, Hi' Ya Sailor, So's Your Uncle* 43 | *In Society, Moon Over Las Vegas*, On Stage Everybody, South Of Dixie*, Twilight On The Prairie*, Weekend Pass* 44 | *The Naughty Nineties* 45 | *Cuban Pete* 46 | *Holiday In Havana* 49 | *Casa Manana* 51. Also produced films marked *.

Yellen, Jack. Bn 6 July 1892, Poland. Lyricist often in collaboration with Milton Ager (*q.v.*). Broadway writer who went West under contract to Fox; wrote for musicals *Honky Tonk* 29 | *Chasing Rainbows,* **The King Of Jazz,** *They Learned About Women* 30 | **George White's Scandals** 34 | *Captain January,* **The King Of Burlesque,** *George White's Scandals Of 1935* 35 | *Pigskin Parade,* **Sing Baby Sing** 36 | *Happy Landing, Rebecca Of Sunnybrook Farm* 38 | *George White's Scandals Of 1945* 45.

Youmans, Vincent. Bn 27 September 1898, New York. Composer. Produced own stage shows, some of which were filmed **Hit The Deck** 29, 55 | *No, No, Nanette* 30, 40; retitled **Tea For Two** 50 | *The Song Of The West* 30 | **Take A Chance** 33. Only original film score was **Flying Down To Rio** 33. Retired that year through illness, spent rest of his life in and out of sanatoriums and died 5 April 1946.

Young, Alan. Bn 19 November 1920, North

Shields, England. Comedian long known on radio and TV, seen in several films including musicals *Margie* (début) 46 | *Aaron Slick From Punkin Crick* 51 | **Gentlemen Marry Brunettes** 55 | *Tom Thumb* 58.

Young, Victor. Bn 8 August 1900, Chicago. Composer-conductor. Violinist with Warsaw Philharmonic at thirteen. Back in USA played in theatre orchestras and conducted on radio. Musical director with Paramount from 1936, wrote scores and conducted hundreds of films including musicals *Frankie And Johnny* 35 | *Rhythm On The Range, Klondike Annie, College Holiday,* **Anything Goes** 36 | *Champagne Waltz, Double Or Nothing, Hideaway Girl* 37 | *Breaking The Ice* 38 | *Fisherman's Wharf, Gulliver's Travels, Man About Town, Way Down South* 39 | **The Road To Singapore,** *Dancing On A Dime, Love Thy Neighbour, Moon Over Burma, Rhythm On The River* 40 | *Glamour Boy, Kiss The Boys Goodbye, Las Vegas Nights, The Road To Zanzibar* 41 | **The Fleet's In,** *The Road To Morocco, Beyond The Blue Horizon, True To The Army, Priorities On Parade, Sweater Girl* 42 | *Riding High, Salute For Three, True To Life* 43 | **And The Angels Sing** 44 | *Masquerade In Mexico, Out Of This World* 45 | *California* 46 | **A Connecticut Yankee At King Arthur's Court,** *Dream Girl, The Emperor Waltz, The Paleface* 48 | *Riding High* 50 | *Honey Chile* 51 | *The Stars Are Singing* 53 | *Knock On Wood* 54 | **The Vagabond King** 56. Died 11 November 1956 and received posthumous Academy Award for his last score, *Around The World In 80 Days*.

Zinnemann, Fred. Bn 29 April 1907, Vienna. Director who started as an extra, directed shorts, crime second features and, latterly, big dramas and epic Westerns like *High Noon*. One musical **Oklahoma!** 55.

Zorina, Vera. Bn 2 January 1917, Berlin, Germany, as Eva Brigitta Hartwig. Prima ballerina who went 'pop' in **On Your Toes** 39, repeating her 1937 stage success. Other musicals **The Goldwyn Follies** 38 | *Louisiana Purchase* 41 | **Star Spangled Rhythm** 42 | *Follow The Boys* 44. Retired from dancing, but has been recording as a narrator of classical works.

Index of Songs

The completely effective index of songs from
screen musicals would list composers, performers, film titles and dates. It would also
warrant a complete volume to itself. For the
present we must be content with this listing
of some 2,750 of the best-known film songs,
cross-referenced to the index of titles.

From the 275 films in the selective filmo-
graphy (listed in bold type in the main index)
the reader will be able to follow up composers
and performers. For the rest, space permits
only a reference to title and date, but this in
itself should prove a valuable source of
reference.

In toto, the 1,443 films listed in this book
must have contained somewhere between
eight to ten thousand songs, many of which
served their immediate functional purpose and
vanished from the public memory a few
minutes after the final credits. So it was
obviously pointless to list some obscure
'point' number from a long-forgotten wartime
Monogram 'B' musical. But on the other side
of the coin, Hollywood musicals have contributed to our popular culture some of the most
durable songs by the greatest popular composers of our times. Every film song which
ever became a 'standard' is here, along with
others, less familiar today perhaps, but which
may be still remembered by *aficionados* as
milestones in the career of a favourite performer
or subjects of especially interesting camera or
dance routines.

I have omitted a number of less familiar
title tunes from well-known films; these will
be self-evident. *Where a song has been reprised
in a number of films, bold type indicates the film for
which it was originally written.* If no bold type
appears in these cases, the song may have
originated in a stage show or as a musical
interpolation in a dramatic film or comedy.
Or it may simply be a well-loved standard
indispensable to a particular era, location or
mood; hence the use of *St Louis blues* in no
fewer than nine musicals, or *Pretty baby* and
By the light of the silvery moon, featured in ten
and eight respectively, despite their pre-talkie
origins.

Throughout this index, the numbers following
the song title refer to the numbers preceding
the film titles in the index of titles, pages
261–78.

Debbie Reynolds, Gene Kelly and Donald O'Connor/
Singin' in the Rain

Index of Titles

Film titles are shown in strict alphabetical order, ignoring the definite and indefinite articles. Titles changed for release in Great Britain are shown and cross-referenced accordingly. Dates indicate the year of production rather than of release, as confirmed either by the film companies or the *Film Daily* Year Books. Titles in bold type are the films included in full detail in the section of Select Filmographies.

266

696 Little Old New York *Fox 1940*

697 The Littlest Rebel *Fox 1935*

698 Living In A Big Way *MGM 1946*

699 Living It Up *Paramount 1954*

700 **Look For The Silver Lining** *Warner 1949*

701 Looking For Love *MGM 1964*

702 Lord Byron Of Broadway (GB: What Price Melody) *MGM 1929*

703 Lost In A Harem *MGM 1944*

704 **Lottery Bride** *United Artists 1930*

705 Lottery Lover *Fox 1934*

706 Louisiana Hayride *Columbia 1944*

707 Louisiana Purchase *Paramount 1941*

708 Love Affair *RKO 1939*

709 Love Among The Millionaires *Paramount 1930*

710 **Love And Hisses** *Fox 1937*

711 Love And Learn *Warner 1947*

712 Love Finds Andy Hardy *MGM 1938*

713 Love Happy *RKO 1948*

714 Love In Bloom *Paramount 1935*

Love In Las Vegas *see 1365*

715 Love In The Rough *MGM 1930*

Lovely To Look At (1937) *see 1279*

716 **Lovely To Look At** *MGM 1952*

717 Love Me Forever (GB: On Wings Of Song) *Columbia 1935*

718 **Love Me Or Leave Me** *MGM 1955*

719 Love Me Tender *Fox 1956*

720 **Love Me Tonight** *Paramount 1932*

721 Love On Toast *Paramount 1938*

722 **The Love Parade** *Paramount 1929*

723 Love Thy Neighbour *Paramount 1940*

724 Loving You *Paramount 1957*

725 Lucky Boy *Tiffany-Stahl 1929*

Lucky Days *see 1087*

726 Lucky In Love *Pathe/ RKO 1929*

727 Lucky Me *Warner 1954*

728 Lullaby Of Broadway *1951*

729 Lulu Belle *Bogeaus/ Columbia 1948*

730 Luxury Liner *MGM 1947*

731 Mad About Music *Universal 1938*

The Mad Hatter *see 116*

732 The Magnificent Rebel *Disney 1962*

733 Ma, He's Making Eyes At Me *Universal 1940*

734 The Main Attraction *Seven Arts/MGM 1963*

735 Make A Wish *Principal/ RKO 1937*

736 Make Believe Ballroom *Columbia 1949*

737 Make Mine Laughs *RKO 1949*

738 **Make Mine Music** *Disney/RKO 1946*

739 **Mammy** *Warner/Vita- phone 1930*

740 Man About Town *Paramount 1939*

The Man From The Folies Bergère *see 328*

741 Manhattan Angel *Columbia 1949*

Manhattan Mary *see 335*

742 Manhattan Merry-Go- Round (GB: Manhattan Music Box) *Republic 1937* Manhattan Music Box *see 742*

743 Manhattan Parade *Warner 1932*

744 The Man I Love *Warner 1946*

Man Of The Family *see 1322*

745 Many Happy Returns *Paramount 1934*

Marching Along *see 1172*

746 Mardi Gras *Wald/Fox 1958*

747 Margie *Universal 1940*

748 Margie *Fox 1946*

749 Marianne *MGM 1929*

750 Marie Galante *Fox 1934* Marshmallow Moon *see 1*

751 Mary Lou *Columbia 1948*

752 **Mary Poppins** *Disney/ Buena Vista 1964*

753 Masquerade In Mexico *Paramount 1945*

754 The Mayor Of 44th Street *RKO 1942*

755 Maytime *MGM 1937*

756 Meet Danny Wilson *Universal 1952*

757 Meet Me After The Show *Fox 1951*

758 Meet Me At The Fair *Universal 1952*

759 **Meet Me In Las Vegas** (GB: **Viva Las Vegas**) *MGM 1956*

760 **Meet Me In St Louis** *MGM 1944*

761 Meet Me On Broadway *Columbia 1945*

762 Meet Miss Bobby-Sox *Columbia 1944*

763 Meet The People *MGM 1943*

764 Melba *United Artists 1953*

765 Melody And Moonlight *Republic 1940*

766 Melody Cruise *RKO 1933*

767 Melody For Two *Warner 1937*

Melody Girl *see 1092*

Melody Inn *see 990*

768 Melody In Spring *Paramount 1934*

769 Melody Lane *Universal 1929*

770 Melody Lane *Universal 1941*

771 Melody Maker *RKO 1947*

772 Melody Man *Columbia 1930*

773 Melody Parade *Monogram 1943*

774 Melody Ranch *Republic 1940*

775 Melody Time *Disney/RKO 1948*

Memory For Two *see 561*

776 **Merry Andrew** *MGM 1958*

777 Merry-Go-Round Of 1938 *Universal 1937*

778 The Merry Monahans *Universal 1944*

872 Oh Sailor Behave *Warner 1930*

873 Oh You Beautiful Doll *Fox 1949*

874 **Oklahoma!** *Magna/ Todd AO 1955*

875 The Old Homestead *Liberty 1935*

876 Old Man Rhythm *RKO 1935*

877 **Oliver!** *Romulus/Columbia 1968*

878 **On A Clear Day You Can See Forever** *Paramount 1969*

879 On An Island With You *MGM 1947*

880 The One And Only Genuine Original Family Band *Disney/Buena Vista 1968*

881 One Heavenly Night *Goldwyn/United Artists 1930*

882 **One Hour With You** *Paramount 1932*

883 101 Dalmatians *Disney 1960*

884 **100 Men And A Girl** *Universal 1936*

885 One In A Million *Fox 1937*

886 **One Night In The Tropics** *Universal 1940*

887 **One Night Of Love** *Columbia 1934*

888 One Sunday Afternoon *Warner 1948*

889 **One Touch Of Venus** *Universal 1948*

890 On Moonlight Bay *Warner 1951*

891 On Stage Everybody *Universal 1944*

892 **On The Avenue** *Fox 1937*

893 On The Riviera *Fox 1951*

894 **On The Town** *MGM 1949*

On Wings Of Song *see* 717

895 On With The Show *Warner 1929*

896 **On Your Toes** *1st National 1939*

897 Operator 13 (GB: Spy 13) *MGM 1934*

898 The Opposite Sex *MGM 1956*

899 **Orchestra Wives** *Fox 1942*

900 Our Little Girl *Fox 1935*

901 Out Of This World *Paramount 1945*

902 Outside Of Paradise *Republic 1938*

903 Pack Up Your Troubles (GB: We're In The Army Now) *Fox 1939*

904 Paddy O'Day *Fox 1935*

905 Pagan Love Song *MGM 1950*

906 Page Miss Glory *Cosmopolitan/Warner 1935*

907 Painting The Clouds With Sunshine *Warner 1951*

908 **Paint Your Wagon** *Paramount 1969*

909 **The Pajama Game** *Warner 1957*

910 The Paleface *Paramount 1948*

911 **Pal Joey** *Essex-Sidney/ Columbia 1957*

912 Palm Springs *Paramount 1936*

913 Palm Springs Weekend *Warner 1963*

914 **Palmy Days** *Goldwyn/ United Artists 1931*

915 Palooka *United Artists 1934*

916 Panama Hattie *MGM 1942*

917 Pan-Americana *RKO 1944*

918 Paradise Hawaiian Style *Paramount 1966*

919 **Paramount On Parade** *Paramount 1930*

920 Pardners *Paramount 1956*

921 Pardon My Rhythm *Universal 1944*

922 Pardon My Sarong *Universal 1942*

923 Paris *1st National 1929*

924 Paris Blues *Pennebaker/ United Artists 1961*

925 Paris Honeymoon *Paramount 1939*

926 Paris In The Spring (GB: Paris Love Song) *Paramount 1935*

Pass To Romance *see* 483

927 Patrick The Great *Universal 1944*

928 Pennies From Heaven *Columbia 1936*

929 Penthouse Rhythm *Universal 1945*

930 People Are Funny *Pine-Thomas/Paramount 1954*

931 **Pepe** *Posa/Columbia 1960*

932 The Perils Of Pauline *Paramount 1947*

933 **Pete Kelly's Blues** *Mark VII/Warner 1955*

934 Peter Pan *Disney/RKO 1953*

935 The Petty Girl (*also known as* Girl of The Year) *Columbia 1950*

936 The Phantom Of The Opera *Universal 1943*

937 The Phantom President *Paramount 1932*

938 Pick A Star *Roach/MGM 1937*

939 Pigskin Parade (GB: Harmony Parade) *Fox 1936*

940 Pinocchio *Disney/RKO 1940*

941 Pin-Up Girl *Fox 1944*

942 **The Pirate** *MGM 1947*

943 The Playboy Of Paris *Paramount 1930*

944 Playmates *RKO 1941*

945 The Pleasure Seekers *Fox 1964*

946 Poor Little Rich Girl *Fox 1936*

947 Poppy *Paramount 1936*

948 **Porgy And Bess** *Goldwyn/Columbia 1959*

949 Pot Of Gold (GB: The Golden Hour) *Roosevelt/ United Artists 1940*

950 The Powers Girl (GB: Hello Beautiful) *Rogers/ United Artists 1942*

Present Arms *see* 666

951 Presenting Lily Mars *MGM 1942*

952 Priorities On Parade *Paramount 1942*

953 Private Buckaroo *Universal 1942*

954 Puddin' Head *Republic 1941*

955 Purple Heart Diary (GB: No Time For Tears) *Columbia 1951*

956 Puttin' On The Ritz *United Artists 1930*

957 Queen High *Paramount 1930*

958 Queen Of Burlesque *P.R.C. 1946*

1045 Scatterbrain *Republic 1940*
1046 Second Chorus
Paramount 1940
1047 **Second Fiddle** *Fox 1939*
1048 See My Lawyer
Universal 1945
1049 The Senorita From The
West *Universal 1945*
1050 Sensations Of 1945 *Stone/ United Artists 1944*
Serenade (1939) *see* 138
1051 Serenade *Warner 1955*
1052 **Seven Brides For Seven Brothers** *MGM 1954*
1053 Seven Days Ashore *RKO 1944*
1054 Seven Days Leave *RKO 1942*
1055 The Seven Hills Of Rome
MGM 1958
1056 The Seven Little Foys
Paramount 1955
1057 Seven Sweethearts *MGM 1942*
1058 Shady Lady *RKO 1933*
1059 Shady Lady *Universal 1945*
1060 **Shall We Dance?** *RKO 1937*
1061 She Couldn't Say No
Warner 1930
1062 She Done Him Wrong
Paramount 1933
1063 She Has What It Takes
Columbia 1943
1064 She Learned About
Sailors *Fox 1934*
1065 She Loves Me Not
Paramount 1934
1066 She's A Sweetheart
Columbia 1944
1067 She's Back On Broadway
Warner 1952
1068 She's For Me *Universal 1944*
1069 She's Got Everything
(GB: She's Got That
Swing) *RKO 1938*
She's My Lovely *see* 374
1070 She's Working Her Way
Through College *Warner 1951*
1071 Shine On Harvest Moon
Warner 1944
1072 Ship Ahoy *MGM 1942*
1073 Ship Cafe *Paramount 1935*
1074 Shipmates Forever
Cosmopolitan/Warner 1935

1075 The Shocking Miss
Pilgrim *Fox 1947*
1076 Shoot The Works (GB:
Thank Your Stars)
Paramount 1934
1077 Shopworn Angel
Paramount 1929
1078 **Show Boat** *Universal 1936*
1079 **Show Boat** *MGM 1951*
1080 **Show Business** *RKO 1944*
1081 A Show Girl In Hollywood *1st National 1930*
1082 The Show-Off *MGM 1946*
1083 **The Show Of Shows**
Warner 1929
1084 **Silk Stockings** *MGM 1956*
1085 Silver Skates *Monogram 1943*
1086 Sincerely Yours *Warner 1955*
1087 Sing A Jingle (GB: Lucky
Days) *Universal 1944*
1088 Sing And Be Happy *Fox 1937*
1089 Sing Another Chorus
Universal 1941
1090 **Sing Baby Sing** *Fox 1936*
1091 Sing Boy Sing *Fox 1957*
1092 Sing, Dance, Plenty Hot
(GB: Melody Girl)
Republic 1940
1093 Sing For Your Supper
Columbia 1941
1094 **The Singing Fool**
Warner 1928
1095 Singin' In The Corn (GB:
Give And Take)
Columbia 1946
1096 **Singin' In The Rain**
MGM 1952
1097 **The Singing Kid** *1st National 1936*
1098 The Singing Marine
Warner 1937
1099 The Singing Nun *MGM 1966*
1100 The Singing Sheriff
Universal 1944
1101 **Sing Me A Love Song**
(GB: **Come Up Smiling**)
1st National 1936
1102 Sing While You Dance
Columbia 1946
1103 Sing Your Way Home
RKO 1945

1104 Sing Your Worries Away
RKO 1942
1105 Sing You Sinners
Paramount 1938
1106 Sis Hopkins *Republic 1941*
1107 Sitting On The Moon
Republic 1936
1108 Sitting Pretty *Paramount 1933*
1109 Six Lessons From
Madame La Zonga
Universal 1941
1110 Ski Party *American-
International 1965*
1111 Skirts Ahoy *MGM 1952*
1112 **The Sky's The Limit**
RKO 1943
1113 Sleeping Beauty *Disney/ Buena Vista 1959*
1114 Sleepy Lagoon *Republic 1943*
1115 Sleepytime Gal *Republic 1942*
1116 Slightly French *Columbia 1949*
1117 Slightly Scandalous
Universal 1947
1118 Slightly Terrific
Universal 1944
1119 Small Town Girl *MGM 1953*
1120 The Smartest Girl In
Town *RKO 1936*
1121 Smiling Irish Eyes *1st National 1929*
1122 The Smiling Lieutenant
Paramount 1931
1123 Snow White And The
Seven Dwarfs *Disney/ RKO 1937*
1124 Snow White And The
Three Stooges (GB:
Snow White and the
Three Clowns) *Fox 1961*
1125 So Dear To My Heart
Disney/RKO 1949
1126 So Long Letty *Warner 1930*
1127 Somebody Loves Me
Paramount 1952
1128 Some Like It Hot
Paramount 1939
1129 **Some Like It Hot**
Ashton-Mirisch/United Artists 1959
1130 Something For The Boys
Fox 1944

WESTMAR COLLEGE LIBRARY.